Multidimensional Differential and Integral Calculus

Giorgio Riccardi · Bruno Antonio Cifra ·
Enrico De Bernardis

Multidimensional Differential and Integral Calculus

A Practical Approach

 Springer

Giorgio Riccardi
Dipartimento di Matematica e Fisica
Università della Campania Luigi Vanvitelli
Caserta, Italy

Bruno Antonio Cifra
Facoltà di Ingegneria Civile e Industriale
Sapienza Università di Roma
Latina, Italy

Enrico De Bernardis
Consiglio Nazionale delle Ricerche
Istituto di Ingegneria del Mare
Rome, Italy

ISBN 978-3-031-70325-6   ISBN 978-3-031-70326-3   (eBook)
https://doi.org/10.1007/978-3-031-70326-3

This Springer imprint is published by the registered company Springer Nature Switzerland AG
The registered company address is: Gewerbestrasse 11, 6330 Cham, Switzerland

If disposing of this product, please recycle the paper.

In memory of Massimo Strani, who showed us a way.

In memory of Ulderico Paolo Bulgarelli, who brought us together.

Preface

This book is the result of experience achieved in the field, teaching for several years in Mathematical Analysis courses for students in the first two years of Engineering degree courses.

With the aim of an agile approach to the contents, the text has been complemented with numerous exercises, accompanied by a significant graphic contribution. Furthermore, the main results of the theory—rather than being presented according to the formal sequence of the "proof of the theorem"—are deduced, starting from concrete examples, through a discursive albeit rigorous path, and finally stated in a more traditional form. From this choice also derives the use of colored boxes to highlight, in different ways, some definitions and results of the theory.

We then attempt to historically place the development of the main arguments of the theory, through the exposition of some biographical notes of protagonists of the discipline. The notes are taken by linking to MacTutor archive, created and maintained by Edmund Robertson and John O'Connor of the School of Mathematics and Statistics at the University of St Andrews, Scotland, and hosted by the University.

The suggested bibliographic path is deliberately minimal, as it is aimed at a more formal rereading of the topics proposed in the book.

The authors hope that this textbook will allow the student who is approaching the discipline to access the basic concepts and become familiar with the main tools, develop skills that help him in solving application problems, and prepare for an expansion of the contents through a broader discussion.

Finally, special thanks go to Dr. Cecilia Leotardi of the National Research Council of Italy, Institute of Marine Engineering. The quality of the manuscript has greatly benefited from her suggestions on various aspects of this work.

Latina, Italy
April 2024

Giorgio Riccardi
Bruno Antonio Cifra
Enrico De Bernardis

Contents

Chapter 1
Basic Concepts and Parametrisation of Curves

In this first lesson we start talking about *curves*, that is subsets of a space that can be put in one-to-one correspondence with segments of the real axis. The space has any number n of dimensions, but we will mostly deal with cases $n = 3$ and $n = 2$. In the latter case, the curve is defined as a *plane* curve.

1.1 Introductory Concepts

The most intuitive example of a curve is provided by the kinematics of a material particle. Consider a material particle P of mass m, which is subjected to a force $\boldsymbol{F}$, depending on the position $\boldsymbol{x}$ and, if applicable, the time t. We place P at the point $\boldsymbol{x}_0$ with velocity $\boldsymbol{v}_0$ at the initial time (as usual, we take such a time to be $t = 0$). The motion of the particle in the time interval $[0, T)$ is studied enforcing that the variation over time of the momentum of P (the momentum is given by the product mass $m \times$ speed $\boldsymbol{v} = d\boldsymbol{x}/dt$) equals the force acting on the particle. Translating this principle in formal terms, we can say that the particle motion is a solution of the Cauchy[1] differential problem:

$$\begin{cases} \dfrac{d}{dt}(m\boldsymbol{v}) = \boldsymbol{F} \\[2mm] \boldsymbol{x}(0) = \boldsymbol{x}_0\,, \quad \boldsymbol{v}(0) = \boldsymbol{v}_0\,. \end{cases} \tag{1.1}$$

Note that, due to the presence of second (time) derivatives, two initial conditions are required. They specify, at the initial time, the position and velocity of the material point, respectively.

---

[1] https://mathshistory.st-andrews.ac.uk/Biographies/Cauchy/.

© The Author(s), under exclusive license to Springer Nature Switzerland AG 2024
G. Riccardi et al., *Multidimensional Differential and Integral Calculus*,
https://doi.org/10.1007/978-3-031-70326-3_1

1

The solution of problem (1.1) provides a *vector valued function* (hereafter abbreviated in *vector function*) of time t:

$$\boldsymbol{x} : [0, T) \longrightarrow \mathbb{R}^3$$
$$t \quad \mapsto \quad \boldsymbol{x}(t) . \tag{1.2}$$

The function $\boldsymbol{x}(t)$ defines the trajectory of the particle P.

What is a vector function of time as (1.2)? It generates the correspondence between a segment of the real axis (in the case of our study, the segment $[0, T)$) and a subset of the space. So our particle moves on a curve in space. Let's take a couple of examples.

Suppose that the particle moves in a plane and that $\boldsymbol{F}$ is an elastic force (that is, one generated by a spring), characterised by a spring constant k, *i.e.* consider the force:

$$\boldsymbol{F}(\boldsymbol{x}) = -k\,(\boldsymbol{x} - \boldsymbol{c}) , \tag{1.3}$$

$\boldsymbol{c}$ being the point in the plane where the elastic force is zero, called the *equilibrium position* of P.

Then, the study of the motion results in the solution of the Cauchy problem:

$$\begin{cases} m\,\dfrac{d^2\boldsymbol{x}}{dt^2} = -k\,(\boldsymbol{x} - \boldsymbol{c}) \\[2mm] \boldsymbol{x}(0) = \boldsymbol{x}_0 , \quad \boldsymbol{v}(0) = \boldsymbol{v}_0 . \end{cases} \tag{1.4}$$

Introduced the radian frequency $\omega = \sqrt{k/m}$ (equivalently, we can also define the period $T = 2\pi/\omega$), we know the problem (1.4) has a solution for all times t given by:

$$\begin{aligned} \boldsymbol{x}(t) &= \begin{pmatrix} x(t) \\ y(t) \end{pmatrix} \\[2mm] &= (\boldsymbol{x}_0 - \boldsymbol{c})\cos(\omega t) + \frac{\boldsymbol{v}_0}{\omega}\sin(\omega t) + \boldsymbol{c} \\[2mm] &= \begin{pmatrix} (x_0 - c_x)\cos(\omega t) + v_{x0}\sin(\omega t)/\omega + c_x \\ (y_0 - c_y)\cos(\omega t) + v_{y0}\sin(\omega t)/\omega + c_y \end{pmatrix}, \end{aligned} \tag{1.5}$$

where we reported both the vector and scalar form of the solution (notice that the vector notation is significantly more compact compared to the other). The solution (1.5) establishes a correspondence between instants of time and points on the plane, so defining a plane curve.

Let's look at (1.5) from a different point of view. For a generic vector of the plane $\boldsymbol{x} = (x, y)$ we introduce the corresponding orthogonal vector: $\boldsymbol{x}^\perp = (-y, x)$ (how is orthogonality verified?). Without losing generality, we choose the origin of reference coincident with the equilibrium position $\boldsymbol{c}$ of the particle (*i.e.*, where $\boldsymbol{F} = 0$) and look for the conditions allowing to solve the system in $\cos(\omega t)$, $\sin(\omega t)$ (omit, for brevity, the dependence on t of the position vector $\boldsymbol{x}$):

$$\begin{cases} x_0 \cos(\omega t) + v_{x0} \sin(\omega t)/\omega = x \\ y_0 \cos(\omega t) + v_{y0} \sin(\omega t)/\omega = y \,. \end{cases} \quad (1.6)$$

The determinant Δ of the coefficient matrix:

$$\Delta = \begin{vmatrix} x_0 & v_{x0}/\omega \\ y_0 & v_{y0}/\omega \end{vmatrix} = \frac{x_0^\perp \cdot v_0}{\omega} \quad (1.7)$$

is nonzero for all initial velocity vectors which are not parallel to the initial position vector x_0. In fact, if this condition is not verified and v_0 is parallel to x_0, we will see that the oscillation of the particle occurs along a straight line passing through c and x_0, that is the motion becomes one-dimensional. We assume for the moment $x_0^\perp \cdot v_0 \neq 0$, and solve the system:

$$\cos(\omega t) = \frac{v_{y0}x - v_{x0}y}{\omega \Delta}, \quad \sin(\omega t) = \frac{-y_0\omega x + x_0\omega y}{\omega \Delta}. \quad (1.8)$$

The definition (1.8) of the cosine and sine of ωt suggests a way to obtain the equation of the curve along which the particle moves in the implicit form, *i.e.* rather than parameterised in time. Represented in this new form, the trajectory curve will appear more familiar. Since $\cos^2(\omega t) + \sin^2(\omega t) \equiv 1$, from (1.8) we get:

$$\underbrace{(v_{y0}^2 + \omega^2 y_0^2)}_{\alpha} x^2 + 2 \underbrace{(-v_{x0}v_{y0} - \omega^2 x_0 y_0)}_{\beta} xy + \underbrace{(v_{x0}^2 + \omega^2 x_0^2)}_{\gamma} y^2 = \omega^2 \Delta^2. \quad (1.9)$$

The curve is then a conic one and, since it is $\alpha\gamma > \beta^2$ (check it up) it is an ellipse. We conclude that the trajectory of our particle is an elliptical arc for $t \leq T$. In particular, the ellipse is completely traversed by the particle during a time interval equal to the period T. When v_0 is parallel to x_0, *i.e.* there is a real number q such that $v_0 = q\omega x_0$, the determinant Δ (1.7) vanishes. The trajectory lies in this case on a segment of the straight line passing through the origin and the point x_0 (and the area of the ellipse vanishes), as it is easily recognised from the solution (1.5):

$$x(t) = x_0[\cos(\omega t) + q \sin(\omega t)]. \quad (1.10)$$

Problem 1.1 Calculate the semi-axes and the orientation of the ellipse where the trajectory lies, performing a rotation of the coordinate system (centered at point c) on the solution (1.5).

Solution 1.1 Introducing the new coordinate system (x', y'), rotated by an angle θ with respect to the old (then $x' = x \cos\theta + y \sin\theta$, $y' = -x \sin\theta + y \cos\theta$) we find that the trajectory is written in the new coordinates as follows:

$$\begin{cases} x' = \underbrace{(+x_0 \cos\theta + y_0 \sin\theta)}_{+a\cos\varphi} \cos(\omega t) + \underbrace{(+v_{x0}\cos\theta + v_{y0}\sin\theta)/\omega}_{-a\sin\varphi} \sin(\omega t) \\[4mm] y' = \underbrace{(-x_0 \sin\theta + y_0 \cos\theta)}_{+b\sin\delta} \cos(\omega t) + \underbrace{(-v_{x0}\sin\theta + v_{y0}\cos\theta)/\omega}_{+b\cos\delta} \sin(\omega t) \,. \end{cases}$$

where a and b are two positive quantities to be determined, and the angles φ and δ are to be defined as well. We note that the quantities a, b, φ and δ all depend on the angle θ and the latter is calculated imposing $\tan\varphi = \tan\delta$. So we get the following trigonometric equation in the unknown θ:

$$\frac{1}{2}[(v_{x0}^2 - v_{y0}^2) + \omega^2(x_0^2 - y_0^2)]\sin 2\theta = (v_{x0}v_{y0} + \omega^2 x_0 y_0)\cos 2\theta \,,$$

whose solution allows to easily find a and b:

$$a^2 = \frac{1}{2}\left(+x_0^2 + y_0^2 + \frac{+v_{x0}^2 + v_{y0}^2}{\omega^2}\right) + \frac{1}{2}\left(+x_0^2 - y_0^2 + \frac{+v_{x0}^2 - v_{y0}^2}{\omega^2}\right)\cos(2\theta)$$
$$+ \left(x_0 y_0 + \frac{v_{x0}v_{y0}}{\omega^2}\right)\sin(2\theta)$$

$$b^2 = \frac{1}{2}\left(+x_0^2 + y_0^2 + \frac{+v_{x0}^2 + v_{y0}^2}{\omega^2}\right) + \frac{1}{2}\left(-x_0^2 + y_0^2 + \frac{-v_{x0}^2 + v_{y0}^2}{\omega^2}\right)\cos(2\theta)$$
$$- \left(x_0 y_0 + \frac{v_{x0}v_{y0}}{\omega^2}\right)\sin(2\theta)\,.$$

Once θ, a and b are known, the parametric equations of trajectory in the new coordinates take the well-known form:

$$\begin{cases} x' = a\cos(\omega t + \varphi) \\ y' = b\sin(\omega t + \varphi) \end{cases} \tag{1.11}$$

that is an ellipse with semi-axes a and b. The semi-axis a is rotated by an angle θ with respect to the x-axis and may not be the major semi-axis. The particle moves along this ellipse, starting at point of phase φ with respect to x'-axis at time 0. Three examples of the particle in motion are illustrated in Fig. 1.1: the particle starts, from the same position, with three different initial velocities. The geometrical parameters of the resulting elliptic trajectories are reported in Table 1.1.

Exercise 1.1 Modify the force (1.3) adding the contribution of the gravitational field along the y-axis:

$$F(x) = \begin{pmatrix} -kx \\ -ky - mg \end{pmatrix} \tag{1.12}$$

with given g, and determine the nature of the new trajectory curve.

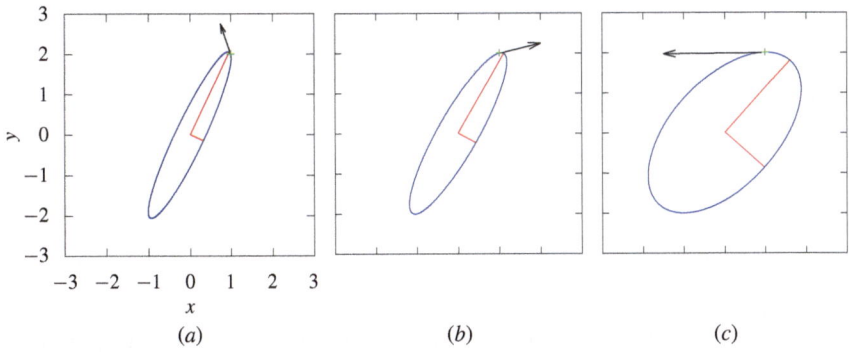

Fig. 1.1 Ellipses supporting the trajectories of a particle starting at $x_0 = (1, 2)$ with three different initial velocities (see Table 1.1). The velocity scale is $1/4$

Table 1.1 Geometric parameters of the ellipse supporting the trajectory of a particle starting at point $x_0 = (1, 2)$ with different initial velocities (see Fig. 1.1)

v_0	θ	a	b	Figure
$(-1, 3)$	$-25°6'54''$	0.35135	2.26492	1.1a
$(4, 1)$	$-29°1'47''$	0.48896	2.27849	1.1b
$(-10, -0.1)$	$-41°42'37''$	1.31449	2.40944	1.1c

A second interesting example is that of the trajectory of a particle subjected to an elastic force including a "*small*" nonlinear contribution as, for example:

$$F(x) = \begin{pmatrix} -kx + \varepsilon y^2 \\ -ky \end{pmatrix}, \tag{1.13}$$

where ε is a *small* dimensional constant. Keeping the same initial position and speed, how is the trajectory of P changed? Does it remain a closed curve, *i.e.* is the motion still periodic? The answer to these questions comes from the integration of the equation of motion (1.1) with the force (1.13):

$$\begin{cases} x(t) = A_1 \cos \omega t + B_1 \sin \omega t + C_1 \cos 2\omega t + D_1 \sin 2\omega t + E_1 \\ y(t) = A_2 \cos \omega t + B_2 \sin \omega t , \end{cases}$$

where the constants are given by the formulae:

$$A_1 = x_0 - \frac{\varepsilon}{3k} y_0^2 - \frac{2\varepsilon}{3k} \frac{v_{y0}^2}{\omega^2}, \quad B_1 = \frac{2\varepsilon}{3k} y_0 \frac{v_{y0}}{\omega} + \frac{v_{x0}}{\omega}, \quad C_1 = -\frac{\varepsilon}{6k} \left(y_0^2 - \frac{v_{y0}^2}{\omega^2} \right)$$

$$D_1 = -\frac{\varepsilon}{3k} y_0 \frac{v_{y0}}{\omega}, \quad E_1 = \frac{\varepsilon}{2k} \left(y_0^2 + \frac{v_{y0}^2}{\omega^2} \right), \quad A_2 = y_0, \quad B_2 = \frac{v_{y0}}{\omega}.$$

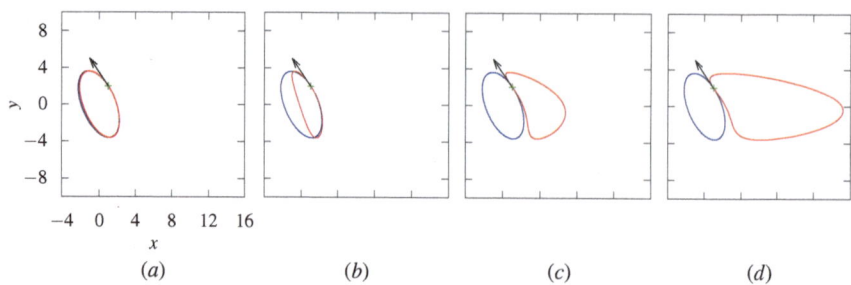

Fig. 1.2 With a red line the trajectories are drawn for a particle starting at point $(1, 2)$ with velocity $(-2, 3)$ and subjected to the force (1.13) with $\varepsilon/k = 0.01$ **a**, 0.1 **b**, 0.5 **c** and 1 **d**. The trajectory obtained with $\varepsilon = 0$ is drawn with a blue line. The corresponding time interval is a period

The resulting motion is still periodic with period $T := 2\pi/\omega$. In Fig. 1.2 the trajectories determined by different values of di ε/k are shown.

From a survey of the solutions obtained with increasing ε, we can argue that the solution exhibits a continuous dependence on the parameter ε and, however large ε is, the particle undergoes a periodic motion with period T. The same conclusions can be derived very easily also from a survey of the above solution.

1.2   Parametrization of a Curve

In all the examples we have proposed so far, the position of the particle along the trajectory was assigned based on time t elapsed from the starting instant. In these cases, the time therefore plays the role of a *parameter* that uniquely identifies the position along the curve. Obviously, the choice of the parameter is not unique: the same curve can be described through a different parameter. Indeed, there are infinitely many possible parameters.

Consider for example an ellipse with semi-axes a and b places along x and y and, for the sake of simplicity, center at the origin. We have seen previously that one way to describe this curve is to adopt the angular parameter $\theta \in [0, 2\pi)$ and write:

$$x = x(\theta) = \begin{pmatrix} a\cos\theta \\ b\sin\theta \end{pmatrix},\qquad(1.14)$$

but any other parameter could be used to describe the same curve. In fact, we consider any positive real number p and the new parameter $\alpha(\theta) = \theta^p$, varying in the range $[0, (2\pi)^p)$ of the real axis. Note that $\alpha(\theta)$ is a *monotonically* increasing function. Reversing the previous relation, we can calculate the angle θ as a function of the new parameter α as $\theta(\alpha) = \alpha^{1/p}$ thus obtaining the new parametrization of the same curve:

$$x = x(\alpha) = \begin{pmatrix} a \cos \alpha^{1/p} \\ b \sin \alpha^{1/p} \end{pmatrix}. \tag{1.15}$$

The relation (1.15) shows that there are infinitely many feasible parameters. In fact, the number p can be chosen arbitrarily. We observe that in both the parameterizations (1.14), (1.15) the curve is traveled counterclockwise for increasing values of θ or α. This is a consequence of the monotonically increasing trend of the angle θ as a function of the new parameter α. On the other hand, one can easily change the path direction choosing a new family of parameters, for example defined by $\beta(\theta) = 1/(1 + \theta^p)$. The new parameter β spans the interval $(1/[1 + (2\pi)^p], 1]$. As θ increases, the new parameter $\beta(\theta)$ decreases: β takes its maximum value 1 at $\theta = 0$ and tends to its minimum value $1/[1 + (2\pi)^p]$ when θ grows to 2π. The function $\beta(\theta)$ is then monotonically *decreasing*: as β is reduced, the corresponding angle θ decreases so that the curve is traveled clockwise (negative direction). Therefore, it is possible to invert the above relation between θ and β and obtain $\theta = (1/\beta - 1)^{1/p}$. As β grows, θ reduces starting from $\theta = 2\pi$ at $\beta = 1/(1 + (2\pi)^p)$ and ending in $\theta = 0$ at $\beta = 1$. In the relation between two parameters of a same curve, the type of trend (increasing or decreasing) determines the direction of path travel, but what is really crucial it seems to be the *monotonic* nature of this function. Why?

To answer this question, let us think of the meaning of a parameter in a general terms. We said at the beginning of the lesson that a curve $\mathscr{C} \subset \mathbb{R}^n$ is obtained by a bijective map of a segment $[a, b)$ of the real axis (From now on this interval will also be indicated by $\mathscr{E}$) to a subset of the n-dimensional space $\mathbb{R}^n$. The variable σ, whose values varying in the interval $[a, b)$ describe the path on the curve, is said to be a *parameter*. So the curve $\mathscr{C}$ is drawn by the map:

$$\begin{aligned} x : [a, b) &\longrightarrow \mathscr{C} \\ \sigma &\longmapsto x(\sigma) \in \mathbb{R}^n \end{aligned} \tag{1.16}$$

that must be *bijective*, which means *injective* ($\forall \sigma_{1,2} \in [a, b) : x(\sigma_1) = x(\sigma_2) \Rightarrow \sigma_1 = \sigma_2$) and *surjective* ($\forall y \in \mathscr{C} : \exists \sigma \in [a, b) \mid x(\sigma) = y$). The curve is said to be closed if $x(a) = x(b)$. Notice that since the map (1.16) is bijective, it can be reversed, that is: given any point y on $\mathscr{C}$, we can find σ such that $x(\sigma) = y$. Consider now a function $\sigma = f(\chi)$, with $\chi \in [c, d)$, in general *non monotonic*: as a sample, given a point $e \in (c, d)$, suppose f is increasing in $[c, e)$, then decreasing in (e, d). Thus, there is a neighborhood of e where two *distinct* values $\chi_{1,2}$ of the parameter χ give rise to a unique value of the original parameter $\sigma = f(\chi_{1,2})$. Therefore, using $f(\chi)$ in place of σ, $x[f(\chi)]$ would associate the same point on the curve to different values of χ ($\chi_{1,2}$), which would cause the new composite map $x[f(\chi)]$ to be no longer injective. This explains the requirement for the function f to be monotonic, in order to perform a change of parameter. Furthermore, if the function f has a derivative f', the above requirement implies that $f'(\chi) \neq 0$ for all $\chi \in [c, d)$.

Finally, we point out that in defining certain curves one gives up, at a finite number of points, the requirement for a bijective parametric representation. The most frequent case is a curve with one or more *double points*. A double point is an intersection of the curve with itself, or a point whose inverse image set is constituted by two distinct values of the parameter.

1.3 Limit on a Curve

Now we are able to make use of some basic tools of Mathematical Analysis to investigate the local properties of the new entity we have just defined. We recall that the domain of a vector function is the intersection of the domains of the individual component functions. In the following, we assume that the parameter σ belongs to the domain of the vector function concerned.

To begin with, we define the concept of *limit* for the map (1.16) whose domain will be denoted by $[a, b)$. We say that x_0 is the limit of $x(\sigma)$ for $\sigma \to \sigma_0 \in [a, b]$ (with σ_0 accumulation point of $\mathscr{E}$, not necessarily belonging to this set), and we write:

$$x_0 = \lim_{\sigma \to \sigma_0} x(\sigma)$$

if given any neighborhood $U(x_0)$ (arbitrarily small) of x_0 we can find a neighborhood $I(\sigma_0) \subset [a, b)$ di σ_0 such that every value of the parameter σ in I correspond to a point $x(\sigma)$ in U. We start observing that U is an open subset of $\mathbb{R}^n$ that contains x_0, whose shape is unimportant. We can think of it as a small sphere centered at x_0 with radius ε conveniently small, or as cube centered at x_0 with sufficiently small side, and so on. Furthermore, I is an open subset containing σ_0 and contained in $\mathscr{E}$: we can imagine it, without loss of generality, as the open interval $(\sigma_0 - p, \sigma_0 + q)$ (with $p, q > 0$ suitably small) of the real axis.

In order to simplify the calculations, one often prefers to think of U as a small ball and take $p = q$ (symmetrical neighborhood), which translates the above statement in the following symbolic form:

$$x_0 = \lim_{\sigma \to \sigma_0} x(\sigma),$$

equivalent to

$$\forall \varepsilon > 0 \, \exists \delta_\varepsilon > 0 \mid \forall \sigma \in (\sigma_0 - \delta_\varepsilon, \sigma_0 + \delta_\varepsilon) \cap [a, b) \Rightarrow |x(\sigma) - x_0| < \varepsilon. \tag{1.17}$$

Examples in the following illustrate the concepts exposed in the present section.

Problem 1.2 Let us prove that:

$$\lim_{\sigma \to 0} \begin{pmatrix} \sigma^2 \\ 1 - \sigma \\ \sigma^3 \end{pmatrix} = \begin{pmatrix} 0 \\ 1 \\ 0 \end{pmatrix}.$$

Solution 1.2 Taken an arbitrarily small positive number ε, we show that one can find a corresponding positive number δ_ε such that, for any $\sigma \in (0 - \delta_\varepsilon, 0 + \delta_\varepsilon)$, the following inequality holds:

$$|x(\sigma) - x_0| = \left\{ (\sigma^2 - 0)^2 + [(1 - \sigma) - 1]^2 + (\sigma^3 - 0)^2 \right\}^{1/2} = (\sigma^4 + \sigma^2 + \sigma^6)^{1/2} < \varepsilon.$$

Assume $\delta_\varepsilon < 1$ (we should only show that such a δ_ε exists) and calculate it enforcing a stronger inequality:

$$(\sigma^4 + \sigma^2 + \sigma^6)^{1/2} < (\delta_\varepsilon^4 + \delta_\varepsilon^2 + \delta_\varepsilon^6)^{1/2} < \sqrt{3}\delta_\varepsilon < \varepsilon,$$

in fact, since $\delta_\varepsilon < 1$ we get $\delta_\varepsilon^4 = \delta_\varepsilon^2 \delta_\varepsilon^2 < \delta_\varepsilon^2$ and further $\delta_\varepsilon^6 = \delta_\varepsilon^2 \delta_\varepsilon^4 < \delta_\varepsilon^2$. It follows that simply taking $\delta_\varepsilon = \varepsilon/2$ fulfills the requirement (1.17).

Problem 1.3 Let us prove that:

$$\lim_{\sigma \to 0} \begin{pmatrix} (\sin \sigma)/\sigma \\ (1 - \cos \sigma)/\sigma^2 \\ \log(1 + \sigma) \end{pmatrix} = \begin{pmatrix} 1 \\ 1/2 \\ 0 \end{pmatrix}.$$

Solution 1.3 In fact, we choose as usual a positive number ε, arbitrarily small, and try to find δ_ε requiring that for every $\sigma \in (0 - \delta_\varepsilon, 0 + \delta_\varepsilon)$ the following inequality be satisfied:

$$\left\{ \left(\frac{\sin \sigma}{\sigma} - 1 \right)^2 + \left(\frac{1 - \cos \sigma}{\sigma^2} - \frac{1}{2} \right)^2 + \left[\log(1 + \sigma) - 0 \right]^2 \right\}^{1/2} < \varepsilon. \quad (1.18)$$

We observe that it must be $\delta_\varepsilon < 1$, since the third component of the vector $x(\sigma)$ does not exist when $\sigma \leq -1$. Then, we can then try $\delta_\varepsilon < 1/2$, for example. We begin to deal with the three addenda under the root separately. As for the first, we observe that $(\sin \sigma)/\sigma \geq 1 - \sigma/\pi$, when $\sigma \in [0, \pi)$. Similarly, $(\sin \sigma)/\sigma \geq 1 + x/\pi$ for $\sigma \in (-\pi, 0]$. It follows that for every $\sigma \in (-\pi, +\pi)$ the following inequality holds:

$$1 - \frac{\sin \sigma}{\sigma} < \frac{|\sigma|}{\pi} =: c_1 |\sigma|.$$

In the same interval, also the following inequality applies to the second term:

$$\frac{1}{2} - \frac{1 - \cos \sigma}{\sigma^2} < \frac{1}{\pi} \left(\frac{1}{2} - \frac{2}{\pi^2} \right) |\sigma| =: c_2 |\sigma|.$$

Observe that $\log(1 + \sigma) \le \sigma$ for $\sigma > 0$, whereas $\log(1 + \sigma) \ge (2 \log 2)\sigma$ for $\sigma \in (-1/2, 0)$.

Also for the third addendum the following kind of relation holds:

$$|\log(1 + \sigma)| < c_3|\sigma|,$$

with $c_3 = 2 \log 2$ (in fact, $2 \log 2 > 1$). Using the above three estimates in the inequality (1.18) we try to satisfy the following stronger inequality:

$$C|\sigma| < C\delta_\varepsilon < \varepsilon,$$

where $C = \sqrt{c_1^2 + c_2^2 + c_3^2}$. It follows that simply put $\delta_\varepsilon = \varepsilon/(C + 1)$ will make the requirement (1.17) satisfied.

Problem 1.4 Finally, let us prove that:

$$\lim_{\sigma \to \pi} \begin{pmatrix} \sin(\sigma/2) \\ \cos \sigma \end{pmatrix} = \begin{pmatrix} 1 \\ -1 \end{pmatrix}.$$

Solution 1.4 We choose a positive number ε, arbitrarily small, let us say less than unity. then try to find, correspondingly, a number δ_ε such that for every $\sigma \in (\pi - \delta_\varepsilon, \pi + \delta_\varepsilon)$ the following inequality holds:

$$\{[\sin(\sigma/2) - 1]^2 + (\cos \sigma + 1)^2\}^{1/2} < \varepsilon. \tag{1.19}$$

We deal with the two addenda separately. First of all, we can definitely assume $\delta_\varepsilon < \pi/2$ and then the following inequality holds:

$$1 - \sin\left(\frac{\sigma}{2}\right) < \frac{|\sigma - \pi|}{\pi},$$

and, under the same conditions, a second inequality holds:

$$1 + \cos \sigma < \frac{2}{\pi}|\sigma - \pi|.$$

It follows from the inequality (1.19):

$$\{[\sin(\sigma/2) - 1]^2 + (\cos \sigma + 1)^2\}^{1/2} < \frac{\sqrt{5}}{\pi}|\sigma - \pi| < \frac{\sqrt{5}}{\pi}\delta_\varepsilon < \varepsilon,$$

so just choose $\delta_\varepsilon = \pi\varepsilon/3$ to make inequality (1.19) satisfied.

Exercise 1.2 Prove the following limits:

$$(a)\ \lim_{\sigma \to 1} \begin{pmatrix} \sigma^4 \\ \sigma - 1 \\ \sigma^2 + \sigma^3 \end{pmatrix} = \begin{pmatrix} 1 \\ 0 \\ 2 \end{pmatrix}, \qquad (b)\ \lim_{\sigma \to 0} \begin{pmatrix} \cosh \sigma \\ \sinh^2 \sigma \\ 1 \end{pmatrix} = \begin{pmatrix} 1 \\ 0 \\ 1 \end{pmatrix},$$

$$(c) \lim_{\sigma \to 2} \left(\frac{e^\sigma}{\log \sigma} \right) = \left(\frac{e^2}{\log 2} \right).$$

In a similar way to what it was done for the scalar functions of a single variable, one can extend to a vector function x of a single variable σ the notion of a divergent limit at a point σ_0. We take a number $M > 0$ arbitrarily large. If we are able to find a corresponding positive number δ_M such that for every $\sigma \in (\sigma_0 - \delta_M, \sigma_0 + \delta_M) \cap \mathscr{E}$ the absolute value $|x(\sigma)|$ is always larger than M, then we say x is divergent at σ_0. Using symbols, the above statement can be formalised as follows:

$$\lim_{\sigma \to \sigma_0} x(\sigma) = \infty,$$

equivalent to

$$\forall M > 0 \exists \delta_m > 0 \mid \forall \sigma \in (\sigma_0 - \delta_M, \sigma_0 + \delta_M) \cap \mathscr{E} \Rightarrow |x(\sigma)| > M. \quad (1.20)$$

We show, as usual, some examples.

Problem 1.5 Verify the following limit:

$$\lim_{\sigma \to 1} \begin{pmatrix} 1/(\sigma^2 - 1) \\ \sigma \\ 1/\sigma^2 \end{pmatrix} = \infty.$$

Solution 1.5 In this case only the first component of x diverges as $\sigma \to 1$. We take an arbitrarily large number M, and try to find $\delta_M > 0$ such that for every $\sigma \in (1 - \delta_M, 1 + \delta_M)$, except σ_0, the inequality

$$\left[\frac{1}{(\sigma^2 - 1)^2} + \sigma^2 + \frac{1}{\sigma^4} \right]^{1/2} > M.$$

holds. Let us take $\delta_M < 1$, then we can require a stronger a stronger inequality to hold:

$$\left[\frac{1}{(\sigma^2 - 1)^2} + \sigma^2 + \frac{1}{\sigma^4} \right]^{1/2} > \frac{1}{3\delta_M} > M,$$

which is verified if we choose $\delta_M = 1/(4M)$.

Problem 1.6 Verify the following limit:

$$\lim_{\sigma \to 0} \begin{pmatrix} \log \sigma \\ 1/\sigma \\ 1/\sin \sigma \end{pmatrix} = \infty.$$

Solution 1.6 In this case all the components of the vector are singular at the point $\sigma_0 = 0$. We take a number $M > 0$ arbitrarily large and try to find a positive number δ_M that we assume less than π such that the inequality:

$$\left(\log^2 \sigma + \frac{1}{\sigma^2} + \frac{1}{\sin^2 \sigma} \right)^{1/2} > M$$

holds for every $\sigma \in (0, \delta_M)$ (notice that this interval is nor symmetric due to the fact that $\mathcal{E} = \{\eta > 0, \, \eta \neq \pi, 2\pi, \ldots\}$). As usual, we require a stronger inequality to hold in order to find δ_M:

$$\left(\log^2 \sigma + \frac{1}{\sigma^2} + \frac{1}{\sin^2 \sigma} \right)^{1/2} > \frac{1}{\delta_M} > M \,,$$

and this is verified when $\delta_M = \min[1/(2M), \pi/2]$ is chosen.

Exercise 1.3 Prove the following limits:

$$(a) \; \lim_{\sigma \to 1} \begin{pmatrix} \sigma \\ 1/(\sigma - 1) \\ 1/(\sigma - 1)^2 \end{pmatrix} = \infty \,, \qquad (b) \; \lim_{\sigma \to 0} \begin{pmatrix} \log \sigma \\ e^\sigma \\ (\sigma + 1)/(\sigma - 1) \end{pmatrix} = \infty \,,$$

$$(c) \; \lim_{\sigma \to -1} \begin{pmatrix} \sigma \\ 1/(\sigma + 1) \end{pmatrix} = \infty \,.$$

Finally, for certain curves the range of the parameter is unbounded. In these cases it makes sense trying to give a meaning to the limit as the parameter tends to infinity, $\lim_{\sigma \to \infty} x = x_0$: if and only if for every (arbitrarily small) neighbourhood U of x_0 is possible to find a neighbourhood of infinity, e.g. an interval $(M, +\infty)$, with M a large, positive number, such that all the vectors $x(\sigma)$, with σ belonging to this neighbourhood, lie in U. This can be written formally as follows:

$$x_0 = \lim_{\sigma \to \infty} x(\sigma) \,,$$

equivalent to

$$\forall \varepsilon > 0 \, \exists M > 0 \, | \, \forall \sigma \in (M, +\infty) \cap \mathcal{E} \Rightarrow |x(\sigma) - x_0| < \varepsilon \,.$$

For example, consider, for $a > 0$, the curve:

$$\boldsymbol{x}(\theta) = \begin{pmatrix} (\cos\theta)/(\theta + a) \\ (\sin\theta/(\theta + a)) \end{pmatrix} \tag{1.21}$$

with $\theta \in [0, +\infty)$ plotted in Fig. 1.2a. It is quite obvious to show that for increasing of θ this curve approaches more and more the origin. And in fact we show that:

$$\lim_{\theta \to +\infty} \boldsymbol{x}(\theta) = 0.$$

We choose a number $\varepsilon > 0$ arbitrarily small and show that we can determine a corresponding positive number M_ε such that for every $\theta > M_\varepsilon$ we get $|\boldsymbol{x}(\theta)| < \varepsilon$. Verification of this inequality is immediate, once considered that $|\boldsymbol{x}(\theta)| = 1/(\theta + a)$ and $M_\varepsilon = 1/\varepsilon$ is chosen.

Exercise 1.4 Prove the following limits:

$$(a) \ \lim_{\theta \to +\infty} \begin{pmatrix} (\cos\theta)/\theta^{1/2} \\ (\sin\theta)/\theta^{1/2} \end{pmatrix} = 0, \qquad (a) \ \lim_{\theta \to 0^+} \begin{pmatrix} (\cos\theta)/\theta^{1/2} \\ (\sin\theta)/\theta^{1/2} \end{pmatrix} = \infty,$$

$$\lim_{\theta \to +\infty} \begin{pmatrix} (\cos\theta)/\theta^{1/4} \\ (\sin\theta)/\theta^{1/4} \end{pmatrix} = 0.$$

The curves studied in the above exercises are depicted in Fig. 1.3b, c.

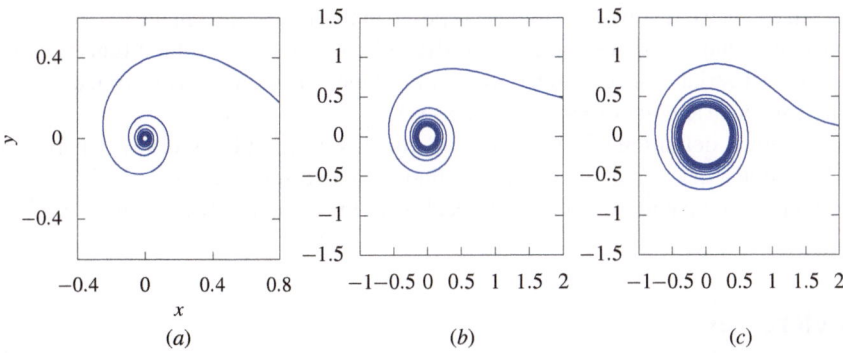

Fig. 1.3 Three different spirals are plotted in the pictures: in **a** the curve (1.21), for $a = 1$; in **b** the curve $x(\theta) = \cos\theta/\theta^{1/2}$, $y(\theta) = \sin\theta/\theta^{1/2}$; in **c** the curve $x(\theta) = \cos\theta/\theta^{1/4}$, $y(\theta) = \sin\theta/\theta^{1/4}$. In all plots, $\theta \in (0, 20\pi)$

1.4   Catalogs of Sample Curves

In order to become familiar with some curves of historical importance, it is interesting to browse some catalogs of curves, many of them available online. Three particularly appealing catalogs of *plane curves* can be found at the following addresses:

https://mathshistory.st-andrews.ac.uk/Curves/

http://xahlee.info/SpecialPlaneCurves_dir/specialPlaneCurves.html

http://progettomatematica.dm.unibo.it/Curve%20celebri/home.html

The first is edited by the university of St. Andrews, England, UK. In the second many historical notes may be found. In the third, edited by the Department of Mathematics of the University of Bologna, Italy, includes many historical notes, as well as the useful details on geometric properties of every curves. Finally, a beautiful catalog of three-dimensional curves is available at the website:

http://www.mathcurve.com/courbes3d/courbes3dit.shtml

where a number of plane curves are also cataloged.

1.5   Suggested Readings

A formal definition of a curve, and of its parametrization and length, is given in Lax and Terrell (2017, Chap. 2, p. 83), whereas the concept of a *smooth curve* (and the related *smooth parametrisation*) is introduced at the beginning of the section where line integrals are introduced (7, p. 279).

An extensive treatment of topics proposed in this lesson (and in the two following ones) is found in Adams and Essex (2010, 11), where vector valued functions of a single real variable are examined from both a kinematic point of view (pp. 621–627) and a geometric point of view (pp. 635–641).

A formal definition of a parametrised curve is given in Lang (1987, Chap. II, p. 51), within a discussion on the differentiation of a vector function.

Curves and paths are defined and dealt with in Apostol (1974, Chap. 6, p. 133).

References

R. A. Adams and C. Essex. *Calculus: Several Variables*. Pearson Education Canada, Toronto, ON, 7th edition, 2010.

T. M. Apostol. *Mathematical Analysis: A Modern Approach to Advanced Calculus*. Pearson Education US, Hoboken, NJ, 2nd edition, 1974.

S. Lang. *Calculus of Several Variables*. Springer, New York, 3rd edition, 1987.

P. D. Lax and M. S. Terrell. *Multivariable Calculus with Applications*. Springer, Cham, CH, 2017.

Chapter 2
Differential and Geometric Properties of Curves

The present lesson investigates the basic differential properties of the curves and defines their principal geometric quantities. The concept of the *limit on a curve, i.e.* its behaviour in small neighbourhoods of assigned points, viewed as a function from $\mathbb{R}$ (the space of the parameter) to $\mathbb{R}^n$ (the space where the curve exists), has been introduced in the previous lesson. It will be now used for building the fundamental tools of the *differential calculus on a curve*.

2.1 Differential and Tangent Vector

For the sake of clarity, consider a *simple curve* $\mathscr{C}$:

$$\begin{aligned} \boldsymbol{x} : [a, b) &\longrightarrow \mathbb{R}^n \\ \sigma &\longmapsto \boldsymbol{x}(\sigma) \end{aligned} \tag{2.1}$$

i.e. the *one-to-one* map between an interval $[a, b)$[1] on the real axis (it can also be unbounded, so that a can be $-\infty$, as well as b can be $+\infty$) and a subset of $\mathbb{R}^n$. Moreover, the curve (2.1) does not have multiple points, *i.e.* it does not intersect itself. In the previous lesson, the concept of curve has been introduced in terms of the vector-valued function giving the positions assumed by a material particle at different times. In the definition (2.1) the time is generalised in a generic parameter σ and the position in the point of $\mathbb{R}^n$.

An observer $O^{(1)}$ moves from the point $\sigma = a$ to the point $\sigma = b$ along the interval $[a, b) \subseteq \mathbb{R}$ in the parameter space. In correspondence to $O^{(1)}$, a second observer $O^{(n)}$ moves along the curve, occupying the point $\boldsymbol{x}(\sigma)$ when $O^{(1)}$ lies in σ. If $O^{(1)}$ moves

---

[1] The choice of a semi-open interval on the real axis prevents the loss of bijectivity in the case of a closed curve.

© The Author(s), under exclusive license to Springer Nature Switzerland AG 2024
G. Riccardi et al., *Multidimensional Differential and Integral Calculus*,
https://doi.org/10.1007/978-3-031-70326-3_2

smoothly from $\sigma = a$ to $\sigma = b$, does the same occur also to $O^{(n)}$? In order to answer to this question, the behaviour of the curve (2.1) about a generic point $x(\sigma_0) =: x_0$ (with $a \leq \sigma_0 < b$) has to be better investigated.

First of all, an *induced topology* on the curve is introduced defining an open set (on the curve) in the following way:

> An open set on the curve $\mathscr{C}$ is given by the intersection between $\mathscr{C}$ and an open set in $\mathbb{R}^n$.

In general, the open set $A \subseteq \mathscr{C}$ is formed by disjointed arcs of the curve. A *neighbourhood* of a point $x_0 \in \mathscr{C}$ on the same curve is an open set containing x_0, but with the restriction that it is formed by only one arc. A simple family of neighbourhoods of the point x_0 is defined intersecting $\mathscr{C}$ with balls centered in that point:

> The neighbourhood $A_\varepsilon(x_0; \mathscr{C})$ of x_0 on $\mathscr{C}$ is the subset of the intersection between the open ball $B_\varepsilon(x_0)$ (having radius ε and center in x_0) and the curve $\mathscr{C}$ that contains x_0.

Hereafter, $A_\varepsilon(x_0; \mathscr{C})$ will be shortened in A_ε, if the point (x_0) and the curve $(\mathscr{C})$ are identified without ambiguity.

In these conditions, a local question appears to be preliminary to the previous one: if the observer $O^{(1)}$ moves near σ_0, does the observer $O^{(n)}$ remain inside the fixed neighbourhood A_ε? In other words, does a neighbourhood $I(\sigma_0)$ of σ_0 in $[a, b)$ exist such that its image through the map (2.1) lies inside A_ε? If it occurs, $O^{(n)}$ remains close to x_0 for σ moving about σ_0:

> If at a point σ_0 of the interval $[a, b)$ it occurs that:
>
> $$\lim_{\sigma \to \sigma_0} x(\sigma) = x(\sigma_0) \qquad (2.2)$$
>
> the curve $\mathscr{C}$ is said to be continuous at its point $x(\sigma_0)$. If the property (2.2) is not verified, the curve is said to be discontinuous in correspondence to the value σ_0 of the parameter.

Obviously, a curve that is continuous in all its points is said to be a *continuous curve*. For example, the curves investigated in lesson 1 are continuous curves. Samples of curves that are not continuous in a point can be easily built considering a function $y = f(x)$ from $\mathbb{R}$ to $\mathbb{R}$, and its usual graph in the cartesian plane (x, y) that is here reread as the curve:

$$x(x) = \begin{pmatrix} x \\ f(x) \end{pmatrix} \tag{2.3}$$

obtained using the abscissa x in place of σ (2.1) as parameter, having in x_0 a finite jump (the limits in x_0 from the left $f(x_0^-)$ and from the right $f(x_0^+)$ exist finite, but they are different) or a vertical asymptote (one or both the previous limits diverge). In these cases, the curve (2.3) is not continuous in correspondence to the value x_0 of the parameter. The values of the parameter where the curve is not continuous can also accumulate in neighbourhoods of finite points: a classical example of a curve that is discontinuous in any value of the parameter is $\mathscr{C} = \{(x, y)$ for $0 \le x \le 1$ with $y = 1$ if x is rational and $y = 0$ otherwise$\}$.

One of the most important consequences of the continuity[2] is that the preimage S of an open set A on the curve $\mathscr{C}$ is open in the parameter domain $[a, b]$, *i.e.* it is the intersection of an open subset of the real axis and the interval $[a, b]$. Indeed, consider an arbitrary $\sigma \in S$, that is mapped in $x(\sigma) \in A$ by the transformation (2.1). Due to the fact that A is open and to the continuity of the map, neighbourhoods B $(\subset A)$ and T of $x(\sigma)$ and σ exist, such that $x(T) \subset B$, *i.e.* the image by means of the map (2.1) of any point $\sigma \in T$ lying inside $B \subset A$. Due to the fact that T is an open subset of S, it has been shown that any point of S possesses a neighbourhood contained in S or, in other words, S is open.

The continuity of the curve $\mathscr{C}$ at its point x_0 assures that for $O^{(1)}$ moving smoothly about σ_0, the point $x(\sigma)$ remains close to $x(\sigma_0) = x_0$. But nothing is known about the smoothness of the motion of $O^{(n)}$. For example, does the motion of $O^{(n)}$ experience abrupt change of direction in correspondence to a certain value of the parameter? In order to gain more information, the concept of *differential* is needed.

The idea is as follows: change the value of the parameter from σ_0 to $\sigma = \sigma_0 + \Delta\sigma \in I(\sigma_0)$, small neighbourhood of σ_0 and observe what happens to the displacement vector $x(\sigma) - x_0$.

If a finite-length vector a exists, such that for any $\sigma \in I(\sigma_0)$ the following relation:

$$x(\sigma) = x_0 + a\,\Delta\sigma + o(\Delta\sigma), \tag{2.4}$$

is valid, with $o(\Delta\sigma)/\Delta\sigma \to 0$ as $\Delta\sigma \to 0$, the curve $\mathscr{C}$ is said to be differentiable (or linearizable) at its point x_0. The vector $a\,\Delta\sigma$ is the differential of the curve in $x(\sigma_0)$ and is indicated as $dx(x_0; \Delta\sigma)$.

The definition (2.4) states that the differential gives the linear (in $\Delta\sigma$) part of the displacement $x(\sigma) - x_0$, and that is also the larger contribution to this difference.

Another important vector associated to the local behaviour of the curve is the *tangent vector*. Its definition and meaning are easily understood introducing the vector incremental ratio. The observer $O^{(n)}$ walking in the neighbourhood A_ε of

---

[2] This property can be generalised to any continuous map.

the point x_0 of a differentiable curve has the feeling of moving along the straight line passing on x_0 and directed along a. This line is the *tangent line* to the curve $\mathscr{C}$ at its point x_0. Indeed, the tangent direction at the point x_0 is obtained evaluating the vector incremental ratio $[x(\sigma_0 + \Delta\sigma) - x(\sigma_0)]/\Delta\sigma$ as $\Delta\sigma \to 0$:

$$t(\sigma_0) = \lim_{\Delta\sigma \to 0} \begin{pmatrix} [x_1(\sigma_0 + \Delta\sigma) - x_1(\sigma_0)]/\Delta\sigma \\ [x_2(\sigma_0 + \Delta\sigma) - x_2(\sigma_0)]/\Delta\sigma \\ \vdots \\ [x_n(\sigma_0 + \Delta\sigma) - x_n(\sigma_0)]/\Delta\sigma \end{pmatrix} = \begin{pmatrix} x_1'(\sigma_0) \\ x_2'(\sigma_0) \\ \vdots \\ x_n'(\sigma_0) \end{pmatrix}, \tag{2.5}$$

that is the derivative of the vector function $x(\sigma)$ with respect to the parameter σ. The vector (2.5) is *tangent* to the curve $\mathscr{C}$ in its point x_0. Moreover, if $x'(\sigma_0) \neq 0$, the tangent unit vector

$$\tau(\sigma_0) = \frac{x'(\sigma_0)}{|x'(\sigma_0)|} \tag{2.6}$$

can be defined. It has the same direction of $t(\sigma_0)$, but unitary modulus. The direction of t is independent of the choice of the parameter on $\mathscr{C}$, whereas its modulus depends on that choice, as it is now verified by a direct calculation. To this aim, the parameter σ is changed in the new one $\tilde{\sigma}$, which is a smooth and monotonic increasing function of the old parameter σ and verifies the condition $\tilde{x}[\tilde{\sigma}(\sigma)] = x(\sigma)$. As a consequence, the function $\tilde{\sigma} = \tilde{\sigma}(\sigma)$ can be derived and its derivative $\tilde{\sigma}'$ is strictly positive. Using the chain rule, the relation between the tangent vector (2.5) and the one corresponding to the new parameter $(\tilde{t})$ is written as:

$$t = \frac{d\tilde{x}}{d\tilde{\sigma}} \frac{d\tilde{\sigma}}{d\sigma} = \tilde{t}\,\tilde{\sigma}'. \tag{2.7}$$

The relation (2.7) shows that the direction of $\tilde{t}$ is the same of t, but its modulus is multiplied by $\tilde{\sigma}'$ and then, in general, different from the one of t. We calculate now this vector for celebrated curves (that are among the ones cited in Sect. 1.4).

Example 2.1 *(Cardioid)* The tangent unit vector on the plane $(n = 2)$ curve *cardioid* is calculated. Using the phase $\theta \in [0, 2\pi)$ and the familiar components x, y in place of x_1, x_2, the curve represented in polar coordinates $x(\theta) = \rho(\theta)\cos\theta$, $y(\theta) = \rho(\theta)\sin\theta$ is:

$$\rho(\theta) = a(1 + \cos\theta), \tag{2.8}$$

where a is a positive number. The derivative of the position (2.5) is specified for the curve (2.8) as:

$$x'(\theta) = a \begin{pmatrix} -\sin\theta - \sin 2\theta \\ \cos\theta + \cos 2\theta \end{pmatrix},$$

and its length is $2a|\cos(\theta/2)|$. It follows that in correspondence to the value $\theta = \pi$ of the parameter ($x = 0$) the vector x' vanishes and the tangent vector cannot be defined. The tangent unit vector τ (2.6) for $\theta \neq \pi$ is:

$$\tau(\theta) = 4\,\text{sign}[\cos(\theta/2)] \begin{pmatrix} -\sin(\theta/2)\,[\cos(\theta/2) - 1/2]\,[\cos(\theta/2) + 1/2] \\ \cos(\theta/2)\,[\cos(\theta/2) - \sqrt{3}/2]\,[\cos(\theta/2) + \sqrt{3}/2] \end{pmatrix},$$

which shows that $\tau_x(\theta)$ is negative for $\theta \in (0, 2\pi/3) \cup (\pi, 4\pi/3)$ and vanishes in $\theta = 0, 2\pi/3$ and $4\pi/3$. The component $\tau_y(\theta)$ is negative for $\theta \in (\pi/3, \pi) \cup (\pi, 5\pi/3)$, whereas it vanishes in $\theta = \pi/3$ and $5\pi/3$. The cardioid (for $a = 1$) is drawn in Fig. 2.1a. On the curve are also drawn (with symbols) the points in which one of the components of τ vanishes.

Example 2.2 (*Lemniscate*) The tangent unit vector (2.6) on the *lemniscate* of Bernoulli[3] is now evaluated. This planar ($n = 2$) curve is written in polar coordinates with $a > 0$ and $\theta \in (-\pi/4, +\pi/4) \cup [3\pi/4, 5\pi/4)$ as:

$$\rho(\theta) = a\sqrt{\cos 2\theta}. \tag{2.9}$$

Due to the above definition of the radial coordinate, the domain of this curve is made up of the union of two disjoint intervals of the real axis, here chosen as $(-\pi/4, +\pi/4)$ and $[3\pi/4, 5\pi/4)$. Note that at the endpoints of these intervals ρ vanishes. The curve intersects itself at the origin! This is a first sample of *double point*. Due to the fact that $\rho' = -a^2 \sin(2\theta)/\rho$, the vector x' is:

$$x'(\theta) = \begin{pmatrix} -a^2 \sin 2\theta \cos\theta/\rho - \rho \sin\theta \\ -a^2 \sin 2\theta \sin\theta/\rho + \rho \cos\theta \end{pmatrix},$$

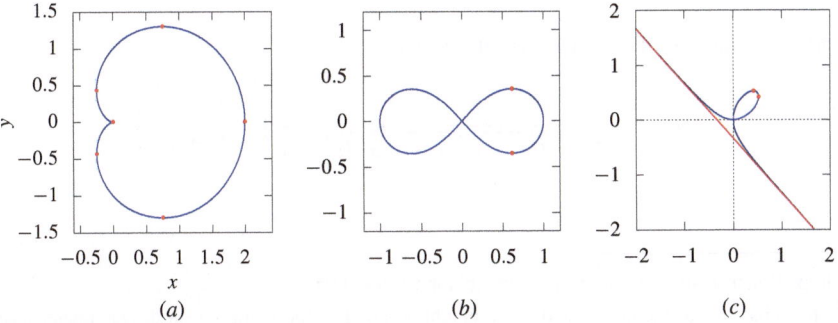

Fig. 2.1 Cardioid (**a**, $a = 1$), *lemniscate* of Bernoulli (**b**, $a = 1$) and *folium* of Descartes (**c**). Red dots are placed at points where one of the components of τ vanishes

---

[3] https://mathshistory.st-andrews.ac.uk/Biographies/Bernoulli_Jacob/.

with $|\boldsymbol{x}'(\theta)| = a^2/\rho(\theta)$. The tangent unit vector (2.6) follows as:

$$\boldsymbol{\tau}(\theta) = \begin{pmatrix} -\sin 3\theta \\ \cos 3\theta \end{pmatrix}.$$

Consider the arc for $\theta \in (-\pi/4, +\pi/4)$ (the other part of the lemniscate is obtained by a reflection across the axis y), the component τ_x is positive for $-\pi/4 \le \theta < 0$ and negative for $0 < \theta \le +\pi/4$. The y-component is positive for $-\pi/6 < \theta < +\pi/6$, vanishes in $\theta = \pm\pi/6$ and is negative otherwise.

Example 2.3 *(Folium of Descartes)* Consider the curve named *folium* of Descartes,[4] given in terms of the parameter $t \ne -1$ by the following formula:

$$\boldsymbol{x}(t) = \frac{1}{t^3 + 1} \begin{pmatrix} t \\ t^2 \end{pmatrix}, \tag{2.10}$$

therefore its domain is made up of the union of the two disjoint and unbounded intervals $(-\infty, -1) \cup (-1, +\infty)$. What happens in a neighbourhood of $t = -1$? This curve is drawn in Fig. 2.1c. In correspondence to the limit value $t \to -\infty$ the point $\boldsymbol{x}(t)$ (2.10) lies on the origin and moves on the lower branch ($x > 0$ and $y < 0$), as t increases, up to the limit position at infinity[5] at $t = -1^-$. In correspondence to the value $t = -1^+$, the point $\boldsymbol{x}(t)$ is at infinity on the upper branch of the curve ($x < 0$, $y > 0$) and moves toward the origin for growing t, up to reach this point for $t = 0$. For increasing positive values of the parameter, the point $\boldsymbol{x}(t)$ moves counterclockwise (why? give an answer) on the closed arc in the first quadrant ($x > 0$, $y > 0$), coming back to the origin for $t \to +\infty$.

The derivative of the vector $\boldsymbol{x}(t)$ (2.10) is written as:

$$\boldsymbol{x}'(t) = \frac{1}{(t^3 + 1)^2} \begin{pmatrix} -2t^3 + 1 \\ -t^4 + 2t \end{pmatrix},$$

and the tangent unit vector (2.6) follows as:

$$\boldsymbol{\tau}(t) = \frac{1}{\sqrt{t^8 + 4t^6 - 4t^5 - 4t^3 + 4t^2 + 1}} \begin{pmatrix} -2t^3 + 1 \\ -t^4 + 2t \end{pmatrix}.$$

---

[4] https://mathshistory.st-andrews.ac.uk/Biographies/Descartes/.

[5] The behaviour of the curve (2.10) in a neighbourhood of the point -1 on the parameter space $(\mathbb{R} - \{-1\})$ is investigated by means of the following asymptotic form of the components of $\boldsymbol{x}(t)$:

$$x(t) = -\frac{1}{3(t + 1)} + O(t + 1), \quad y(t) = +\frac{1}{3(t + 1)} - \frac{1}{3} + O(t + 1),$$

How the asymptot in Fig. 2.1c is obtained?

Note that at $t = 2^{-1/3}$ and $t = 2^{1/3}$ the components x_1' x_2' vanish. The corresponding points on the curve are drawn with symbols in Fig. 2.1c.

Exercise 2.1 Calculate, if it is possible, the tangent unit vector (2.6) on the following curves:

(a) $y = \sin x$ $x \in (0, 2\pi)$

(b) $y = \cos x/(1 + \sin^2 x)$ $x \in (0, 2\pi)$

(c) $x(\sigma) = \sigma/(1 + \sigma^2)$, $y(\sigma) = \sigma^2$ $\sigma \in (-1, +1)$

(d) $x(\sigma) = \cos \sigma$, $y(\sigma) = \sin \sigma^2$ $\sigma \in (0, 2\pi)$.

Once the tangent vector (2.5) has been introduced, it is also possible to specify the vector a in the differential (2.4). If the curve is linearizable in x_0, moving at the left hand side of the relation (2.4) the vector x_0 and dividing bot sides times $\Delta\sigma$, one obtains:

$$\frac{x(\sigma) - x(\sigma_0)}{\Delta\sigma} = a + \frac{o(\Delta\sigma)}{\Delta\sigma}.$$

Accounting for that $o(\Delta\sigma)$ is an infinitesimal vector of order higher than one ($o(\Delta\sigma)/\Delta\sigma \to 0$ as $\Delta\sigma \to 0$), the limit as $\Delta\sigma \to 0$ gives $a = x'(\sigma_0)$. As a consequence, the parameter σ being the independent variable in the vector-valued function $x = x(\sigma)$, its increment ($\Delta\sigma$) is completely given by its differential ($d\sigma$). For this reason the identity $d\sigma = \Delta\sigma$ will be used hereafter. If the curve is linearizable in x_0, its differential at that point becomes:

$$dx(\sigma_0; d\sigma) = x'(\sigma_0)\, d\sigma. \tag{2.11}$$

The relation (2.11) explains the local behaviour of $O^{(n)}$, for $O^{(1)}$ moving in a neighbourhood of σ_0. $O^{(n)}$ runs in the direction given by the vector $x'(\sigma_0)$ (tangent to the curve) and the growing of its distance from the point x_0 depends on $|x'|$. However, from a global point of view the situation is in general more complicated, due to the fact that the curve gets away from its tangent line in x_0 for increasing $|\Delta\sigma|$. In these conditions, how change the direction in which $O^{(n)}$ moves? An answer to this question will be discussed in Sect. 2.3.

2.2 Curvilinear Abscissa

In the previous section the tangent vector t to the curve (2.1) at its point x_0 has been introduced (2.5). Equation (2.7) shows that the direction of t is independent of the choice of the parameter, whereas its modulus depends on this choice. As a consequence, a preferred change of parameter exists, such that the modulus of $\tilde{t}$ is unitary. In this case, the parameter takes the name of *curvilinear abscissa* and it is indicated by s. Its relation with the generic parameter σ is deduced using the Eq. (2.7) with

$\tilde{\sigma} = s$. Taking the modulus of the vectors and enforcing the tangent vector $d\boldsymbol{x}/ds$ to have unitary modulus, the relation:

$$\frac{ds}{d\sigma} = |\boldsymbol{t}| = \left|\frac{d\boldsymbol{x}}{d\sigma}\right|$$

is obtained. Once an origin of the curvilinear abscissa is arbitrarily fixed (for example, $s(\sigma_0) = 0$), this differential equation is integrated and gives s as a function of the generic parameter σ (the name of the parameter is changed inside the integral, in order to avoid confusion with the integral limits):

$$s(\sigma) = \int_{\sigma_0}^{\sigma} d\gamma \left|\frac{d\boldsymbol{x}}{d\gamma}(\gamma)\right|. \tag{2.12}$$

It is left to the reader to verify that the definition (2.12) of curvilinear abscissa does not depend on the curve parametrization and that s is a monotonic increasing function of σ. Due to the fact that the modulus of the tangent vector obtained evaluating the derivative $d\boldsymbol{x}/ds$ has been enforced unitary, it follows that $d\boldsymbol{x}/ds = \boldsymbol{\tau}$, $\boldsymbol{\tau}$ being defined in terms of a generic parameter σ in Eq. (2.6). It is left to the reader to verify this property directly starting from the definition of s given in Eq. (2.12).

The meaning of the curvilinear abscissa (2.12) is lighten by the following observations, based on the definition of the integral at the right hand side. Assume that the length $l(\sigma_0, \sigma)$ of the arc of the curve $\mathscr{C}$ corresponding to the interval $[\sigma_0, \sigma]$ in the parameter space has to be measured. To this aim, this interval is partitioned in a large number (n) of subintervals $[\sigma_{i-1}, \sigma_i]$ for $i = 1, 2, \ldots, n$ ($\sigma_n = \sigma$). As well known, the degree of refinement of the partition is measured in terms of its norm δ, defined as the maximum length of the subintervals. An approximation of the length $l(\sigma_0, \sigma)$ is obtained adding the lengths of the n line segments joining $\boldsymbol{x}(\sigma_{i-1})$ to $\boldsymbol{x}(\sigma_i)$ for $i = 1, 2, \ldots, n$:

$$l(\sigma_0, \sigma) \simeq \sum_{i=1}^{n} |\boldsymbol{x}(\sigma_i) - \boldsymbol{x}(\sigma_{i-1})| \tag{2.13}$$

this equation being exact only in the limit as $\delta \to 0$. Choosing (in an arbitrary way) a point η_k inside the kth interval $[\sigma_{k-1}, \sigma_k]$, the length $|\boldsymbol{x}(\sigma_i) - \boldsymbol{x}(\sigma_{i-1})|$ in the above sum is approximated with $|\boldsymbol{x}'(\eta_i)|(\sigma_i - \sigma_{i-1})$, making an error at least of the order $(\sigma_i - \sigma_{i-1})^2$. The right hand side of Eq. (2.13) becomes:

$$\sum_{i=1}^{n} |\boldsymbol{x}(\sigma_i) - \boldsymbol{x}(\sigma_{i-1})| = \sum_{i=1}^{n} \left[\left|\frac{d\boldsymbol{x}}{d\sigma}(\eta_i)\right|(\sigma_i - \sigma_{i-1}) + O(\sigma_i - \sigma_{i-1})^2\right]$$

in which the sum of the linear contributions are the *Riemann sum* for the function $|\boldsymbol{x}'|$ integrated on the interval $[\sigma_0, \sigma]$, whereas the sum of the contributions of $O(\sigma_i - \sigma_{i-1})^2$ vanishes as $\delta \to 0$. Hence the Eq. (2.13) gives in that limit:

$$l(\sigma_0, \sigma) = \int_{\sigma_0}^{\sigma} d\gamma \left| \frac{d\boldsymbol{x}}{d\gamma}(\gamma) \right| = s(\sigma),$$

by the definition (2.12) of s. As a consequence, the curvilinear abscissa $s(\sigma)$ acquires the important meaning of length of the arc of the curve $\mathscr{C}$ corresponding to the interval $[\sigma_0, \sigma]$ of the parameter space.

Problem 2.1 Evaluate the curvilinear abscissa on the parabola $y = x^2/2$.

Solution 2.1 It is a planar curve, so that $n = 2$. The abscissa x is chosen as parameter on that curve and the origin of the curvilinear abscissa is placed on the origin of the coordinates, *i.e.* $s(0) = 0$ is assumed. The tangent vector:

$$\frac{d\boldsymbol{x}}{dx} = \begin{pmatrix} 1 \\ x \end{pmatrix}$$

follows, and the length of the arc of the parabola corresponding to the interval $[0, x]$ in the parameter space is obtained specializing the definition (2.12) for the above tangent vector:

$$s(x) = \int_0^x dt \sqrt{1 + t^2} = \frac{1}{2} \log(x + \sqrt{x^2 + 1}) + \frac{1}{2} x \sqrt{x^2 + 1}.$$

Problem 2.2 Evaluate the curvilinear abscissa along the curve $y = \log x$, assuming $x \geq 1$ and $s(1) = 0$.

Solution 2.2 It is a planar curve, so that $n = 2$. Using the abscissa x as a parameter on that curve the tangent vector is written as:

$$\frac{d\boldsymbol{x}}{dx} = \begin{pmatrix} 1 \\ 1/x \end{pmatrix}$$

and the curvilinear abscissa (2.12) follows as:

$$s(x) = \int_1^x dt \, \frac{\sqrt{t^2 + 1}}{t}.$$

The integration is performed changing the variable from t to ξ with: $t = \sinh \xi$, $\xi = \log(t + \sqrt{t^2 + 1})$. In terms of the new variable, the above integral becomes:

$$s(x) = \int_{\xi(1)}^{\xi(x)} d\xi \, \cosh \xi \, \frac{\cosh \xi}{\sinh \xi} = \int_{\xi(1)}^{\xi(x)} d\xi \, \sinh \xi + \int_{\xi(1)}^{\xi(x)} \frac{d\xi}{\sinh \xi},$$

in which the first integral is elementary, whereas the second is evaluated by means of the new change of variable $w = \exp(\xi)$. The curvilinear abscissa:

$$s(x) = \sqrt{x^2 + 1} - \sqrt{2} + \log(1 + \sqrt{2}) + \log \frac{x - 1 + \sqrt{x^2 + 1}}{x + 1 + \sqrt{x^2 + 1}}$$

follows.

2.3 Normal Unit Vector and Curvature

During its motion on $\mathscr{C}$, the point $O^{(n)}$ changes direction, if the curve is not a straight line. In the present section, the changes of tangent unit vector $\boldsymbol{\tau}$ vs. the curvilinear abscissa s is investigated. Due to the fact that the modulus of $\boldsymbol{\tau}$ is constant in s, its derivative in s is orthogonal to $\boldsymbol{\tau}$:

$$0 = \frac{d}{ds} |\boldsymbol{\tau}|^2 = \frac{d}{ds} (\boldsymbol{\tau} \cdot \boldsymbol{\tau}) = 2 \frac{d\boldsymbol{\tau}}{ds} \cdot \boldsymbol{\tau},$$

so that the derivative of $\boldsymbol{\tau}$ identifies *a direction* normal to the curve (if $n > 2$, infinitely many normal vectors exist). Therefore the *normal unit vector* $\boldsymbol{v}$ is defined as:

$$v = \frac{\dfrac{d\boldsymbol{\tau}}{ds}}{\left| \dfrac{d\boldsymbol{\tau}}{ds} \right|}. \tag{2.14}$$

The denominator, *i.e.* the modulus of the derivative of the tangent unit vector with respect to the curvilinear abscissa, is the (absolute) *curvature* $k(s)$ of the curve $\mathscr{C}$ at the point $\boldsymbol{x}(s)$. It gives a measure of the rapidity of the change of $\boldsymbol{\tau}$ with the curvilinear abscissa. From the definition of the normal unit vector (2.14) and of the curvature, the derivative in s of the tangent unit vector is written as:

$$\frac{d\boldsymbol{\tau}}{ds}(s) = k(s) \, \boldsymbol{v}(s). \tag{2.15}$$

The above formulae for the normal unit vector (2.14) and for the curvature (2.15) are now specified in terms of a generic parameter σ in two ($n = 2$) or three ($n = 3$) dimensions. Denoting with an apex the derivative with respect to σ and recalling the definition (2.12) of the curvilinear abscissa, the modulus of $\boldsymbol{x}'$ is also the derivative in σ of s:

$$\boldsymbol{\tau} = \frac{\boldsymbol{x}'}{s'}. \tag{2.16}$$

Due to the fact that $|\boldsymbol{x}'| = (x_1'^2 + \cdots + x_n'^2)^{1/2}$, the second derivative in σ of s is written as:

$$s'' = \frac{\boldsymbol{x}' \cdot \boldsymbol{x}''}{|\boldsymbol{x}'|}, \tag{2.17}$$

from which it follows that the derivative in s of $\boldsymbol{\tau}$ (2.16) is rewritten in terms of the parametrization in σ as:

$$\frac{d\boldsymbol{\tau}}{ds} = \frac{1}{s'}\frac{d\boldsymbol{\tau}}{d\sigma} = \frac{s'\boldsymbol{x}'' - s''\boldsymbol{x}'}{s'^3}. \tag{2.18}$$

The modulus of the vector (2.18) is evaluated using the second derivative of s (2.17) and gives the curvature:

$$k = \frac{1}{s'^3}\sqrt{|\boldsymbol{x}'|^2|\boldsymbol{x}''|^2 - (\boldsymbol{x}'\cdot\boldsymbol{x}'')^2}.$$

It can be easily shown that the relation: $|\boldsymbol{x}'|^2|\boldsymbol{x}''|^2 - (\boldsymbol{x}'\cdot\boldsymbol{x}'')^2 = |\boldsymbol{x}'\times\boldsymbol{x}''|^2$ holds. As a consequence, the curvature can be rewritten as:

$$k = \frac{|\boldsymbol{x}'\times\boldsymbol{x}''|}{|\boldsymbol{x}'|^3}. \tag{2.19}$$

Once the curvature has been evaluated, the normal unit vector is also obtained:

$$\boldsymbol{v} = \frac{1}{k}\frac{d\boldsymbol{\tau}}{ds} = \frac{|\boldsymbol{x}'|^2\boldsymbol{x}'' - \boldsymbol{x}'\cdot\boldsymbol{x}''\,\boldsymbol{x}'}{|\boldsymbol{x}'|\,|\boldsymbol{x}'\times\boldsymbol{x}''|}. \tag{2.20}$$

Problem 2.3 Write the expression of curvature for the curve represented by Eq. (2.8).

Solution 2.3 The curvature of the cardioid is written using the result $|\boldsymbol{x}'\times\boldsymbol{x}''| = 6a^2\cos^2(\theta/2)$, that leads to the result:

$$k(\theta) = \frac{3}{4a}\frac{1}{|\cos(\theta/2)|}.$$

Note that $k(\theta) \to \infty$ as $\theta \to \pi$. The normal unit vector takes the form:

$$\boldsymbol{v} = \frac{1}{4|\cos(\theta/2)|^3}\begin{pmatrix} 1 - 3\cos^2\theta - 2\cos^3\theta \\ -\sin\theta(1 + 3\cos\theta + 2\cos^3\theta) \end{pmatrix}.$$

The curvilinear abscissa (2.12) with origin on the point $(2a, 0)$ is:

$$s(\theta) = \int_0^\theta d\eta\, 2a\left|\cos\frac{\eta}{2}\right| = \begin{cases} 4a\sin(\theta/2) & \text{for } 0 \le \theta \le \pi \\ 4a[2 - \sin(\theta/2)] & \text{for } \pi < \theta < 2\pi \end{cases}$$

Problem 2.4 Write the expression of curvature for the curve represented by Eq. (2.9).

Solution 2.4 The curvature of the lemniscate of Bernoulli follows as:

$$k(\theta) = 3\,\frac{\rho}{a^2}$$

and the normal unit vector is:

$$\boldsymbol{v}(\theta) = -\begin{pmatrix} \cos 3\theta \\ \sin 3\theta \end{pmatrix}.$$

The calculation of the curvilinear abscissa is quite complicated. Using the change of variable $\beta = \theta + \pi/4$, where $\theta \in (-\pi/4, +\pi/4)$ so that $0 < \beta < \pi$, the curvilinear abscissa becomes:

$$s(\theta) = a \int_{-\pi/4}^{\theta} \frac{d\alpha}{\sqrt{\cos 2\alpha}} = a \int_{0}^{\theta+\pi/4} \frac{d\beta}{\sqrt{\sin 2\beta}}.$$

The change of variable $t = \tan\beta$ leads to:

$$s(\theta) = \sqrt{2}a \int_{0}^{\tan(\theta+\pi/4)} \frac{d\sqrt{t}}{\sqrt{1+t^2}},$$

which can be rewritten with the other change of variable $\xi = \sqrt{t}$ as:

$$s(\theta) = \sqrt{2}a \int_{0}^{[\tan(\theta+\pi/4)]^{1/2}} \frac{d\xi}{\sqrt{1+\xi^4}}.$$

Finally, the change of the integration variable from ξ to φ with:

$$\cos\varphi = \frac{\xi^2 - 1}{\xi^2 + 1}, \quad \xi^2 = \frac{1 + \cos\varphi}{1 - \cos\varphi}, \quad d\xi = -\frac{d\varphi}{1 - \cos\varphi}$$

leads to the following form of the integral:

$$s(\theta) = \frac{\sqrt{2}}{2}\,a \int_{\arccos(\tan\theta)}^{\pi} \frac{d\varphi}{\sqrt{1 - (\sqrt{2}/2)^2 \sin^2\varphi}}. \tag{2.21}$$

This result can be rewritten by means of the new function:

$$F(x; k) := \int_{0}^{x} \frac{dy}{\sqrt{1 - k^2 \sin^2 y}}, \tag{2.22}$$

that is named *elliptic integral of the first kind* of modulus k. Once k is set to $\sqrt{2}/2$, the curvilinear abscissa (2.21) becomes:

$$s(\theta) = \frac{\sqrt{2}}{2} a \left\{ F(\pi; \sqrt{2}/2) - F[\arccos(\tan\theta); \sqrt{2}/2] \right\}.$$

The new function (2.22) has to be handled as the other transcendental functions (as $\sin x$, $\log x$, ...). Indeed, library functions exist that evaluate the elliptic integral (2.22) with the desired accuracy. However, the numerical calculation of this function will be also discussed below, by means of the *series of functions*.

2.4 Suggested Readings

For a concise analysis of topics introduced in this lesson one can see Adams and Essex (2010, Chap. 11, pp. 642–645). A formal definition of the *unit tangent vector* to a curve is given in Lax and Terrell (2017, Chap. 7, p. 285).

References

R. A. Adams and C. Essex. *Calculus: Several Variables*. Pearson Education Canada, Toronto, ON, 7th edition, 2010.
P. D. Lax and M. S. Terrell. *Multivariable Calculus with Applications*. Springer, Cham, CH, 2017.

Chapter 3
Curves in Space: The Frenet Frame

In order to complete the study of differential properties of curves in space, another vector is introduced. This describes locally an important characteristic of curves in $\mathbb{R}^3$, is the third vector of the frame defined on the curve itself to represent its geometric behaviour.

3.1 Binormal Unit Vector, Torsion

In the past lectures we have defined the unit tangent vector $\boldsymbol{\tau}$ and the unit normal vector $\boldsymbol{v}$ to a curve $\mathscr{C}$ in $\mathbb{R}^n$ ($n = 2, 3$). If the curve is a plane one, these two vectors are sufficient to define a local base, whereas if the curve is three-dimensional, a third base vector is required. In the latter case, the third vector is called the unit *binormal* vector:

$$\boldsymbol{b} := \boldsymbol{\tau} \times \boldsymbol{v}, \tag{3.1}$$

that is always normal to the plane of the curve, if the curve is a planar one.

We are able to calculate the derivative of unit tangent vector with respect to the arc length parameter:

$$\frac{d\boldsymbol{\tau}}{ds} = k\boldsymbol{v}, \tag{3.2}$$

let us try now to calculate the derivative, with respect to same parameter, of the unit normal vector. Since the vector thus obtained must be orthogonal to $\boldsymbol{v}$ (as a consequence of the fact that the modulus of $\boldsymbol{v}$ is independent of s), there must be two scalar constants μ, χ such that:

$$\frac{d\boldsymbol{v}}{ds} = -\mu\boldsymbol{\tau} - \chi\boldsymbol{b}. \tag{3.3}$$

G. Riccardi et al., *Multidimensional Differential and Integral Calculus*,
https://doi.org/10.1007/978-3-031-70326-3_3

We observe that, substituting (3.3) in the derivative of b with respect to s (3.1) one gets the following relation:

$$\frac{db}{ds} = \frac{d\tau}{ds} \times v + \tau \times \frac{dv}{ds}$$
$$= \tau \times (-\mu\tau - \chi b)$$
$$= \chi v , \tag{3.4}$$

where the first term in the right-hand side ($d\tau/ds \times v$) vanishes as a consequence of (3.2). The nonvanishing vector db/ds indicates that the curve departs from the plane defined by the vectors τ and v, called the *osculating plane*. The speed at which this departure occurs is defined by the scalar χ, that is a real number (with sign) called *torsion* of the curve. Finally, we define the scalar μ. From the dot product of the relation (3.3) by τ, taking into account that:

$$\tau \cdot \frac{dv}{ds} = -v \cdot \frac{d\tau}{ds} = -k$$

we are led to the relation $\mu = k$. Thus, we have obtained the derivatives with respect to the arc length (3.2), (3.3), (3.4) of the three unit vectors τ, v, b, that are a local base for the curve $\mathscr{C}$. The above relations are called Frenet[1]-Serret[2] equations.

In the last lecture we have derived the expressions of the unit tangent vector, of the curvature and of the unit normal vector with an arbitrary parametrization of the curve. We report them here for the sake of clarity:

$$\tau = \frac{x'}{|x'|} , \quad k = \frac{|x' \times x''|}{|x'|^3} , \quad v = \frac{|x'|^2 x'' - x' \cdot x'' x'}{|x'| \, |x' \times x''|} . \tag{3.5}$$

Now we extend this procedure to the calculation of the torsion and the unit binormal vector. From the vector product between τ and v written in the form (3.5) we immediately get:

$$b = \frac{x' \times x''}{|x' \times x''|} . \tag{3.6}$$

As observed above, if the curve is plane, the unit binormal vector simply turns out to be orthogonal to the plane of the curve, since x' and x'' lie in the same plane. In order to calculate the torsion, we exploit the above form of the unit binormal vector:

$$\chi = v \cdot \frac{db}{ds}$$
$$= v \cdot \frac{1}{|x'|} \frac{db}{d\sigma}$$

---

[1] https://mathshistory.st-andrews.ac.uk/Biographies/Frenet/.
[2] https://mathshistory.st-andrews.ac.uk/Biographies/Serret/.

$$= \left(\underbrace{\frac{\boldsymbol{x}''}{|\boldsymbol{x}' \times \boldsymbol{x}''|}}_{\boldsymbol{v}_1} - \underbrace{\frac{\boldsymbol{x}' \cdot \boldsymbol{x}''}{|\boldsymbol{x}'|^2 \, |\boldsymbol{x}' \times \boldsymbol{x}''|} \boldsymbol{x}'}_{\boldsymbol{v}_2} \right) \cdot \left(\underbrace{\frac{\boldsymbol{x}' \times \boldsymbol{x}'''}{|\boldsymbol{x}' \times \boldsymbol{x}''|}}_{\boldsymbol{b}'_1} + \underbrace{\boldsymbol{x}' \times \boldsymbol{x}'' \frac{d}{d\sigma} \frac{1}{|\boldsymbol{x}' \times \boldsymbol{x}''|}}_{\boldsymbol{b}'_2} \right).$$

Since in the derivative $d\boldsymbol{b}/d\sigma$ both vectors $\boldsymbol{b}'_{1,2}$ are orthogonal to $\boldsymbol{x}'$, the dot product between $\boldsymbol{v}_2$ and that derivative vanishes. Moreover, the dot product between $\boldsymbol{v}_1$ and $\boldsymbol{b}'_2$ also vanish, so that there is no need for calculating the derivative of $1/|\boldsymbol{x}' \times \boldsymbol{x}''|$. The only nonzero dot product is $\boldsymbol{v}_1 \cdot \boldsymbol{b}'_1$, where the quantity:

$$\boldsymbol{x}'' \cdot \boldsymbol{x}' \times \boldsymbol{x}''' = \boldsymbol{x}' \cdot \boldsymbol{x}''' \times \boldsymbol{x}'' = -\boldsymbol{x}' \cdot \boldsymbol{x}'' \times \boldsymbol{x}''',$$

appears; thus the torsion is given by the following formula:

$$\chi = -\frac{\boldsymbol{x}' \cdot \boldsymbol{x}'' \times \boldsymbol{x}'''}{|\boldsymbol{x}' \times \boldsymbol{x}''|^2}. \tag{3.7}$$

We observe that if the curve is plane, the three vectors $\boldsymbol{x}'$, $\boldsymbol{x}''$ and $\boldsymbol{x}'''$ all lie in the same plane, so that their scalar triple product vanishes, as does χ.

The relations (3.6), (3.7) accomplish (3.5) and allow the calculation of all relevant quantities, using any parametrization. In order to carry out simple dimensional checks, we clarify two important aspects of the above formulas. The first is the presence of derivatives in the parameter σ (whose physical dimension is arbitrary). The total number of derivatives with respect to σ appearing in the numerator of each quantity should always be the same as the total number of those derivatives appearing in the denominator. For example, for the unit normal vector we have:

$$\boldsymbol{v} : \frac{\text{number of derivative in } \sigma = 4}{\text{number of derivative in } \sigma = 4},$$

so that the physical dimension of σ vanishes in each quantity. Furthermore, comparing the number of appearances of vector $\boldsymbol{x}$ (whose physical dimension is length, so it is homogeneous to s) in the numerator with the corresponding number in the denominator, one must find them to be the same for unit vectors (these being nondimensional), whereas the number in the denominator must exceed of one that in the numerator for k and χ, since the physical dimension of these scalar quantities is inverse length. For example:

$$\boldsymbol{v} : \frac{\text{number of vectors } \boldsymbol{x} = 3}{\text{number of vectors } \boldsymbol{x} = 3}, \quad k : \frac{\text{number of vectors } \boldsymbol{x} = 2}{\text{number of vectors } \boldsymbol{x} = 3}.$$

Example 3.1 Let us consider the *circular helix* with constant pitch, described by the parameter $\theta \in \mathbb{R}$:

$$\begin{cases} x(\theta) = r \cos \theta \\ y(\theta) = r \sin \theta \\ z(\theta) = p\theta/(2\pi) \end{cases} \tag{3.8}$$

where r is the radius of circle projected onto the (x, y)-plane and $p > 0$ is the *pitch*, that is the increase in z at each lap of the projected circle. Let p' be the quantity $p/(2\pi)$. Now calculate the derivatives of position vector:

$$x' = \begin{pmatrix} -r\sin\theta \\ r\cos\theta \\ p' \end{pmatrix}, \quad x'' = \begin{pmatrix} -r\cos\theta \\ -r\sin\theta \\ 0 \end{pmatrix}, \quad x''' = \begin{pmatrix} r\sin\theta \\ -r\cos\theta \\ 0 \end{pmatrix},$$

since $x' \equiv \sqrt{r^2 + p'^2}$, the arc length is proportional to θ, whereas the unit tangent vector is:

$$\tau(\theta) = \frac{1}{\sqrt{r^2 + p'^2}} \begin{pmatrix} -r\sin\theta \\ r\cos\theta \\ p' \end{pmatrix}.$$

Once the vector product

$$x' \times x'' = \begin{pmatrix} rp'\sin\theta \\ -rp'\cos\theta \\ r^2 \end{pmatrix}$$

is calculated, the curvature immediately follows:

$$k = \frac{r}{r^2 + p'^2},$$

whereas, considering that $x' \cdot x'' \equiv 0$, the unit normal vector is:

$$\nu = \begin{pmatrix} -\cos\theta \\ -\sin\theta \\ 0 \end{pmatrix}.$$

The unit binormal vector follows from the relation (3.6):

$$b = \frac{1}{\sqrt{r^2 + p'^2}} \begin{pmatrix} p'\sin\theta \\ -p'\cos\theta \\ r \end{pmatrix}.$$

Finally, the torsion can be written based on the relation (3.7):

$$\chi = -\frac{p'}{p'^2 + r^2}.$$

Therefore, at any point of the circular helix we find the same curvature and torsion.

Example 3.2 Consider the three-dimensional curve defined in parametric form as follows:

$$x(\sigma) = \begin{pmatrix} a_1\sigma^3 + b_1\sigma^2 + c_1\sigma + d_1 \\ a_2\sigma^3 + b_2\sigma^2 + c_2\sigma + d_2 \\ a_3\sigma^3 + b_3\sigma^2 + c_3\sigma + d_3 \end{pmatrix} \tag{3.9}$$

with $a_{1,2,3}$, $b_{1,2,3}$, $c_{1,2,3}$ and $d_{1,2,3}$ given constants and $\sigma \in (-\infty, +\infty)$. In the following, it is worth thinking these quantities as the components of four vectors a, b, c and d in the space $\mathbb{R}^3$. In this way, the definition (3.9) of the curve can be rewritten in a more concise form using a vector notation:

$$x(\sigma) = a\sigma^3 + b\sigma^2 + c\sigma + d . \tag{3.10}$$

Let us determine the triple of unit vectors (τ, v, b), the curvature k and the torsion χ. We begin with writing the derivatives of vectors:

$$x' = 3a\sigma^2 + 2b\sigma + c , \quad x'' = 2(3a\sigma + b) , \quad x''' \equiv 6a .$$

Let us calculate the modulus of vector x':

$$|x'| = \sqrt{9|a|^2\sigma^4 + 12a \cdot b\sigma^3 + 2(3a \cdot c + 2|b|^2)\sigma^2 + 4b \cdot c\sigma + |c|^2} = f(\sigma)$$

and then the unit tangent vector:

$$\tau(\sigma) = \frac{3a\sigma^2 + 2b\sigma + c}{f(\sigma)} .$$

The curvature is easily written, once the vector product $x' \times x'' = -2(3a \times b\sigma^2 + 3a \times c\sigma + b \times c)$ has been calculated. The modulus of this can be written:[3]

$$|x' \times x''| = 2\Big[9|a \times b|^2\sigma^4 + 18a \times b \cdot a \times c\sigma^3 + 3(2a \times b \cdot b \times c + 3|a \times c|^2)\sigma^2$$

$$+6a \times c \cdot b \times c\,\sigma + |b \times c|^2\Big]^{1/2}$$

$$=: g(\sigma) .$$

---

[3] To avoid the calculation of vector products, one can use the following formulas:

$$|a \times b|^2 = |a|^2|b|^2 - (a \cdot b)^2 \qquad a \times b \cdot a \times c = |a|^2 b \cdot c - a \cdot ba \cdot c$$

$$|a \times c|^2 = |a|^2|c|^2 - (a \cdot c)^2 \qquad a \times b \cdot b \times c = a \cdot bb \cdot c - a \cdot c|b|^2$$

$$|b \times c|^2 = |b|^2|c|^2 - (b \cdot c)^2 \qquad a \times c \cdot b \times c = a \cdot b|c|^2 - a \cdot cb \cdot c$$

The curvature is now written:

$$k = \frac{g}{f^3},$$

whereas the unit normal vector is:

$$v = \frac{2}{fg}\{ \ 9(a \cdot b \, a - |a|^2 b)\sigma^4 + 3\left[(2|b|^2 + 3a \cdot c)a - 2a \cdot b \, b - 3|a|^2 c\right]\sigma^3$$
$$+9(b \cdot c \, a - a \cdot b \, c)\sigma^2 + \left[3|c|^2 a + 2b \cdot c \, b - (2|b|^2 + 3a \cdot c)c\right]\sigma$$
$$+(|c|^2 b - b \cdot c \, b)\} \, .$$

The unit binormal vector is written as:

$$b = -2 \, \frac{3a \times b \, \sigma^2 + 3a \times c \, \sigma + b \times c}{g},$$

whereas the torsion takes the form:

$$\chi = 2 \, \frac{a \cdot b \times c}{g^2} \, .$$

Example 3.3 *(Horse foot of Eudoxus)* Consider the curve generated by the inter-section of a sphere (to make it simple, centered at the origin) of radius R and a right circular cylinder whose cross sections have their centers lying on a straight line parallel to the z-axis and passing through the point $(a, 0, 0)$, with $a \in (0, R)$. The radius r of cross sections of the cylinder is taken as $R - a$, so as to make cylinder and sphere be tangent at point $(R, 0, 0)$. The curve is thus obtained from the solution of the following system:

$$\begin{cases} x^2 + y^2 + z^2 = R^2 \\ (x - a)^2 + y^2 = r^2 \, , \end{cases}$$

and has the parametric form:

$$\begin{cases} x = a + r \cos\theta \\ y = r \sin\theta \\ z = 2\sqrt{ar} \sin(\theta/2) \, . \end{cases} \tag{3.11}$$

The curve (3.11) is called *horse foot* of Eudoxus.[4] If one makes $a = R/2$ a particular curve is obtained, which is generally called *curve* of Viviani.[5] The geometry of this curve will then be defined for a given value of the ratio a/R, which we will call α.

---

[4] https://mathshistory.st-andrews.ac.uk/Biographies/Eudoxus/.
[5] https://mathshistory.st-andrews.ac.uk/Biographies/Viviani/.

Once the derivatives of x have been calculated:

$$x' = \begin{pmatrix} -r\sin\theta \\ r\cos\theta \\ \sqrt{ar}\,\cos\dfrac{\theta}{2} \end{pmatrix}, \quad x'' = \begin{pmatrix} -r\cos\theta \\ -r\sin\theta \\ -\dfrac{\sqrt{ar}}{2}\sin\dfrac{\theta}{2} \end{pmatrix}, \quad x''' = \begin{pmatrix} r\sin\theta \\ -r\cos\theta \\ -\dfrac{\sqrt{ar}}{4}\cos\dfrac{\theta}{2} \end{pmatrix},$$

since $|x'| = r^{1/2}[R - a\sin^2(\theta/2)]^{1/2}$, the unit tangent vector is written as:

$$\tau = \frac{1}{[1 - \alpha\sin^2(\theta/2)]^{1/2}} \begin{pmatrix} -\sqrt{1-\alpha}\,\sin\theta \\ \sqrt{1-\alpha}\,\cos\theta \\ \sqrt{\alpha}\,\cos\dfrac{\theta}{2} \end{pmatrix}.$$

Moreover, since $|x' \times x''| = r^{3/2}[R - 3a\sin^2(\theta/2)/4]^{1/2}$, the curvature is:

$$kR = \frac{\left(1 - \dfrac{3}{4}\alpha\sin^2\dfrac{\theta}{2}\right)^{1/2}}{\left(1 - \alpha\sin^2\dfrac{\theta}{2}\right)^{3/2}}.$$

The unit normal vector is written as:

$$v = -\frac{1}{4}\frac{1}{\left(1 - \alpha\sin^2\dfrac{\theta}{2}\right)^{1/2}\left(1 - \dfrac{3}{4}\alpha\sin^2\dfrac{\theta}{2}\right)^{1/2}} \begin{pmatrix} \alpha\cos^2\theta + 2(2-\alpha)\cos\theta + \alpha \\ \sin\theta\,[2(2-\alpha) + \alpha\cos\theta] \\ 2\sqrt{\alpha(1-\alpha)}\,\sin\dfrac{\theta}{2} \end{pmatrix},$$

whereas the unit binormal vector takes the form:

$$b = \frac{1}{\left(1 - \dfrac{3}{4}\alpha\sin^2\dfrac{\theta}{2}\right)^{1/2}} \begin{pmatrix} \dfrac{\sqrt{\alpha}}{2}\left(\sin\theta\,\cos\dfrac{\theta}{2} + \sin\dfrac{\theta}{2}\right) \\ -\dfrac{\sqrt{\alpha}}{2}\left(\cos\theta\,\cos\dfrac{\theta}{2} + \cos\dfrac{\theta}{2}\right) \\ \sqrt{1-\alpha} \end{pmatrix}.$$

Finally, the torsion is:

$$\chi R = -\frac{3}{4}\sqrt{\frac{\alpha}{1-\alpha}}\,\frac{\cos\dfrac{\theta}{2}}{1 - \dfrac{3}{4}\alpha\sin^2\dfrac{\theta}{2}}.$$

In Fig. 3.1a the curve is drawn, whereas in (b) some intrinsic frames are drawn along the same curve. The curvature kR and the torsion χR are plotted in Fig. 3.2 as functions of θ.

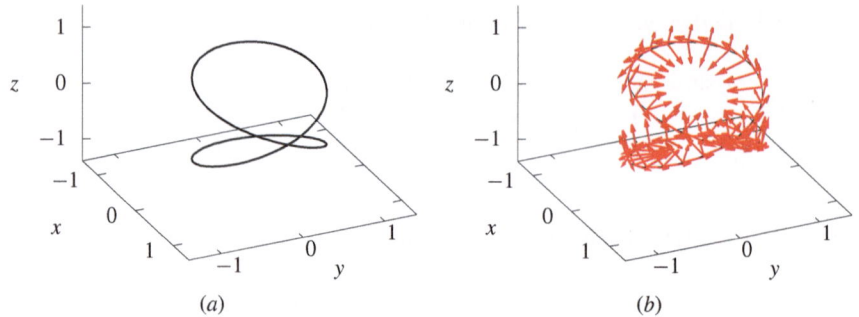

Fig. 3.1 Horse foot of Eudoxus with $R = 1$ and $\alpha = 1/4$ **a**, and some intrinsic frames on this curve (**b**, Vector scale: 0.4)

Fig. 3.2 Curvature (black line) and torsion (red) of the curve horse foot of Eudoxo per $R = 1$ ed $\alpha = 1/4$ as a function of the parameter θ (radians)

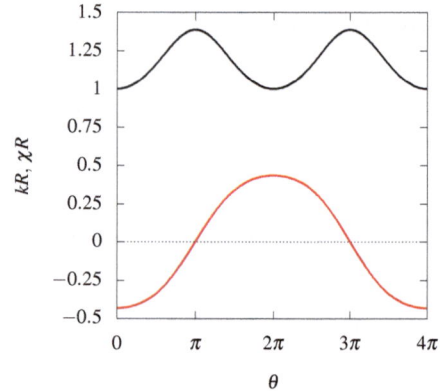

A wide catalogue of three-dimensional curves can be found at:
http://www.mathcurve.com/courbes3d/courbes3dit.shtml

Problem 3.1 Calculate the intrinsic frame, curvature and torsion for a *conic spiral* of Pappus:[6]

$$\begin{cases} x = a\,\theta \cos\theta \\ y = a\,\theta \sin\theta \\ z = b\,\theta\,, \end{cases} \tag{3.12}$$

with $a, b > 0$ and $-\infty < \theta < +\infty$ (vedi Figs. 3.3 and 3.5a).

---

[6] https://mathshistory.st-andrews.ac.uk/Biographies/Pappus/.

Solution 3.1

$$\boldsymbol{\tau} = \frac{1}{A(\theta)} \begin{pmatrix} a(\cos\theta - \theta\sin\theta) \\ a(\sin\theta + \theta\cos\theta) \\ b \end{pmatrix},$$

$$\boldsymbol{v} = \frac{1}{A(\theta)B(\theta)} \begin{pmatrix} -[(2+\theta^2)a^2 + 2b^2]\sin\theta - [(2+\theta^2)a^2 + b^2]\theta\cos\theta \\ [(2+\theta^2)a^2 + 2b^2]\cos\theta - [(2+\theta^2)a^2 + b^2]\theta\sin\theta \\ -ab\theta \end{pmatrix},$$

$$\boldsymbol{b} = \frac{1}{B(\theta)} \begin{pmatrix} -b(2\cos\theta - \theta\sin\theta) \\ -b(2\sin\theta + \theta\cos\theta) \\ a(2+\theta^2) \end{pmatrix},$$

$$k = \frac{aB(\theta)}{A^3(\theta)}, \qquad \chi = -\frac{(6+\theta^2)b}{B^2(\theta)},$$

where: $A(\theta) = [(1+\theta^2)a^2 + b^2]^{1/2}$ and $B(\theta) = [(2+\theta^2)^2a^2 + (4+\theta^2)b^2]^{1/2}$.

Problem 3.2 Calculate the intrinsic frame, curvature and torsion for the curve represented by parametric equations:

$$\begin{cases} x = 2\cos\theta \\ y = \sin\theta \\ z = 2\sin 2\theta, \end{cases} \tag{3.13}$$

with $0 \le \theta < 2\pi$ (see Figs. 3.4 and 3.5b).

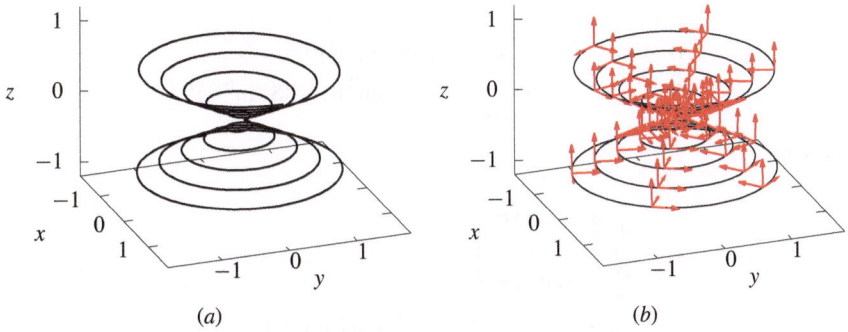

(a) (b)

Fig. 3.3 Conical spiral of Pappus (3.12), **a** with $a = 0.05$, $b = 0.025$, $-10\pi < \theta < +10\pi$ and some intrinsic frames on this curve (scale of vectors: 0.4)

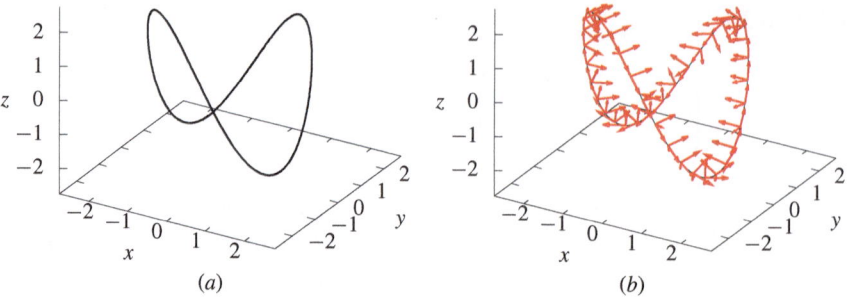

Fig. 3.4 Curve (3.13) **a** and some intrinsic frames on it (**b**, vector scale: 0.5)

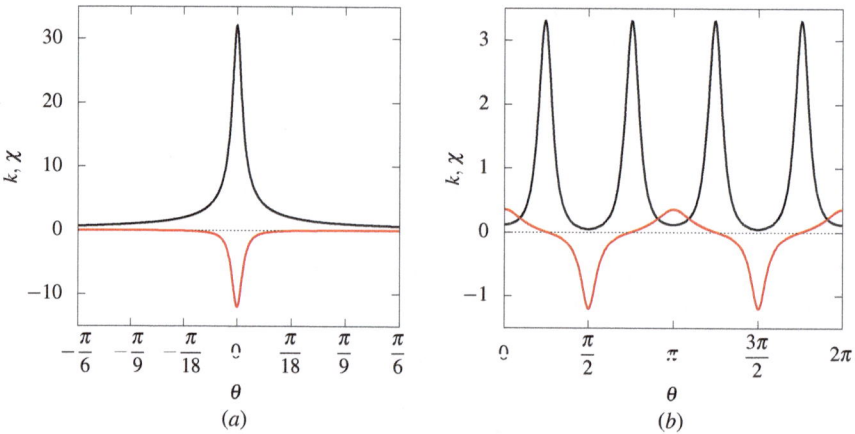

Fig. 3.5 Curvature (black line) and torsion (red line) of **a** the conic spiral of Pappus (3.12) and **b** the curve (3.13) as functions of the parameter θ (radians)

Solution 3.2 Setting:

$$A(\theta) = (17 - 61 \sin^2 \theta + 64 \sin^4 \theta)^{1/2}$$
$$B(\theta) = (17 + 84 \sin^2 \theta - 48 \sin^4 \theta - 48 \sin^6 \theta)^{1/2},$$

one gets:

$$\boldsymbol{\tau} = \frac{1}{A} \begin{pmatrix} -2 \sin \theta \\ \cos \theta \\ 4 \cos 2\theta \end{pmatrix},$$

$$\boldsymbol{\nu} = \frac{1}{AB} \begin{pmatrix} \cos \theta (-17 + 64 \sin^4 \theta) \\ 2 \sin \theta (11 - 32 \sin^2 \theta + 16 \sin^4 \theta) \\ -\sin 2\theta (7 + 6 \sin^2 \theta) \end{pmatrix},$$

$$
\boldsymbol{b} = \frac{1}{B} \begin{pmatrix} -2\sin\theta\,(3 - 2\sin^2\theta) \\ -4\cos\theta\,(1 + 2\sin^2\theta) \\ 1 \end{pmatrix},
$$

$$
k = \frac{2B}{A^3}, \qquad \chi = \frac{6\cos 2\theta}{B^2}.
$$

3.2 Suggested Readings

A comprehensive analysis of issues treated in this lesson can be found in Adams and Essex (2010, Chap. 11, pp. 646–648).

Reference

R. A. Adams and C. Essex. *Calculus: Several Variables*. Pearson Education Canada, Toronto, ON, 7th edition, 2010.

Chapter 4
Functions of a Vector Variable

Some properties, in particular differential properties of curves, *i.e.*, vector functions of a scalar variable (parameter), were explored in the previous lectures. Now the dual situation will be faced, that is to study *scalar functions of several scalar variables* or, as commonly said, *functions of a vector variable*.

4.1   Scalar Functions of Several Scalar Variables

A number of such functions is well known: for example, if $x = (x, y, z)$ is a vector in $\mathbb{R}^3$ its modulus

$$|x| = \sqrt{x^2 + y^2 + z^2} = f(x, y, z)$$

is a correspondence between a vector (*i.e.*, the three scalar quantities x, y and z) and a real (in this case, non-negative) number. Notice that it is possible to calculate the modulus of any vector: This fact can be more properly expressed just saying that the function $|x|$ *exists* throughout $\mathbb{R}^3$.

More well-known important examples come from the *implicit* representation of plane curves and surfaces in three dimensional space. For example, a conic may be seen as the intersection between the $z = 0$ plane and the surface of $\mathbb{R}^3$ explicitly represented by the equation $z = p_2(x, y)$, where p_2 is a second degree polynomial in the variables x and y:

$$p_2(x, y) = \alpha x^2 + 2\beta xy + \gamma y^2 + \delta x + \mu y + \omega = 0, \qquad (4.1)$$

where α, β, γ, δ, μ and ω are given real numbers. Notice that the function p_2 also *exists* everywhere in the (x, y) plane. As typical examples, in Figs. 4.1 and 4.2 (in Fig. 4.1 degenerate cases are drawn, whereas in Fig. 4.2 two degenerate cases and the circumference are shown) a red line draws the surfaces $z = p_2(x, y)$ in a proper

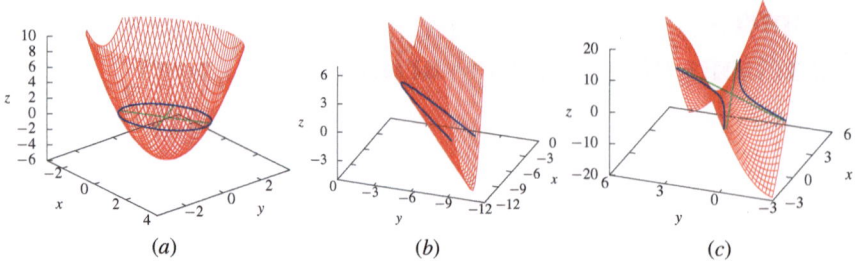

Fig. 4.1 Conics and related surfaces $z = p_2(x, y)$. Coefficients of p_2 are listed in Table 4.1. The ellipse **a** is centered in $x_0 = (0.35, -0.05)$, and has semi-axes $a \simeq 1.20890$, $b \simeq 1.95603$. The semi-axis a forms an angle of $121° \, 43' \, 3''$ with the positive x axis. The parabola **b** $y' = ax'^2$ has the vertex at point $x_0 \simeq (-3.01563, -3.32813)$ and the x' axis forms an angolo of $135°$ with the positive direction x axis. The coefficient **a** is about 11.31371. Finally, the hyperbola **c** has the center at $x_0 = (1.25, 0.875)$. In normal form $x'^2/a^2 - y'^2/b^2 = 1$ has coefficients $a \simeq 0.69468$, $b \simeq 1.74949$. The x' axis forms an angle of $127° \, 58' \, 55''$ with the positive x axis

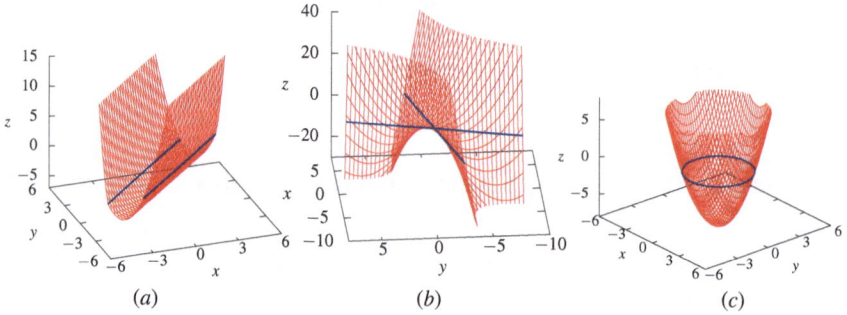

Fig. 4.2 Conics and related surfaces $z = p_2(x, y)$. Coefficients of p_2 are listed in Table 4.2. The conic in **a** degenerate in two parallel lines of equations $px + qy + r_{1,2} = 0$, with $p = -q \simeq 1.41421$ and $r_1 \simeq 2.12132, r_2 \simeq -1.41421$. Conversely, the conic in **b** degenerates in two intersecting lines at point $x_0 = (0, -0.5)$ of equations $y' = \pm mx'$, with $m \simeq 1.73205$. The x' axis forms an angle of $135°$ with the positive x axis. Finally, in **c** the conic is a circumference centered at the point $x_0 = (1.5, -1)$ with radius $R \simeq 2.69258$

Table 4.1 Coefficients of conics drawn in Fig. 4.1

α	β	γ	δ	μ	ω	Type of conic
2	−1	3	−1.5	1	−5	Ellipse 4.1a
2	−2	2	1.5	−1	1	Parabola 4.1b
1	−2	2	1	1.5	−3	Hyperbola 4.1c

region of the (x, y) plane, whereas a blue line draws the curves generated by the intersection of such surfaces with the $z = 0$ plane, that are indeed curves in the (x, y) plane, called conics.

Table 4.2 Coefficients of conics drawn in Fig. 4.2

α	β	γ	δ	μ	ω	Type of conic
2	−1	3	−1.5	1	−5	Parallel lines 4.2a
1	−2	1	−2	1	0.25	Intersecting lines 4.2b
1	0	1	−3	2	−4	Circumference 4.2c

In principle, it is not difficult to extend the concept of a function to a set of n real variables $(x_1, x_2, \ldots, x_n)$, that is:

$$f : \mathbb{R}^n \longrightarrow \mathbb{R}^1$$
$$x \mapsto f(x).$$

(4.2)

Writing (4.2) means one has a map f that, combining the n real numbers $x_1, x_2, \ldots,$ x_n (for the sake of clarity, one may think of them as the n components of vector x), results in another real number. In order to make it clear that this latter depends on the n numbers $x_1, x_2, \ldots, x_n$, the notation $f(x_1, x_2, \ldots, x_n)$ or, more synthetically, $f(x)$ is used. The natural number n can be arbitrarily large: for example, in Fluid Mechanics the numerical simulation of flow fields deals with functions of thousands of variables.

4.2 Field of Existence

When studying a function of several variables (4.2), one must first identify the *field of existence* $\mathcal{E}$, that is to specify the subset (open, closed or none of them) of $\mathbb{R}^n$ where the function f (4.2) is defined. In the previous two examples (the function $f(x) = |x|$ and the polynomials $p_2(x, y)$) the field of existence was the entire space ($\mathbb{R}^3$ and $\mathbb{R}^2$, respectively), but it is not so in general. Some examples, are shown in the following, in the form of worked problems. For the sake of clarity of the graphical representation, the analysis will be limited to functions of two variables.

Problem 4.1 Define the field of existence of the function:

$$f(x, y) = \frac{x + y}{x(x^2 - y^2 + 2)}.$$

(4.3)

Solution 4.1 The function (4.3) exists in the (x, y) plane wherever the denominator is not zero (in general, existence is to be checked at points where the denominator is zero, when the numerator vanishes as well. In this example the origin is such a point). The field of existence is then the open set of points exterior to the curves $x = 0$, $y = \pm\sqrt{x^2 + 2}$ drawn with a red line in Fig. 4.3a.

Now one has to explain why the function (4.3) does not exist at the point denoted by a red square in Fig. 4.3a, that is at the origin. In fact, this is a point where both the numerator and the denominator of f (4.3) vanish. If one represents the vector x in a

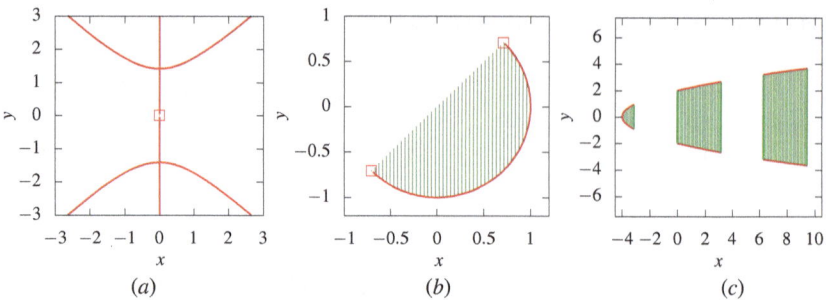

Fig. 4.3 Fields of existence for the functions (4.3) **a**, (4.4) **b** and (4.5) **c**. Solid red lines denote curves outside the field, whereas dashed green lines denote the interior of the field

neighbourhood of the origin using polar coordinates ($x = \rho \cos \theta$, $y = \rho \sin \theta$ with ρ very small) the function is will be written in the form:

$$\frac{x + y}{x(x^2 - y^2 + 2)} \equiv \frac{1}{x^2 - y^2 + 2} \frac{x + y}{x} \simeq \frac{1}{2}(1 + \tan \theta) \ .$$

This representation (terms of order ρ^2 are neglected) shows how the function could take any real value (including $\pm \infty$) in an arbitrarily small neighbourhood of the origin for varying θ. Thus, the function (4.3) does not exist at point $x = 0$.

Problem 4.2 Define the field of existence of the function:

$$f(x, y) = \frac{\sqrt{x - y}}{\sqrt{1 - x^2 - y^2}} \ . \tag{4.4}$$

Solution 4.2 The field of existence is determined by the intersection of the following subsets of the (x, y) plane: $I: x \geq y$; $II: |x| \leq 1$; $III: |x| \neq 1$. The set I guarantees the expression under the square root in the numerator to result in a nonnegative number, the set II ensures the same condition is met for the square root in the denominator, whereas the set III prevents the denominator to vanish. The field of existence is the domain (not open nor closed) where the above conditions are met, and it is represented in Fig. 4.3b. Notice that the open bisector line segment joining the two points $(-\sqrt{2}/2, -\sqrt{2}/2)$ and $(+\sqrt{2}/2, +\sqrt{2}/2)$ (drawn in green) is included in the field of existence (the function (4.4) is 0 in this set), whereas the arc of the unit circle between the same points drawn in red is not included. Few remarks are in order on the existence of the function (4.4) at points denoted by the red square in Fig. 4.3b. It must be kept in mind that at such points both the numerator and the denominator vanish. The function (4.4) is now studied in a neighbourhood of the point $(\sqrt{2}/2, \sqrt{2}/2)$ (analogous discussion can be conducted on the point $(-\sqrt{2}/2, -\sqrt{2}/2)$), setting:

$$x = \frac{\sqrt{2}}{2} + \rho \cos \theta \ , \qquad y = \frac{\sqrt{2}}{2} + \rho \sin \theta$$

with $5\pi/4 \leq \theta < 7\pi/4$ and ρ very small. Furthermore, consider that, in this range of θ, $\sin\theta$ is negative and its absolute value (*i.e.* $-\sin\theta$) is larger than $\sqrt{2}/2$ so that, in particular, it is larger than the corresponding value of $\cos\theta$. It follows that the function (4.4) takes values, in a very small neighbourhood of the point $(\sqrt{2}/2, \sqrt{2}/2)$, given by:

$$\frac{\sqrt{x-y}}{\sqrt{1-x^2-y^2}} \simeq \frac{\sqrt{-\sin\theta+\cos\theta}}{2^{1/4}\sqrt{-\sin\theta-\cos\theta}},$$

which still depends on θ. One can derive from this that the function does not exist at point $(\sqrt{2}/2, \sqrt{2}/2)$. Analogous discussion leads to the same result at point $(-\sqrt{2}/2, -\sqrt{2}/2)$.

Problem 4.3 Define the field of existence of the function:

$$f(x, y) = \sqrt{\sin x} \log(x - y^2 + 4).$$ (4.5)

Solution 4.3 Again, the field of existence is obtained as the intersection of two sets. *I*: the verticali strips where $\sin x \geq 0$ (that is $2k\pi \leq x \leq (2k+1)\pi$ for any integer k); *II*: the concavity of the parabola with horizontal axis $x = y^2 - 4$. Part of such a field is depicted in Fig. 4.3c.

Problem 4.4 Identify the field of existence of the function:

$$f(x, y, z) = \log(1 - x^2 - y^2)\sqrt{4 - x^2 - y^2 - z^2}.$$ (4.6)

Solution 4.4 The field $\mathscr{E}$ is obtained as the intersection of the field of existence of the logarithm ($\mathscr{A}$) and the field of existence of the square root ($\mathscr{B}$). Since the argument of the logarithm has to be positive, $\mathscr{A}$ is the straight circular cylinder with unit radius, having the z axis as the symmetry axis. Observe that the surface of this cylinder *is not part* of the field of existence. Furthermore, since the argument of the square root must be nonnegative, the set $\mathscr{B}$ is given by the closed sphere $B_2(\mathbf{0})$.

4.3 Limit of a Function of Several Scalar Variables

In order to investigate the local properties of the function (4.2), we introduce now the required tools. One says that the limit of f as the vector $\mathbf{x} \in \mathbb{R}^n$ tends to the vector $\mathbf{x}_0$ is $\varphi \in \mathbb{R}^1$ and one writes:

$$\lim_{\mathbf{x}\to\mathbf{x}_0} f(\mathbf{x}) = \varphi$$ (4.7)

when given any neighbourhood of φ, $I(\varphi)$, (for example a symmetric interval, centered at φ: $(\varphi - \varepsilon, \varphi + \varepsilon)$, having taken $\varepsilon > 0$) one can find a neighbourhood of $\mathbf{x}_0$, not necessarily including $\mathbf{x}_0$, $U(\mathbf{x}_0)$ (for example the intersection of a sphere $B_{\delta_\varepsilon}(\mathbf{x}_0)$ centered at $\mathbf{x}_0$ with radius δ_ε and the field of existence $\mathscr{E}$ of the function) such that every point $\mathbf{x} \in U(\mathbf{x}_0)$ is mapped by the function f into a point of the given interval

$I(\varphi)$. When translating into symbols, one prefers to use symmetric neighbourhoods (symmetric intervals and open spheres centered at point). So, the statement (4.7) is equivalent to

$$\forall \varepsilon > 0 \exists \delta_\varepsilon > 0 \mid \forall x \in B_{\delta_\varepsilon}(x_0) \cap \mathcal{E} \Rightarrow |f(x) - \varphi| < \varepsilon. \qquad (4.8)$$

Problem 4.5 Prove that:

$$\lim_{x \to (-1,0)^\mathsf{T}} \frac{x}{x^2 + y^2} = -1. \qquad (4.9)$$

Solution 4.5 The function $f(x, y) = x/(x^2 + y^2)$ exists at every point in the plane ($\mathcal{E} = \mathbb{R}^2 - \{\mathbf{0}\}$) and then the point $(-1, 0)$ is an inner point of the field of existence. Let ε be a positive number; for the sake of simplicity, it will be taken less than 1. Points in a spherical neighbourhood of the point $(-1, 0)$ with radius δ_ε are now represented in polar form, that is $x = -1 + \rho \cos \theta$, $y = \rho \sin \theta$ with $\rho \in (0, \delta_\varepsilon)$. It is searched δ_ε such that $f[B_{\delta_\varepsilon}(x_0)] \subset (-1 - \varepsilon, -1 + \varepsilon)$, being $f(x, y) = x/(x^2 + y^2)$. In order to show that $-1 - \varepsilon < f(x, y) < -1 + \varepsilon$ for points $(x, y) \in B_{\delta_\varepsilon}(x_0)$ one considers the quantity:

$$|1 + f| = \rho \frac{|\rho - \cos \theta|}{1 - 2\rho \cos \theta + \rho^2},$$

which must result less than ε for a properly selected δ_ε. It is assumed $\delta_\varepsilon < 1$ and a quantity larger than $|1 + f|$ is found as follows:

$$|1 + f| < \frac{\rho(1 + \rho)}{(1 - \rho)^2} := g(\rho),$$

where it is easily seen that the new function $g(\rho)$ is monotonically increasing in the interval $(0, 1)$. It follows that, if $g(\delta_\varepsilon) < \varepsilon$ is required, $|1 + f| < \varepsilon$ for every $\rho \in (0, \delta_\varepsilon)$ and the limit (4.9) is proved. Solving (with respect to δ_ε) the quadratic inequality $g(\delta_\varepsilon) < \varepsilon$ one finds δ_ε as:

$$\delta_\varepsilon = \frac{\sqrt{1 + 8\varepsilon} - (1 + 2\varepsilon)}{2(1 - \varepsilon)},$$

which allows to verify the limit (4.9).

Problem 4.6 Prove that:

$$\lim_{x \to (1,0,-1)^\mathsf{T}} (x + y - z) \cos(x^2 - z^2) = 2. \qquad (4.10)$$

Solution 4.6 The function $f(x, y, z) = (x + y - z) \cos(x^2 - z^2)$ exists at every point of the space ($\mathcal{E} = \mathbb{R}^3$), and then the point $(1, 0, -1)$ is an inner point to the

the field of existence. In this case it is worth considering a cubic neighborhood of the point x_0, since there is no possible combination of variables x, y and z that could lead to $|x|$. Fixed $\varepsilon > 0$ arbitrarily small one search for $\delta_\varepsilon > 0$ such that at each point $x \in (1 - \delta_\varepsilon, 1 + \delta_\varepsilon) \times (-\delta_\varepsilon, +\delta_\varepsilon) \times (-1 - \delta_\varepsilon, -1 + \delta_\varepsilon)$, written as $(1 + \xi, \eta, -1 + \zeta)$ the following relation holds:

$$|(x + y - z)\cos(x^2 - z^2) - 2| = |(2 + \xi + \eta - \zeta)\cos[(\xi + \zeta)(2 + \xi - \zeta)] - 2| < \varepsilon.$$

Since it has to be ε arbitrarily small, one can certainly assume $\varepsilon < 1$. It is requested a stronger inequality than the above, but much simpler:

$$|(2 + \xi + \eta - \zeta)\cos[(\xi + \zeta)(2 + \xi - \zeta)] - 2|$$

$$< 2\{1 - \cos[(\xi + \zeta)(2 + \xi - \zeta)]\} + |\xi + \eta - \zeta||\cos[(\xi + \zeta)(2 + \xi - \zeta)]| < \varepsilon.$$

Observe that taking $\delta_\varepsilon < \pi/16$ one gets the absolute value of the argument of cosine, that is certainly less than $4\delta_\varepsilon(1 + \delta_\varepsilon) < \pi/2$. Thus, the following inequalities hold:

$$1 - \cos[(\xi + \zeta)(2 + \xi - \zeta)] \leq \frac{2}{\pi}|\xi + \zeta||2 + \xi - \zeta| < \frac{2}{\pi}4\delta_\varepsilon(1 + \delta_\varepsilon) < 5\delta_\varepsilon$$

$$|\xi + \eta - \zeta||\cos[(\xi + \zeta)(2 + \xi - \zeta)]| \leq 3\delta_\varepsilon.$$

So, requiring $13\delta_\varepsilon < \varepsilon$ the limit (4.10) has been proven. Then, one can take $\delta_\varepsilon = \varepsilon/14$.

Problem 4.7 Verify the following limit:

$$\lim_{x \to (0,0)^\mathsf{T}} \sqrt{1 - x^2 - y^2}\cos(x^2 + y^2) = 1. \tag{4.11}$$

Solution 4.7 The function $f(x, y) = \sqrt{1 - x^2 - y^2}\cos(x^2 + y^2)$ exists inside and on the unit circle (circle of radius 1 and centered at the origin) and then the origin is an inner point to the field of existence. Take an arbitrarily small positive number ε and search for the radius δ_ε of a spherical neighbourhood $B_{\delta_\varepsilon}(0)$ of the origin such that for each $x \in B_{\delta_\varepsilon}(0)$ one has $1 - \varepsilon < f(x, y) < 1 + \varepsilon$. Of course, one takes $\delta_\varepsilon < 1$. Represent in polar form the point x ($x = \rho\cos\theta$, $y = \rho\sin\theta$ with $0 < \rho < \delta_\varepsilon$ and $0 \leq \theta < 2\pi$) and consider the inequality:

$$1 - \sqrt{1 - x^2 - y^2}\cos(x^2 + y^2) = 1 - \sqrt{1 - \rho^2}\cos(\rho^2)$$

$$\equiv \left(1 - \sqrt{1 - \rho^2}\right) + \sqrt{1 - \rho^2}\left[1 - \cos(\rho^2)\right] < \varepsilon.$$

Since in the interval $0 < \rho < 1$ the curve $z = \sqrt{1 - \rho^2}$ lies above the straight line $z = 1 - \rho$, one has: $1 - \sqrt{1 - \rho^2} < \rho$. Analogously, the curve $z = \cos(\rho^2)$ in the interval $0 < \rho < \sqrt{\pi/2}$ lies above the straight line $z = 1 - \sqrt{2/\pi}\rho$ and then one can write: $1 - \cos(\rho^2) < \sqrt{2/\pi}\rho$. Thus, requiring the stronger inequality:

$$\left(1 - \sqrt{1 - \rho^2}\right) + \sqrt{1 - \rho^2}\left[1 - \cos(\rho^2)\right] < \rho + \sqrt{\frac{2}{\pi}}\,\rho < 2\delta_\varepsilon < \varepsilon,$$

that is taking $\delta_\varepsilon = \varepsilon/3$, the limit (4.11) is proven.

Given a function (4.2) and a point $x_0 \in \mathcal{E}$, one says that f is *continuous at* x_0 if it happens that:

$$\lim_{x \to x_0} f(x) = f(x_0). \tag{4.12}$$

Exercise 4.1 Determine the fields of existence and study the continuity of the following functions at the given points:

(a) $\quad f(x, y) = \sqrt{\dfrac{x^2 + y^2}{1 + x + y}}$ $\qquad\qquad$ in $(1, 0), (-1/2, -1/2), (0, 0)$

(b) $\quad f(x, y) = \cos(x)\log(1 - y)$ $\qquad\qquad$ in $(0, 0), (-\pi, 1), (\pi, 0)$

(c) $\quad f(x, y) = \arccos(x + y)$ $\qquad\qquad\quad$ in $(1, 0), (-1, 1), (0, 0)$

(d) $\quad f(x, y) = \log\left(1 + \dfrac{1 - 2xy}{x^2 + y^2}\right)$ $\qquad$ in $(0, 1), (0, 0), (-1, 1)$.

The idea of limit extends to case of a function (4.2) diverging at a given point x_0 of the space $\mathbb{R}^n$. In this case one writes:

$$\lim_{x \to x_0} f(x) = +\infty, \tag{4.13}$$

if the function diverges to $+\infty$, whereas an analogous writing is used if the function diverges to $-\infty$. Referring to the case $+\infty$, the meaning of such a writing is as follows. However one takes an arbitrarily large positive number M, it is possible to find a neighbourhood of the point x_0 (for example, given by the intersection of a sphere centered in x_0 of radius δ_M and $\mathcal{E}$) such that the values the function takes at every point in this neighbourhood are greater than M. Translated in symbols, the statement (4.13) is equivalent to

$$\forall M > 0 \,\exists \delta_M > 0 \mid \forall x \in B_{\delta_M}(x_0) \cap \mathscr{E} \Rightarrow f(x) > M \,. \qquad (4.14)$$

Finally, the definition of limit may be extended to the case $x \to \infty$. For example:

$$\lim_{x \to \infty} f(x) = \varphi \,,$$

with φ finite, means that however one takes a neighbourhood of the point φ (for example, a symmetric one with arbitrarily small amplitude ε) it is possible to find a number M_ε large enough to guarantee that, for each x such that $|x| > M_\varepsilon$ (therefore, belonging to a neighbourhood of $+\infty$), the inequality: $|f(x) - \varphi| < \varepsilon$ holds. Examples and verifications will be illustrated in the following lecture.

4.4 Suggested Readings

For introductory concepts on functions of several variables, see Apostol (1969 Chap. 8, pp. 243–252), and Lax and Terrell (2017 Chap. 2, pp. 63–78).

References

T. M. Apostol. *Calculus, Volume II*. Wiley, Hoboken, NJ, 2nd edition, 1969.
P. D. Lax and M. S. Terrell. *Multivariable Calculus with Applications*. Springer, Cham, CH, 2017.

Chapter 5
Continuity and Differentiability of Functions of a Vector Variable

In the previous lesson the definition of the limit for a function f of the n variables x_1, $x_2, \ldots, x_n$ (for short, they can be considered as the components of the vector $x \in \mathbb{R}^n$) has been given. In particular, the limit for a divergent function has been introduced. Examples of limits of such kind are discussed in the following. Moreover, some topological consequences of the continuity and the differentiability of functions of several scalar variables are presented.

5.1 Divergent Limits and Limits at Infinity

Example 5.1 Consider the function of the two variables x and y:

$$f(x, y) = \log(x^2 + y^2). \tag{5.1}$$

Its existence field is $\mathscr{E} = \mathbb{R}^2 - \{0\}$, *i.e.* the real plane deprived of the origin. In correspondence to the origin, f behaves as specified by the limit:

$$\lim_{x \to 0} f(x) = -\infty. \tag{5.2}$$

It diverges to $-\infty$. This fact is proved in the following way. First of all, a neighbourhood of the "point" $-\infty$ is fixed, choosing a very large positive number M and considering the values of f below the threshold $-M$. It is now showed that in correspondence to the above neighbourhood of $-\infty$, it is possible to find in the existence field $\mathbb{R}^2 - \{0\}$ a neighbourhood of the origin so small that all the points x belonging to it are mapped inside the fixed neighbourhood of $-\infty$. In order to simplify the discussion, the shape of the neighbourhood of the origin is chosen as circular, with radius δ_M depending on the arbitrary number M. Using a polar representation for the generic point of $B_{\delta_M}(0)$ ($x = \rho \cos \theta$, $y = \rho \sin \theta$ with $0 < \rho < \delta_M$

© The Author(s), under exclusive license to Springer Nature Switzerland AG 2024
G. Riccardi et al., *Multidimensional Differential and Integral Calculus*,
https://doi.org/10.1007/978-3-031-70326-3_5

and $0 \le \theta < 2\pi$), one obtains: $f(x, y) = 2 \log \rho < 2 \log \delta_M$, so that, if it is possible to enforce the inequality $2 \log \delta_M < -M$, the limit (5.2) is verified. Taking the exponential (remember that the exponential e^ξ is a growing function of ξ and then the direction of the inequality does not change) of both sides, one obtains: $\delta_M^2 < \exp(-M)$. It follows that one can choose (as an example) $\delta_M = \exp(-M/2)/2$ and this choice implies that all the points with $0 < \rho < \delta_M$ satisfy $f(x, y) < -M$, i.e. the image of the circular neighbourhood of the origin in the domain $\mathbb{R}^2 - \{0\}$ lies inside the fixed neighbourhood of $-\infty$.

Example 5.2 Verify that the function of the three variables x, y and z:

$$f(x, y, z) = \frac{\cos x \cos y \cos z}{(x^2 + y^2 + z^2)^{3/2}} \tag{5.3}$$

diverges to $+\infty$ as $x \to 0$:

$$\lim_{x \to 0} f(x) = +\infty . \tag{5.4}$$

As before, the function (5.3) is not defined at the point $x = 0$, its existence field being $\mathscr{E} = \mathbb{R}^3 - \{0\}$. In order to prove the limit (5.4), an arbitrary neighbourhood of the point $+\infty$ is fixed, choosing a large positive number M and considering the values of f above the threshold $+M$. It is now shown that in correspondence to this number is possible to define a neighbourhood of the origin in the existence field $\mathbb{R}^3$ such that the images of its points through the function f (5.3) lie inside the above neighbourhood of $+\infty$. In order to simplify the algebraic details, a spherical neighbourhood of the origin $B_{\delta_M}(0)$ is considered, having the radius $\delta_M < \pi/2$ and depending on the above (arbitrary) number M. Note that, due to the behaviour of the function $\cos \xi$ for $\xi \in (0, \pi/2)$, the important inequality $f(x, y, z) > [(\cos \delta_M)/\delta_M]^3$ is everywhere verified in $B_{\delta_M}(0)$, so that if δ_M is selected in such a way that $\cos \delta_M/\delta_M > M^{1/3}$, therefore the limit (5.4) is proved. However, this inequality is still too complicated to be directly satisfied. In order to obtain a simpler (but stronger) inequality, the lower bound is the cosine. So, $\cos \xi > 1 - 2\xi/\pi$ for $\xi \in (0, \pi/2)$ is used and the chain of inequalities: $(\cos \delta_M)/\delta_M > (1 - 2\delta_M/\pi)/\delta_M > M^{1/3}$ is obtained. The latter is simpler than the original one and leads to the choice $\delta_M = 1/(M^{1/3} + 1)$, which demonstrates the limit (5.4).

Example 5.3 The function of the variables x, y and z:

$$f(x, y, z) = \frac{\sqrt{4 - x^2} + \sqrt{-z}}{(x - 1)^2 + y^2 + (z + 1)^2} \tag{5.5}$$

diverges to $+\infty$ as $x \to (1, 0, -1)^{\mathsf{T}} =: x_0$. Once it is translated in the symbolic language, this behaviour of the function f (5.5) becomes:

$$\lim_{x \to x_0} f(x) = +\infty . \tag{5.6}$$

The existence field of the function f (5.5) is obtained by the intersection of three sets. I : (for the existence of the first root at numerator) the strip included between the planes $x = -2$ and $x = +2$; II : (for the existence of the second root at numerator) the half-space $z \leq 0$; III: (to enforce that the denominator does not vanish) the three-dimensional space deprived of the point x_0 $\mathbb{R}^3 - \{x_0\}$. As before, $i.e.$, a large positive number M is arbitrarily fixed, such that the corresponding neighbourhood of the point $+\infty$ is also defined by the inequality $f > +M$. If the limit (5.6) is true, a neighbourhood of the point x_0 exists, such that it is mapped inside the previous neighbourhood of $+\infty$ by the function f (5.5). In order to simplify the algebraic handling of the inequalities, the spherical neighbourhood $B_{\delta_M}(x_0)$ of the point x_0 is considered. Moreover, the radius δ_M of $B_{\delta_M}(x_0)$ is assumed as $\delta_M < 1$. In this way, the set $B_{\delta_M}(x_0) - \{x_0\}$ lies inside the existence field $\mathscr{E}$. The three components of any point $(x, y, z) \in B_{\delta_M}(x_0)$ verify the restrictions: $1 - \delta_M < x < 1 + \delta_M, -\delta_M < y < +\delta_M$ and $-1 - \delta_M < z < -1 + \delta_M$. It follows the inequality:

$$f(x, y, z) > \frac{\sqrt{4 - (1 + \delta_M)^2} + \sqrt{1 - \delta_M}}{\delta_M^2}, \tag{5.7}$$

which gives the possibility to find a δ_M discussing a simpler inequality. Indeed, for any $\xi \in (0, 1)$ the following inequalities:

$$4 - (1 + \xi)^2 > 3(1 - \xi), \quad \sqrt{1 - \xi} > 1 - \xi$$

hold. They are used in order to obtain simpler (but stronger) inequalities from the one (5.7):

$$\frac{\sqrt{4 - (1 + \delta_M)^2} + \sqrt{1 - \delta_M}}{\delta_M^2} > \frac{(3^{1/2} + 1)\sqrt{1 - \delta_M}}{\delta_M^2} > \frac{1 - \delta_M}{\delta_M^2}.$$

At this point, the simplest inequality $(1 - \delta_M)/\delta_M^2 > M$ is enforced, leading to the choice $\delta_M = (\sqrt{4M + 1} - 1)/(3M)$. The limit (5.6) is then proved.

Example 5.4 A limit as $x \to \infty$ is now verified. The rational function:

$$f(x, y) = \frac{x - y}{x^2 + 2y^2} \tag{5.8}$$

vanishes in this limit, $i.e.$ in the symbolic language:

$$\lim_{x \to \infty} f(x) = 0. \tag{5.9}$$

The property (5.9) is proved showing that for any neighbourhood of the point 0 in $\mathbb{R}^1$, it is possible to find a neighbourhood of infinity in $\mathbb{R}^n$ the image of which through the map f (5.8) lies inside the neighbourhood of 0. As before, in order to simplify the algebraic handling of the involved inequalities, the above neighbourhoods are chosen as spherical. Once the radius ε of the neighbourhood of the origin $B_\varepsilon(0)$ in $\mathbb{R}^1$ is fixed in an arbitrary way, the radius M_ε of the neighbourhood of infinity $B_{M_\varepsilon}(\infty) = \{x \in \mathbb{R}^n \mid |x| > M_\varepsilon\}$ is searched in order to satisfy the request $f(B_{M_\varepsilon}(\infty)) \subset B_\varepsilon(0)$. Representing a point $x \in B_{M_\varepsilon}(\infty)$ in polar form ($x = \rho \cos\theta$, $y = \rho \sin\theta$ with $\rho > M_\varepsilon$ and $0 \le \theta < 2\pi$), the value of the function f (5.8) at the point x satisfies the relation: $|f(x)| = |\cos\theta - \sin\theta|/[(1 + \sin^2\theta)\rho]$, in which the trigonometric factor is smaller than $3/2$ for any θ. As a consequence, it becomes possible to satisfy a stronger, but much more simpler, inequality:

$$|f(x)| < \frac{3}{2\rho} < \frac{3}{2M_\varepsilon} < \varepsilon,$$

from which it follows that the choice $M_\varepsilon = 2/\varepsilon$ proves the limit (5.9).

Example 5.5 The function:

$$f(x, y, z) = \tanh[x^2 + y^2 + (z - 1)^2] \tag{5.10}$$

goes to 1 for $x \to \infty$:

$$\lim_{x \to \infty} f(x) = 1. \tag{5.11}$$

In order to prove the limit (5.11), an arbitrary neighbourhood of the point 1 in $\mathbb{R}^1$ is fixed and a neighbourhood of infinity that is mapped inside the first one by the function f (5.10) is searched. As before, the spherical neighbourhoods are chosen, with radii ε and M_ε. For technical reasons, it is also required that $\varepsilon < 2$ and $M_\varepsilon > 1$, without loss of generality. The proof consists in showing that in correspondence to the fixed ε a positive number M_ε (diverging to $+\infty$ as $\varepsilon \to 0$) can be found, such that for any $x \in B_{M_\varepsilon}(\infty)$, the following inequality:

$$1 - f(x) = \frac{2}{1 + \exp\{2[x^2 + y^2 + (z - 1)^2]\}} < \varepsilon$$

is verified. In $B_{M_\varepsilon}(\infty)$ a spherical coordinate system is used, so that a point $x \in B_{M_\varepsilon}(\infty)$ has coordinates: $x = \rho \cos\theta \cos\varphi$, $y = \rho \sin\theta \cos\varphi$ and $z = \rho \sin\varphi$, with $0 \le \theta < 2\pi$, $-\pi/2 \le \varphi \le +\pi/2$ and $\rho > M_\varepsilon$. The argument of the hyperbolic tangent $x^2 + y^2 + (z - 1)^2$ becomes: $\rho^2 \cos^2\varphi + (\rho \sin\varphi - 1)^2 = \rho^2 - 2\rho \sin\varphi + 1 \ge (\rho - 1)^2$. It follows that the exponential $\exp\{2[x^2 + y^2 + (z - 1)^2]\}$ is not smaller than $\exp[2(\rho - 1)^2] > \exp[2(M_\varepsilon - 1)^2]$ and it becomes possible to satisfy a stronger (but simpler) inequality:

$$\frac{2}{1 + \exp\{2[x^2 + y^2 + (z - 1)^2]\}} < \frac{2}{1 + \exp[2(M_\varepsilon - 1)^2]} < \varepsilon.$$

A possible choice of the radius of the neighbourhood of the infinity is then the following one:

$$M_\varepsilon = 1 + \sqrt{\log\left(\frac{2}{\varepsilon}\right)},$$

which proves the limit (5.11).

Exercise 5.1 Show that the following limits are correct:

(a) $\displaystyle\lim_{x \to 0} \frac{\cos x \cos y}{(x^2 + y^2)^{3/2}} = +\infty$

(b) $\displaystyle\lim_{x \to (-1,0,+1)^\mathsf{T}} \log\left[(x + 1)^2 + y^2 + (z - 1)^2\right] = -\infty$

(c) $\displaystyle\lim_{x \to (0,0,1)^\mathsf{T}} \frac{1}{\arcsin\left[x^2 + y^2 + (z - 1)^2\right]} = +\infty$

(d) $\displaystyle\lim_{x \to \infty} \frac{x^2 - y^2}{x^4 + y^4} = 0$

(e) $\displaystyle\lim_{x \to \infty} \exp\left[-(x^4 + y^4 + z^4)\right] = 0$

(f) $\displaystyle\lim_{x \to \infty} \sqrt{1 - \frac{z}{x^2 + y^2 + z^2}} = 1.$

5.2 Topological Properties of $\mathbb{R}^n$

A summary of the elementary topological properties about the subsets of $\mathbb{R}^n$ is now presented, with the aim of using these important issues in the successive developments.

– $A \subseteq \mathbb{R}^n$ is *open* if and only if for any $x \in A$ a sphere having center on this point and lying within A exists;
– $C \subseteq \mathbb{R}^n$ is *closed* if its complementary set C' is open;
– $B \subset \mathbb{R}^n$. The *interior* of B (indicated with $B°$) is the set of the points of B that are centers of spheres lying within B;
– $B \subset \mathbb{R}^n$. The *boundary* of B (indicated with ∂B) is the set of points of B that are not centers of spheres lying within B or B'. In other words, if $x \in \partial B$ each sphere having center on x intersects both B and B'.

The elementary properties of the open sets in $\mathbb{R}^n$ are now deduced. The arbitrary union of open sets is an open set. Indeed, consider the arbitrary (in general, uncountable) family of open sets:

$$\mathscr{F}^n = \{A_\lambda \subseteq \mathbb{R}^n, \ A_\lambda \text{ open}\}$$

the index λ running on a subset $\Lambda \subseteq \mathbb{R}$. The union of the sets of the family $\mathscr{F}^n$ is indicated with A:

$$A = \bigcup_{\lambda \in \Lambda} A_\lambda$$

and it is now shown that A is open. This property is proved taking an arbitrary point $x \in A$ and considering that it must belong to some A_λ, that is open. As a consequence, a sphere centered on x and lying within A_λ exists, but $A_\lambda \subseteq A$ so that the sphere is also inside A. For this reason, A is open. Consider now the intersection of a *finite number* (say, m) of sets of the family $\mathscr{F}^n$:

$$B = \bigcap_{k=1}^{m} A_k .$$

An arbitrary point $x \in B$ belongs to any set A_k for $k = 1, 2, \ldots, m$. As a consequence, m spheres having their centers superimposed to the point x and radii $r_1, r_2, \ldots, r_m$ exist such that $B_{r_k}(x) \subset A_k$ (for $k = 1, 2, \ldots, m$). Named with r the minimum of these radii (it is not a vanishing number, because the number of radii m is finite), the sphere $B_r(x) \subseteq B_{r_k}(x)$ lies within any A_k (for $k = 1, 2, \ldots, m$). For this reason it is inside B and therefore this set is open. As it can be foreseen from the above discussion (the minimum radius r can vanish, if the number of radii is infinite), the intersection of infinitely many sets of the family $\mathscr{F}^n$ behaves in a much more complicated way and could lead to a set that is not open, nor closed. In order to show such a behaviour, the (countable) family of open subsets of $\mathbb{R}^1$:

$$\mathscr{F}_1^1 = \{(-1/k, 1 + 1/k) \text{ with } k \in \mathbb{N}^+\}$$

($\mathbb{N}^+$ is the set of natural and positive numbers) is considered. The intersection of the sets of $\mathscr{F}_1^1$ is closed:

$$\bigcap_{k \in \mathbb{N}^+} \left(-\frac{1}{k}, 1 + \frac{1}{k}\right) = [0, 1],$$

the points 0 and 1 belonging to all the sets of $\mathscr{F}_1^1$. As another sample case, consider the (countable) family of open subsets of $\mathbb{R}^1$:

$$\mathscr{F}_2^1 = \{(0, 1 + 1/k) \text{ with } k \in \mathbb{N}^+\}.$$

The intersection of the sets of $\mathscr{F}_2^1$ is the interval $(0, 1]$, that is not open, nor closed. Another key-observation concerns the nature of the entire space ($\mathbb{R}^n$). By definition, it is an open set. But note that the boundary of $\mathbb{R}^n$ is empty, then $\mathbb{R}^n$ coincides with the union of itself and its boundary. In other words, $\mathbb{R}^n$ is also closed. As a consequence, the entire space is at the same time open and closed! Taking the complement sets, it is easy to show that the same thing occurs to the empty set $\emptyset$, too. The properties discussed above (*I*: the union of open sets is open; *II*: the finite intersection of open sets is open; *III*: the empty set and the entire space are open and closed at the same time) are the basic ones for building the so-called *topology* on the space $\mathbb{R}^n$. The same building can be made on an arbitrary space, once it is equipped with a family of its (by definition, open) subsets verifying the properties *I*, *II* and *III*.

Other consequences of the above definitions and properties are deduced using de Morgan's[1] rules. Given two subsets (D and E) of $\mathbb{R}^n$, the rules are:

$$(D \cup E)' = D' \cap E', \quad (D \cap E)' = D' \cup E' \tag{5.12}$$

(one follows from the other one, both are easily proved). The same rules (5.12) can be applied to the union or to the intersection of arbitrarily many sets. The first important consequence of the above rules is that the intersection of arbitrary many closed sets is closed, whereas nothing is known about their union. This fact is used in defining the closure of a set $B \subseteq \mathbb{R}^n$: the intersection of all the closed sets including B is the *closure* of B and is indicated by $\overline{B}$. Obviously, if B is closed its closure coincides with B, *i.e.* $\overline{B} = B$, whereas it is left to the Reader to verify that the closure of B is also obtained as the union between this set and its boundary, *i.e.* $\overline{B} = B \cup \partial B$.

All the definitions and the properties discussed above in $\mathbb{R}^n$ can be exported to its subsets. Indeed, consider a subset B of $\mathbb{R}^n$ (for example, B it is not open, nor closed) and think to B as the whole available space. An open set in B is now defined as the intersection between an open set $A \subseteq \mathbb{R}^n$ and the ambient space B. In general, the resulting set is not open nor closed in $\mathbb{R}^n$ (unless B is open). The collection of these open sets forms the so-called *induced topology* The closed sets are defined in an analogous way.

5.3 Consequences of the Continuity

The concept of continuous function has been introduced and discussed in the previous lesson, but now it becomes possible to examine some properties of such functions, as laws of transformation from $\mathbb{R}^n$ to $\mathbb{R}^1$. From this new point of view, the continuity takes an important geometric meaning, that will be discussed below.

Assume that a function $f : \mathbb{R}^n \to \mathbb{R}$ transform the subset $A \subseteq \mathbb{R}^n$ in the open set $I = (a, b) \subseteq \mathbb{R}$, being continuous in A. As it is well known, I is the image of the set A through the map f and A is the preimage of the set I by means of the same function. Take an arbitrary point $y \in A$, which is mapped in the number $\varphi \in I$ through the map f, *i.e.* $f(y) = \varphi \in (a, b)$. Due to the fact that the image I is an open set, a neighbourhood of the point φ exists such that $B(\varphi) \subseteq I$. Moreover, the function f being continuous in y, it is possible to find a neighbourhood $B(y)$ of the point y the image through f of which is inside the above neighbourhood $B(\varphi)$. As a consequence, given a point $y \in A$ a neighbourhood of it lying inside A is found, thanks to the continuity of the function f. The point $y \in A$ being arbitrary, the subset

---

[1] https://mathshistory.st-andrews.ac.uk/Biographies/De_Morgan/.

A of $\mathbb{R}^n$ results to be open, *i.e.* the preimage of an open set through a continuous function is still an open set.

Assume now that the function f is continuous in its domain $\mathbb{R}^n$. Its image $f(\mathbb{R}^n)$ is a subset of $\mathbb{R}$, say F, which is, in general, not open, nor closed. Consider a point $\varphi \in F$. Due to the fact that the set $\mathbb{R} - \{\varphi\}$ is open in $\mathbb{R}$, the set $F - \{\varphi\} = F \cap (\mathbb{R} - \{\varphi\})$ is open in F, too. As a consequence, the preimage of $F - \{\varphi\}$ is open in $\mathbb{R}^n$, *i.e.* if f is continuous the set:

$$\{x \in \mathbb{R}^n \mid f(x) \neq \varphi\} \tag{5.13}$$

is open. Taking the complementary sets, it follows also that the set:

$$\{x \in \mathbb{R}^n \mid f(x) = \varphi\} \tag{5.14}$$

is closed $\mathbb{R}^n$. In an analogous way, due to the fact that the set $(-\infty, \varphi)$ is open in $\mathbb{R}$, the intersection $F \cap (-\infty, \varphi)$ is also open in F and then the set:

$$\{x \in \mathbb{R}^n \mid f(x) < \varphi\} \tag{5.15}$$

is open in $\mathbb{R}^n$. Taking the complementary sets, it follows that the set:

$$\{x \in \mathbb{R}^n \mid f(x) \geq \varphi\} \tag{5.16}$$

is closed in $\mathbb{R}^n$. Obviously, also sets as in Eq. (5.15), with the inequality $f(x) > \varphi$, or (5.16), with $f(x) \leq \varphi$, are open and closed, respectively.

5.4 Differential of a Function $f : \mathbb{R}^n \to \mathbb{R}$

The continuity of a function $f : \mathbb{R}^n \to \mathbb{R}$ at an interior point x_0 of its existence field means that for any point x near x_0, the corresponding image lies near $f(x_0)$, too. But nothing is known about the smoothness of the behaviour of $f(x)$ for any x lying in a neighbourhood of x_0.

Assume that the function f is continuous in a neighbourhood of the point x_0 belonging to its existence domain, where the function takes the (finite) value $f(x_0)$. The behaviour of f near x_0 is investigated considering another point x close to x_0 and comparing the corresponding value of f with $f(x_0)$. In order to highlight the fact that x lies in a "small" neighbourhood of x_0, it is rewritten summing to x_0 the (vector) small increment Δx, *i.e.* $x =: x_0 + \Delta x$. Naming $\Delta f(x, \Delta x)$ the (algebraic, *i.e.* it may be negative) corresponding increment experienced by the function f:

$$\Delta f(x_0, \Delta x) := f(x_0 + \Delta x) - f(x_0), \tag{5.17}$$

the continuity of f assures that the quantity (5.17) vanishes for $\Delta x \to 0$. How behaves Δf in terms of Δx?

A considerable knowledge is achieved analyzing the part of Δf which is linear in Δx, that is the *differential of f*. It is named by df and, as the entire increment of f (5.17), it depends on x_0 and on its increment Δx. Hence, the following definitions are set on the function f and on its increment Δf (5.17).

Consider a function $f : \mathbb{R}^n \to \mathbb{R}$ and an interior point x_0 of its existence field. Assume that f is continuous in x_0 and examine what happens in points $x = x_0 + \Delta x$ belonging to a small neighbourhood of x_0. If exists a vector a depending on x_0 *but not* on Δx such that:

$$\Delta f(x_0, \Delta x) = a(x_0) \cdot \Delta x + o(\Delta x), \qquad (5.18)$$

the function f is said to be *differentiable* in x_0 and its differential is given by:

$$df(x_0, \Delta x) := a(x_0) \cdot \Delta x, \qquad (5.19)$$

i.e. by the part of the increment (5.17) that behaves linearly with respect to Δx. The remaining part of the increment (5.17) is indicated by the symbol $o(\Delta x)$ in the structural decomposition (5.18), because it depends on powers higher than 1 of the modulus $|\Delta x|$ and, as a consequence, vanishes more rapidly than the differential (5.19) as $\Delta x \to 0$. For this reason it is said to be an infinitesimal of higher order with respect to Δx, in symbols: $o(\Delta x)/|\Delta x| \to 0$ as $\Delta x \to 0$. It is worth noting that the scalar product $a \cdot \Delta x$ is needed in the definition (5.19) due to the fact that Δx is an n-dimensional vector whereas df is a scalar quantity.

A graphical representation of the meaning of the differential is shown in Fig. 5.1 for the case $n = 2$. A sample of function $f(x, y)$ of the two variables x and y (hereafter grouped in the vector $x = (x, y)$) is represented giving the value $f(x)$ to the third coordinate z. In this way, as it will be discussed in the next lessons, the function f appears as a *surface* in $\mathbb{R}^3$. Once a point x_0 in the (x, y)-plane is fixed, several

Fig. 5.1 Increments of $x_0 = (0.55, 0.37)$ (red vectors), surface $z = f(x)$ (green) and graphical representation of the decomposition of the corresponding increments of f (5.17) in the sum (5.18) of differentials ((5.19), red lines) and higher order terms (blue). These increments are drawn starting from the plane $z = f(x_0)$ and ending on the surface $z = f(x)$

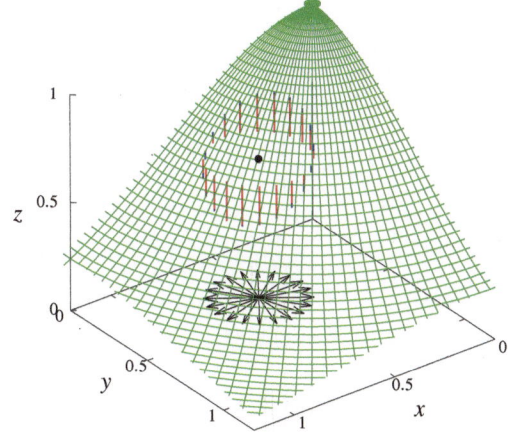

increments (for the sake of simplicity, having the same modulus) Δx are drawn. The value of f on x_0 is used in order to define the point drawn with a black bullet on the surface $z = f(x)$. The increments of f (5.17) corresponding to those of x are now drawn starting from the plane $z = f(x_0)$, for graphical convenience. They are decomposed in the sum of the differentials (5.19) (drawn with red lines) and the higher order terms (blue lines), as in Eq. (5.18). Hence, the sum of red and blue contributions defines a point that still lies on the surface $z = f(x)$. Figure 5.1 shows the important role played by the differential (5.19): it gives a clear representation of the behaviour of f in a "small" neighbourhood of the point x_0.

5.5 Suggested Readings

The concept of continuity for functions of a vector variable, along its main consequences, is illustrated in Lax and Terrell (2017, Chap. 2, pp. 78–94), Adams and Essex (2010, Chap. 12, pp. 677–680). The definition of a differentiable function is given in Apostol (1969, Chap. 8, p. 258), Lang (1987, Chap. III, pp. 77–82) and Lax and Terrell (2017, Chap. 3, p. 104), where the relation with continuity is immediately shown.

References

R. A. Adams and C. Essex. *Calculus: Several Variables*. Pearson Education Canada, Toronto, ON, 7th edition, 2010.

T. M. Apostol. *Calculus, Volume II*. Wiley, Hoboken, NJ, 2nd edition, 1969.

S. Lang. *Calculus of Several Variables*. Springer, New York, 3rd edition, 1987.

P. D. Lax and M. S. Terrell. *Multivariable Calculus with Applications*. Springer, Cham, CH, 2017.

Chapter 6
Partial Derivatives

In the previous lesson, the idea of the *differential* of a function $f : \mathbb{R}^n \to \mathbb{R}$ has been discussed, showing that a function which is *differentiable* in a point x_0 of its existence field is *linearizable* in a neighbourhood $B(x_0)$ of that point. It means that a vector $a \in \mathbb{R}^n$ depending on x_0 exists such that the values of $f(x)$ for $x \in B(x_0)$ are very close to their linear approximation $f(x_0) + a \cdot (x - x_0)$. To underline the local character of such a property, the increment of the independent variable $\Delta x = x - x_0$ and of the function $\Delta f(x_0, \Delta x) = f(x_0 + \Delta x) - f(x_0)$ are often introduced and the differential df is identified as the part of the increment Δf which is linear in Δx. The remaining part of this increment depends on powers higher than 1 of Δx, or, equivalently, is an infinitesimal of higher order with respect to Δx: $\Delta f - df = o(\Delta x)$.

In the present lesson the tool of the derivative is extended to functions of several variables, giving a way to evaluate the vector a in the differential df.

6.1 Partial Derivatives

The increment Δx is now chosen along a coordinate direction, for example along the ith direction (specified by the unit vector e_i): $\Delta x = \Delta x_i e_i$. The corresponding increment of the function f: $f(x_1, \ldots, x_{i-1}, x_i + \Delta x_i, x_{i+1}, \ldots, x_n) - f(x_1, \ldots, x_{i-1}, x_i, x_{i+1}, \ldots, x_n)$ will be indicated with $\Delta_{e_i} f$, for shortness. The incremental ratio along the coordinate direction e_i follows as:

$$\frac{\Delta_{e_i} f(x, \Delta x_i e_i)}{\Delta x_i} =$$
$$= \frac{f(x_1, \ldots, x_{i-1}, x_i + \Delta x_i, x_{i+1}, \ldots, x_n) - f(x_1, \ldots, x_{i-1}, x_i, x_{i+1}, \ldots, x_n)}{\Delta x_i}$$

$$(6.1)$$

G. Riccardi et al., *Multidimensional Differential and Integral Calculus*,
https://doi.org/10.1007/978-3-031-70326-3_6

and the *partial derivative* in the coordinate x_i is defined as the limit for $\Delta x_i \to 0$ of the above incremental ratio, if it exists finite. This limit is indicated with the following symbol:

$$\frac{\partial f}{\partial x_i}(x) = \lim_{\Delta x_i \to 0} \frac{\Delta_{e_i} f(x, \Delta x_i e_i)}{\Delta x_i} , \tag{6.2}$$

or, shortly, with $\partial_{x_i} f$, or $\partial_i f$. The above definition suggests a way to calculate the partial derivative $\partial_{x_i} f$: the components of the vector x different from the ith, *i.e.* x_1, ..., $x_{i-1}, x_{i+1}, ..., x_n$, are kept constant and the usual derivative with respect to the ith coordinate x_i is evaluated. For this reason, the well known rules to evaluate the derivatives remain valid for the partial derivatives. As sample cases, the derivative of a linear combination of the functions $f, g : \mathbb{R}^n \to \mathbb{R}$, *i.e.* $\alpha f + \beta g$ where α and β are constants, is the linear combination of the derivatives, *i.e.* $\alpha \partial_{x_i} f + \beta \partial_{x_i} g$, and the derivative of the product $f g$ is $\partial_{x_i} f g + f \partial_{x_i} g$. Moreover, if $f : \mathbb{R} \to \mathbb{R}$ and $g : \mathbb{R}^n \to \mathbb{R}$, the derivative with respect to x_i of the composite function $f[g(x)]$ is given by the product of the derivatives $f'[g(x)]$ (ordinary derivative of the function $f(y)$ of the variable y, evaluated at the point $y = g(x)$) and $\partial_i g(x)$, *i.e.* $\partial_i f[g(x)] = f'[g(x)] \partial_i g(x)$. In order to acquire a certain familiarity with the partial derivatives, it is important to do several exercises.

Exercise 6.1 Evaluate the following partial derivatives:

$(a) \quad \partial_x \sqrt{x^2 + y^2 + z^2}$

$(b) \quad \partial_y \dfrac{xyz}{x^4 + y^4 + z^4}$

$(c) \quad \partial_z \cos \dfrac{x^4}{z^2}$

$(d) \quad \partial_y \log(x^2 y^2 z^2 + 1)$

$(e) \quad \partial_z \sqrt{1 - x^4 - z^4}$

$(f) \quad \partial_x \exp[1 + \cos(xyz)]$.

A function f for which the limit (6.2) exists (at point x) is said to be *partially differentiable with respect to x_i* (at x), whereas if f possesses all the partial derivatives (at x) it is said to be *partially differentiable* (at x). Note that the existence of all the partial derivatives *does not implies that f is differentiable*.

For a function $f : \mathbb{R}^n \to \mathbb{R}$ that is partially differentiable in x, it is possible to build the vector of its partial derivatives:

$$\begin{pmatrix} \partial_{x_1} f \\ \partial_{x_1} f \\ \vdots \\ \partial_{x_n} f \end{pmatrix} = \nabla f , \tag{6.3}$$

which is named *gradient vector of* f. In this case the partial derivative $\partial_i f$ can be also interpreted as the ith component of the gradient vector of f. The gradient vector has an important meaning for the functions f having their domain in $\mathbb{R}^3$, *i.e.* in the case $n = 3$.

Before discussing the geometrical meaning of the vector ∇f, the definition of *surface* in $\mathbb{R}^3$ is needed. A *surface* $S \subset \mathbb{R}^3$ is a set of points the coordinates of which $(x, y$ and $z)$ are functions of two *parameters* u and v:

$$\begin{cases} x = x(u, v) \\ y = y(u, v) \\ z = z(u, v) \end{cases} \tag{6.4}$$

that verify the following conditions: (*I*) they have a common domain $D \subseteq \mathbb{R}^2$; (*II*) they are continuous and partially differentiable; (*III*) curves (lying on S) corresponding to the coordinate lines $u \equiv u_0$ and $v \equiv v_0$ in D are not parallel (the tangent vectors for $v = v_0$ and $u = u_0$, respectively, are not parallel).

Several observations on this definition of the surface S are now briefly discussed. First of all, the coordinates x, y and z are often grouped in a vector x, as well as the parameters u and v in u, so that the laws (6.4) are a first example of a vector function $(x \in \mathbb{R}^3)$ of a vector variable $(u \in \mathbb{R}^2)$. Moreover, as it will be discussed later, the condition *III* assures that the vectors $\partial_u x(u, v_0)$ and $\partial_v x(u_0, v)$, tangent to the curves $x = x(u, v_0)$ and $x = x(u_0, v)$ on S, are also local basis vectors for the tangent plane in $x(u_0, v_0) \in S$. In this way their vector product $\partial_u x \times \partial_v x$ is non-vanishing and identifies the normal direction at the point $x(u_0, v_0)$ of S. Its modulus is written in terms of the quantities:

$$E = |\partial_u x|^2, \qquad G = |\partial_v x|^2, \qquad F = \partial_u x \cdot \partial_v x \tag{6.5}$$

in the following form:

$$|\partial_u x \times \partial_v x| = \sqrt{EG - F^2} \tag{6.6}$$

(a proof of the relation (6.6) is left to the Reader, as exercise). In order to satisfy the condition *III*, the function of the parameter vector u at the right hand side of the above equation has to be non-vanishing for any $u \in D$. It plays an important role in the calculation of the *area of the surface* S, see below. Finally, as it occurs also for the curves, the choice of the parameters is not unique. In principle, the same surface S can be defined by means of infinite pairs of parameters.

The definition of the surface S by means of the parameters u and v and of the functions (6.4) is called *parametric representation* of S. It is not the only way to define this surface. Indeed, if it is possible to find a continuous and partially differentiable function $f : \mathbb{R}^3 \to \mathbb{R}$ such that: $f[x(u)] \equiv 0$ for any $u \in D$, the equation:

$$f(x) = 0 \tag{6.7}$$

gives a second way to define the same surface S: the *implicit representation*. Also in this case, the representation is not unique. For example, it is sufficient to multiply the function f at the left hand side of Eq. (6.7) times any non-vanishing, continuous and partially differentiable function of x, for generating another implicit representation of the same surface S.

Assume now that, for a given surface S, both a parametric representation (6.4) and an implicit one (6.7) are given. Inserting the functions (6.4) inside f, the identity $f(x(u)) \equiv 0$ is obtained. It is verified in any $u \in D$, so that its partial derivatives with respect to u and v vanish:

$$0 \equiv \partial_x f \partial_u x + \partial_y f \partial_u y + \partial_z f \partial_u z = \nabla f \cdot \partial_u x$$

$$0 \equiv \partial_x f \partial_v x + \partial_y f \partial_v y + \partial_z f \partial_v z = \nabla f \cdot \partial_v x \,.$$

The above relations show that the gradient vector ∇f is orthogonal to both vectors $\partial_u x$ and $\partial_v x$. Due to the fact that they are tangent to S in its point $x(u)$, it follows that ∇f is normal to the surface $f(x) = 0$ in the same point. This important property of any implicit representation (6.6) is extensively used in optimization, to find maxima (or minima) of a given f.

A third way to represent the surface S is often possible writing one of the coordinate as a continuous and partially differentiable function of the other ones. For example, if the x-coordinate can be written as $x = g_x(y, z)$, with g_x continuous and partially differentiable, one obtains the *explicit representation* of the surface S with respect to x. In an analogous way two other representations can be obtained *i.e.* $y = g_y(x, z)$ and $z = g_z(x, y)$. Note that an explicit representation, if it exists, is unique.

It is worth noting that, in general, the above representations (parametric, implicit and explicit) of S are not able to cover the entire surface and several representations (of the same kind) are needed to describe S.

Exercise 6.2 Define the existence fields and evaluate the gradient vectors of the following functions (drawn in Fig. 6.1):

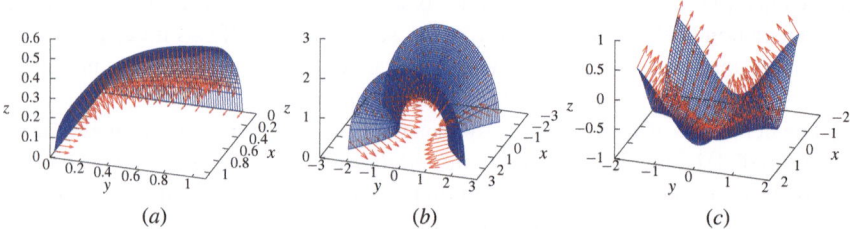

(a) (b) (c)

Fig. 6.1 Surfaces in the Exercise 6.2 and gradient vectors (rescaled times 0.2, 0.1 and 0.2)

$$(a) \quad f(x, y, z) = \frac{xy}{1 + x^2 + y^2} - z^2$$

$$(b) \quad f(x, y, z) = \sqrt{1 + x^2 - y^2} - z$$

$$(c) \quad f(x, y, z) = \cos(xy) + z\sqrt{1 + \sin^2 x + \sin^2 y}.$$

Give parametric and explicit representations of the surfaces $f = 0$.

6.2 Directional Derivative

For the functions of several variables, a new kind of derivative is introduced. Indeed, consider an arbitrary unit vector $\boldsymbol{d}$ (in general, it is not directed along any coordinate direction) and introduce a scalar increment Δs. In this way, it becomes possible to introduce the increment of $\boldsymbol{x}$ along the direction $\boldsymbol{d}$ as $\Delta \boldsymbol{x} = \boldsymbol{d}\,\Delta s$ and the incremental ratio along the same direction:

$$\Delta_{\boldsymbol{d}} f(\boldsymbol{x}_0, \Delta s) = \frac{f(\boldsymbol{x}_0 + \boldsymbol{d}\,\Delta s) - f(\boldsymbol{x}_0)}{\Delta s}. \tag{6.8}$$

If the limit as $\Delta s \to 0$ of the ratio (6.8) exists, it will be named as *derivative of f along the direction $\boldsymbol{d}$*:

$$\frac{\partial f}{\partial \boldsymbol{d}}(\boldsymbol{x}_0) \equiv \partial_{\boldsymbol{d}} f(\boldsymbol{x}_0) = \lim_{\Delta s \to 0} \frac{f(\boldsymbol{x}_0 + \boldsymbol{d}\,\Delta s) - f(\boldsymbol{x}_0)}{\Delta s}. \tag{6.9}$$

The derivative of f along $\boldsymbol{d}$ (6.9) can be rewritten as a linear combination of the partial derivatives of f. Assume that all the components of $\boldsymbol{d}$ $(d_1, d_2, \ldots, d_n)$ are non-vanishing and rewrite the increment of f corresponding to the increment $\Delta s\,\boldsymbol{d}$ of $\boldsymbol{x}$ in a slightly different way:

$$
\begin{aligned}
\Delta_{\boldsymbol{d}} f(\boldsymbol{x}_0, \Delta s) \equiv \quad & d_1 \frac{f(\boldsymbol{x}_0 + \Delta s\,\boldsymbol{d}) - f(x_1, x_2 + d_2\Delta s, \ldots, x_n + d_n\Delta s)}{d_1 \Delta s} \\
+ & d_2 \frac{f(x_1, x_2 + d_2\Delta s, \ldots, x_n + d_n\Delta s) - f(x_1, x_2, \ldots, x_n + d_n\Delta s)}{d_2 \Delta s} \\
+ & \ldots \\
+ & d_n \frac{f(x_1, x_2, \ldots, x_n + d_n\Delta s) - f(x_1, x_2, \ldots, x_n)}{d_n \Delta s},
\end{aligned}
$$

which leads for $\Delta s \to 0$ to the following relation:

$$\partial_{\boldsymbol{d}} f = d_1\,\partial_1 f + d_2\,\partial_2 f + \ldots + d_n\,\partial_n f = \boldsymbol{d} \cdot \nabla f, \tag{6.10}$$

which holds even though one or more components of $\boldsymbol{d}$ vanish. Moreover, if $\boldsymbol{d}$ is directed along the coordinate direction $\boldsymbol{e}_i$ the directional derivative (6.10) reduces to the partial derivative along x_i, i.e. $\partial_{\boldsymbol{e}_i} f = \partial_i f$.

6.3   ... Again on the Differential

The definition (5.19) of the differential of a function $f : \mathbb{R}^n \to \mathbb{R}$ in a point x_0 of its existence field requires to find a vector a, in general depending on x_0 but not on the increment Δx. A rule for obtaining a is now given.

Assume that f is differentiable in x_0, so that, according to the formula (5.18), the increment of the function corresponding to an increment Δx of the independent variable x is the sum of a contribution linear in Δx, that is just the differential $df(x_0, \Delta x)$, and of higher order terms represented by $o(\Delta x)$ in the formula (5.18). The differential is written in Eq. (5.19) as the scalar product of the increment times a certain vector a. It is evaluated calling with Δs the modulus of Δx:

$$\Delta s = |\Delta x| = \sqrt{\Delta x_1^2 + \Delta x_2^2 + \ldots + \Delta x_n^2}$$

and with d (a unitary vector) its direction, so that $\Delta x = \Delta s\, d$, and rewriting the increment of the function (5.18) divided by Δs as:

$$\frac{f(x_0 + d\,\Delta s) - f(x_0)}{\Delta s} = a(x_0) \cdot d + O(\Delta s),$$

where the symbol $O(\Delta s)$ stays for terms which vanish as $\Delta s \to 0$. Using the result (6.10) at the left-hand side of the above relation, this limit implies:

$$[\nabla f(x_0) - a(x_0)] \cdot d = 0. \tag{6.11}$$

Since the increment Δx is an arbitrary (small) vector, also the unit vector d is arbitrary and the relation (6.11) implies $a(x_0) = \nabla f(x_0)$. This fact leads to an important geometrical meaning of the gradient vector. Indeed, it is now possible to rewrite the increment of f (5.18) in the more significant way:

$$\Delta f(x_0, \Delta x) = \nabla f(x_0) \cdot \Delta x + o(\Delta x). \tag{6.12}$$

Suppose now to set the length Δs of the increment Δx at a certain value Δs_0 and to change its direction d. The relation (6.12) says that the maximum (minimum) increment of f is obtained for Δx directed as $+\nabla f$ $(-\nabla f)$. As a consequence, $+\nabla f$ $(-\nabla f)$ specifies the direction of maximum (minimum) increase of the function f. However, in general this meaning of the gradient vector works only locally, i.e. on points $x_0 + \Delta x$ belonging to a (small) neighbourhood of x_0.

It is now possible to give a sufficient condition for the function f to be differentiable. Assume that f is continuous in x_0 and partially derivable, i.e. its partial derivatives exist, in a neighbourhood of this point. In order to fix the ideas, start with a function of two variables ($n = 2$) and reconsider the increment of the function $f : \mathbb{R}^2 \to \mathbb{R}$ in passing from the point $x_0 = (x_0, y_0)$ to the one $x_0 + \Delta x = (x_0 + \Delta x, y_0 + \Delta y)$: $\Delta f(x; \Delta x) = f(x + \Delta x) - f(x)$. This increment is now rewritten so that the increment ratios for the partial derivatives of f

$$\frac{f(x' + \Delta x, y') - f(x', y')}{\Delta x} = \partial_x f(x') + O(\Delta x)$$

$$\frac{f(x'', y'' + \Delta y) - f(x'', y'')}{\Delta y} = \partial_y f(x'') + O(\Delta y)$$

(6.13)

appear. The relations (6.13) follow only from the existence of the partial derivatives, the symbol $O(\Delta x)$ (or $O(\Delta y)$) stays for terms that vanish for $\Delta x \to 0$ (or for $\Delta y \to 0$) and the points x', x'' lie in a neighbourhood of x_0 (they will be specified later). Assume that both the components of the increment Δx are non-vanishing (if one is zero, the corresponding incremental ratio must be omitted), indicate with Δ their order of magnitude and rewrite the increment of f as follows:

$$\Delta f(x_0, \Delta x) \equiv$$

$$\equiv \ [f(x_0 + \Delta x, y_0 + \Delta y) - f(x_0 + \Delta x, y_0)] + [f(x_0 + \Delta x, y_0) - f(x_0, y_0)]$$

$$\equiv \ \Delta y \left[\frac{f(x_0 + \Delta x, y_0 + \Delta y) - f(x_0 + \Delta x, y_0)}{\Delta y} - \partial_y f(x_0 + \Delta x, y_0) \right]$$

$$+ \Delta y \, \partial_y f(x_0 + \Delta x, y_0)$$

$$+ \Delta x \left[\frac{f(x_0 + \Delta x, y_0) - f(x_0, y_0)}{\Delta x} - \partial_x f(x_0, y_0) \right] + \Delta x \, \partial_x f(x_0, y_0)$$

$$= \ \Delta x \, \partial_x f(x_0, y_0) + \Delta y \, \partial_y f(x_0, y_0)$$

$$+ \Delta y \, [\partial_y f(x_0 + \Delta x, y_0) - \partial_y f(x_0, y_0)] + O(\Delta^2),$$

(6.14)

having used the first relation (6.13) in $x' = x_0$, $y' = y_0$ and the second one in $x'' = x_0 + \Delta x$, $y'' = y_0$. As shown by the decomposition (6.14), if the partial derivative $\partial_y f$ is continuous in x_0, the increment $\Delta f(x_0, \Delta x)$ differs from the differential $\Delta x \, \partial_x f(x_0) + \Delta y \, \partial_y f(x_0)$ for terms of order higher than Δ. The same result is obtained using the other decomposition of the increment of f:

$$\Delta f(x_0, \Delta x) \equiv [f(x_0 + \Delta x, y_0 + \Delta y) - f(x_0, y_0 + \Delta y)] + [f(x_0, y_0 + \Delta y) - f(x_0, y_0)]$$

and assuming continuous in x_0 the derivative $\partial_x f$. This result can be generalised easily to any n.

> If $f : \mathbb{R}^n \to \mathbb{R}$ is continuous and partially differentiable at the point x_0, then a sufficient condition to be f differentiable at x_0 is that $n - 1$ partial derivatives are continuous in this point.

Note that the converse is not true and a function can be differentiable in a point, even if its partial derivatives are discontinuous.

6.4   Successive Partial Derivatives

Up to now it has been discussed how build, given a function $f : \mathbb{R}^n \to \mathbb{R}$, its partial derivative in the coordinate direction x_i. If it is possible to define this derivative in a certain subset $D_i \subseteq \mathscr{E}$ of the existence field of the function f, the partial derivative in the coordinate direction x_j (in general, $j \neq i$) of the new function $g_i := \partial_{x_i} f : \mathbb{R}^n \to \mathbb{R}$ can be evaluated and so on. If it is possible to perform this calculation in some points of D_i, the *second partial derivative* of the function f has been built. It is made deriving firstly in x_i and then in x_j, as a consequence the second partial derivative is indicated by $\partial^2_{x_j x_i} f$, or in a more concise way as $\partial^2_{ji} f$.

Once the successive partial derivatives of a function f have been defined, the *Laplacian* $\nabla^2 f$ of f is introduced:

$$\partial^2_{x_1 x_1} f + \partial^2_{x_2 x_2} f + \ldots + \partial^2_{x_3 x_3} f = \nabla^2 f , \tag{6.15}$$

that is the sum of the second derivatives in each coordinate direction x_i. The Laplacian plays an important role in many physical applications. In particular, it often appears in differential equations involving functions of several variables.

Exercise 6.3   Prove that: $\nabla^2 f = \nabla \cdot (\nabla f)$.

Exercise 6.4   Evaluate the Laplacian of the following functions:

$$(a) \quad f(x, y) = \frac{x}{x^2 + y^2}$$

$$(b) \quad f(x, y) = |\boldsymbol{x}| = (x^2 + y^2)^{1/2}$$

$$(c) \quad f(x, y) = \log |\boldsymbol{x}|$$

$$(d) \quad f(x, y, z) = xyz$$

$$(e) \quad f(x, y, z) = \frac{1}{|\boldsymbol{x}|}$$

$$(f) \quad f(x, y, z) = \sqrt{x^2 + y + z^2} .$$

In evaluating the second partial derivatives, a new issue appears: what happens if the second derivative $\partial^2_{x_j x_i} f$ for $i \neq j$ is evaluated firstly deriving along x_j and then along x_i? Does exist a relation between the derivative $\partial^2_{x_j x_i} f$ and the new one, *i.e.* $\partial^2_{x_i x_j} f$, obtained by inverting the order of the derivatives? It is now shown that, if the above second derivatives $\partial^2_{x_j x_i} f(\boldsymbol{x}_0)$ and $\partial^2_{x_i x_j} f(\boldsymbol{x}_0)$ are continuous in a neighbourhood of the point $\boldsymbol{x}_0$, they are equal, so that does not matter the order in which the derivatives are calculated. As before, it is convenient to analyse this property for a function $f : \mathbb{R}^2 \to \mathbb{R}$ of two variables x and y ($n = 2$). Assume that both the second derivatives $\partial_{xy}^2 f$ and $\partial_{yx}^2 f$ exist at the point $\boldsymbol{x}_0 = (x, y)$ of the existence field $\mathcal{E}$ of f and consider an increment $\Delta \boldsymbol{x}$ sufficiently small to give a new point $\boldsymbol{x}_0 + \Delta \boldsymbol{x}$ which still lies inside $\mathcal{E}$. In these conditions, define the *double increment* of f as follows:

$$
\begin{aligned}
\Delta^2 f(\boldsymbol{x}_0, \Delta \boldsymbol{x}) &= \\
&= f(x_0 + \Delta x, y_0 + \Delta y) - [f(x_0, y_0 + \Delta y) + f(x_0 + \Delta x, y_0)] + f(x_0, y_0) \\
&\equiv \underbrace{[f(x_0 + \Delta x, y_0 + \Delta y) - f(x_0, y_0 + \Delta y)]}_{G(y_0 + \Delta y)} - \underbrace{[f(x_0 + \Delta x, y_0) - f(x_0, y_0)]}_{G(y_0)} \\
&\equiv \underbrace{[f(x_0 + \Delta x, y_0 + \Delta y) - f(x_0 + \Delta x, y_0)]}_{H(x_0 + \Delta x)} - \underbrace{[f(x_0, y_0 + \Delta y) - f(x_0, y_0)]}_{H(x_0)} .
\end{aligned}
$$

The double increment of f defined in the first line of the above equation can be seen (second line) as the increment of the function of y alone: $f(x_0 + \Delta x, y) - f(x_0, y) =: G(y)$ in passing from y_0 to $y_0 + \Delta y$, or (third line) as the increment of the function of x alone: $f(x, y_0 + \Delta y) - f(x, y_0) =: H(x)$ in passing from x_0 to $x_0 + \Delta x_0$. In the first case, due to the fact that the derivative $\partial_y f$ is continuous in a neighbourhood of $\boldsymbol{x}_0$, a point $y' \in (y_0, y_0 + \Delta y)$ exists such that:

$$
\Delta^2 f(\boldsymbol{x}_0, \Delta \boldsymbol{x}) = \frac{dG}{dy}(y') \, \Delta y = [\partial_y f(x_0 + \Delta x, y') - \partial_y f(x_0, y')] \, \Delta y , \quad (6.16)
$$

whereas in the second case, the derivative $\partial_x f$ being continuous in a neighbourhood of $\boldsymbol{x}_0$, a point $x'' \in (x_0, x_0 + \Delta x)$ can be found such that the following relation:

$$
\Delta^2 f(\boldsymbol{x}_0, \Delta \boldsymbol{x}) = \frac{dH}{dx}(x'') \, \Delta x = [\partial_x f(x'', y_0 + \Delta y) - \partial_x f(x'', y_0)] \, \Delta x \quad (6.17)
$$

is valid. Due to the fact that also $\partial_{xy}^2 f$ is continuous in a neighbourhood of $\boldsymbol{x}_0$, from the relation (6.16) it follows that a point $x' \in (x_0, x_0 + \Delta x)$ exists such that:

$$
\Delta^2 f(\boldsymbol{x}_0, \Delta \boldsymbol{x}) = \partial_{xy}^2 f(x', y') \, \Delta y \, \Delta x \quad (6.18)
$$

and in an analogous way, the derivative $\partial_{yx}^2 f$ being continuous in a neighbourhood of $\boldsymbol{x}_0$, a point $y'' \in (y_0, y_0 + \Delta y)$ can be found where the relation:

$$
\Delta^2 f(\boldsymbol{x}_0, \Delta \boldsymbol{x}) = \partial_{yx}^2 f(x'', y'') \, \Delta x \, \Delta y \quad (6.19)
$$

holds. The forms (6.18), (6.19) of the double increment of f imply that two points $\boldsymbol{x}' = (x', y')$ and $\boldsymbol{x}'' = (x'', y'')$ exist, which go to $\boldsymbol{x}_0$ as $\Delta \boldsymbol{x} \to \boldsymbol{0}$, such that:

$$
\partial_{xy}^2 f(x', y') = \partial_{yx}^2 f(x'', y'') .
$$

From the continuity of both the above second derivatives, the limit as $\Delta x \rightarrow 0$ implies that they take the same value at x_0. This result is known as *Schwarz*[1] *theorem* in literature. An identical result is valid for any number of dimensions of the vector x.

6.5 Suggested Readings

The definition of (first) partial derivative of a function is given in Lax and Terrell (2017, Chap. 3, p. 105), Adams and Essex (2010, Chap. 12, p. 681), Lang (1987, Chap. III, pp. 70–74), Apostol (1974, Chap. 12, p. 345) and Apostol (1969, Chap. 8, pp. 254–255) (introduced, in the latter two, as a particular directional derivative). Higher order derivatives are successively introduced in Lax and Terrell (2017, Chap. 4, pp. 161), Adams and Essex (2010, Chap. 12, p. 688), Lang (1987, Chap. III, pp. 82–85), Apostol (1974, Chap. 12, pp. 358–360), Apostol (1969, Chap. 8, p. 255).

References

R. A. Adams and C. Essex. *Calculus: Several Variables*. Pearson Education Canada, Toronto, ON, 7th edition, 2010.

T. M. Apostol. *Calculus, Volume II*. Wiley, Hoboken, NJ, 2nd edition, 1969.

T. M. Apostol. *Mathematical Analysis: A Modern Approach to Advanced Calculus*. Pearson Education US, Hoboken, NJ, 2nd edition, 1974.

S. Lang. *Calculus of Several Variables*. Springer, New York, 3rd edition, 1987.

P. D. Lax and M. S. Terrell. *Multivariable Calculus with Applications*. Springer, Cham, CH, 2017.

---

[1] https://mathshistory.st-andrews.ac.uk/Biographies/Schwarz/.

Chapter 7
Sequences of Functions

In the present lesson, as well as in the following one, crucial theoretical issues are introduced. Unlike the rest of the book, the structure of these two lessons follows a more traditional scheme in the presentation of contents. This is because topics proposed here are intended to provide tools required for most of the applications presented in the following lessons, rather than be themselves a target issue.

7.1 Pointwise Convergence

Consider the Cauchy differential problem

$$\begin{cases} y'(x) = y(x) & 0 < x \le 1 \\ y(0) = 1. \end{cases} \tag{7.1}$$

For every $n \ge 1$ and $x \in [0, 1]$ consider the function $f_n(x) = (1 + x/n)^n$. The initial condition $f_n(0) = 1$ is satisfied for every n, but $f'_n(x) = (1 + x/n)^{n-1} \ne f_n(x)$, even though, actually, for increasing n, the difference between f and f' decreases rapidly, as it can be seen from the relation

$$|f_n(x) - f'_n(x)| = \left(1 + \frac{x}{n}\right)^n \left[1 - \frac{1}{(1 + x/n)}\right]$$

$$\le \left(1 + \frac{1}{n}\right)^n \left[1 - \frac{1}{(1 + 1/n)}\right] \le 3 \left[1 - \frac{1}{(1 + 1/n)}\right] = \frac{3}{n+1}.$$

This suggests a couple of very significant comments:

(a) functions $f_n(x)$ are approximate solutions of problem (7.1), whose accuracy increases with n,

© The Author(s), under exclusive license to Springer Nature Switzerland AG 2024
G. Riccardi et al., *Multidimensional Differential and Integral Calculus*,
https://doi.org/10.1007/978-3-031-70326-3_7

(b) it seems to be natural expressing the solution $f(x)$ of problem (7.1) in the form
$f(x) = \lim_{n\to\infty} f_n(x)$, for every $x \in [0, 1]$.

However, a number of issues must be considered:

(1) does the limit $\lim_{n\to\infty} f_n(x)$ exist?..., and, if yes, for which x?
(2) does the limit $\lim_{n\to\infty} f'_n(x)$ exist?..., and, if yes, for which x?
(3) if one assumes that $f(x) = \lim_{n\to\infty} f_n(x)$ exists, can one say that $f'_n(x)$ exist?...,
and for which x?
(4) if one further assumes that $g(x) = \lim_{n\to\infty} f'_n(x)$ exists, can one say that $f'_n(x) = g(x)$?

For the problem addressed the answers to the above questions are affirmative, and
$\forall x \in \mathbb{R}$ one has:

$$\lim_{n\to\infty} f_n(x) = \lim_{n\to\infty} \left(1 + \frac{x}{n}\right)^n = e^x, \qquad \lim_{n\to\infty} f'_n(x) = \lim_{n\to\infty} \left(1 + \frac{x}{n}\right)^{n-1} = e^x.$$

As a consequence $f(x) = e^x$ is a solution of (7.1). However, answers to questions
similar to the ones above are not, in general, affirmative. Then, a question emerges:
what are the crucial properties determining the behaviour of the *family* of functions
$\{f_n\}$ or, more precisely, the *sequence* of functions $\{f_n\}$, defined as follows?

> Consider a family F of functions, all defined in a non empty set $E \subseteq \mathbb{R}$, and
> let $\tilde{N}$ be an infinite subset of $\mathbb{N}$. A map $n \in \tilde{N} \to f_n \in F$ that let each natural
> number correspond to a function in F is said to be a *sequence of functions*
> with *domain* E, and is denoted by $\{f_n\}$.

It is observed that, in general, the single functions in the countable collection mentioned above may be defined on different sets (*e.g.*, depending on their expressions).
In order to operate with the sequence, it is required that the entire family is defined
on a common set. This is called the *context domain*.

Example 7.1 Consider, for $n > 0$, the family

$$\{f_n\} = \left\{\sqrt{x - \frac{1}{n}}\right\} = \left\{\sqrt{x - 1}, \sqrt{x - \frac{1}{2}}, \sqrt{x - \frac{1}{3}}, \ldots\right\}.$$

It is a sequence of functions with context domain $E = [1, +\infty)$. Actually, for each
$i \in \mathbb{N}^+$, one has:

$$f_i(x) = \sqrt{x - \frac{1}{i}} \quad \text{whose domain is } E_i = [1/i, +\infty), \qquad i = 1, 2, \ldots, n, \ldots$$

The context domain E is defined as the intersection of all E_n:

$$E = \bigcap_{i=1}^{+\infty} E_i = [1, +\infty).$$

Example 7.2 Consider the family

$$\{f_n\} = \{\log(x - n)\} = \{\log(x - 1), \log(x - 2), \log(x - 3), \ldots\}.$$

It is not a sequence of functions, since its context domain is empty. In fact the function $f_n(x) = \log(x - n)$ has the domain $E_n = (n, +\infty)$, which implies

$$E = \bigcap_{i=1}^{+\infty} E_i = \emptyset.$$

Example 7.3 Consider the family

$$\{f_n\} = \{\log((n + 2)x - n - 1)\} = \{\log(2x - 1), \log(3x - 2), \log(4x - 3), \ldots\}.$$

The function $f_n(x) = \log((n + 2)x - n - 1)$ has the domain

$$E_n = \left(\frac{n + 1}{n + 2}, +\infty\right) = \left(1 - \frac{1}{n + 2}, +\infty\right),$$

whence the context domain

$$E = \bigcap_{n=1}^{+\infty} \left(\frac{n + 1}{n + 2}, +\infty\right) = \bigcap_{n=1}^{+\infty} \left(1 - \frac{1}{n + 2}, +\infty\right) = [1, +\infty).$$

Example 7.4 Consider, for $n > 0$, the family

$$\{f_n\} = \left\{\sqrt{e^{\frac{x}{n}} - \frac{1}{n}}\right\}.$$

The function $f_n(x)$ exist if $e^{\frac{x}{n}} - 1/n \geq 0$, which means $x \geq n\log(1/n)$, and

$$E_n = \left[n\log\frac{1}{n}, +\infty\right)$$

Since $n\log(1/n) \leq 0$, the context domain is

$$E_n = \bigcap_{n=1}^{+\infty} \left[n\log\frac{1}{n}, +\infty\right) = [0, +\infty).$$

Consider the sequence $\{f_n(x)\} = \{x^n\} = \{x, x^2, x^3, \ldots, x^n, \ldots\}$. It is a sequence with a context domain $E = \mathbb{R}$. As x changes its value, the sequence changes its behaviour. As an example, if $x = 2$, $\{f_n(x)\}$ generates the sequence of real numbers $\{2^n\}$, which is a monotonically increasing sequence, diverging to $+\infty$. Otherwise, if $x = 1/2$, the sequence generated is $\{(1/2)^n\}$, which is a monotonically decreasing sequence, bounded and convergent to 0. Furthermore, if $x = -1$, the sequence obtained is $\{(-1)^n\}$, which is bounded and oscillating. If one takes $x = -2$, then one gets the sequence of real numbers $\{(-2)^n\}$, unbounded and oscillating. Thus, the above observations can be synthesised in:

$$\lim_{n \to +\infty} f_n(x) = \begin{cases} 0 & \text{if } |x| < 1 \\ +\infty & \text{if } x > 1 \\ 1 & \text{if } x = 1 \\ \text{does not exist} & \text{if } x \leq -1. \end{cases}$$

If the context domain is not empty, it does make sense, at each x, to calculate the limit as $n \to +\infty$ of $f_n(x)$. If such a limit exists, one says that the sequence $\{f_n\}$ converges pointwise. To conduct a pointwise analysis of a sequence of functions $\{f_n(x)\}$ is equivalent to extract from it a collection of sequences of real numbers and assess their character with respect to convergence.

Let $\{f_n(x)\}$ be a sequence of functions with context domain E, $D \subseteq E$ a nonempty set, and let $f : D \to \mathbb{R}$ be a function with domain D.
A sequence of functions $\{f_n(x)\}$ converges *pointwise* to f in D (in symbols $f_n \to f$) if, $\forall x \in D$, the sequence of numbers $f_n(x)$ converges to $f(x)$. D is said to be the *domain of pointwise convergence*.

Example 7.5 Consider the sequence of functions

$$\{f_n\} = \left\{\frac{n^2 x + 1}{x^2 + n}\right\}$$

in its context domain $E = \mathbb{R}$. Given x, observe that

$$\lim_{n \to +\infty} f_n(x) = \begin{cases} 0 & \text{if } x = 0 \\ +\infty & \text{if } x > 0 \\ -\infty & \text{if } x < 0. \end{cases}$$

The sequence is regular everywhere in E but its domain of convergence reduces to $D = \{0\}$.

Example 7.6 Consider the sequence of functions

$$\{f_n\} = \left\{\frac{n^2x + 1}{x^2 + n^2}\right\}$$

in its context domain $E = \mathbb{R}$. Observe that, for each given $x \in E$, $\lim_{n \to +\infty} f_n(x) = x$
The sequence converges everywhere. The domain of pointwise convergence of $\{f_n\}$
coincides with its context domain: $D = E$.

Example 7.7 Consider the sequence of functions

$$\{f_n\} = \left\{\frac{nx^n + 2}{1 + x^2n}\right\}$$

in its context domain $E = \mathbb{R}$. Taken $x \neq 0$, with $|x| < 1$, one has

$$\lim_{n \to +\infty} \frac{nx^n + 2}{1 + x^2n} = \lim_{n \to +\infty} \frac{n(x^n + 2/n)}{n(1/n + x^2)} = \lim_{n \to +\infty} \frac{x^n + 2/n}{1/n + x^2} = 0.$$

If $x = 0$ the sequence equals 2 for each n, then it converges to 2. If $x = 1$ one has

$$\lim_{n \to +\infty} f_n(1) = \lim_{n \to +\infty} \frac{n + 2}{1 + n} = 1.$$

If $x > 1$ one has

$$\lim_{n \to +\infty} \frac{nx^n + 2}{1 + x^2n} = \lim_{n \to +\infty} \frac{nx^n[1 + 2/(nx^n)]}{n(1/n + x^2)} = \lim_{n \to +\infty} \frac{x^n[1 + 2/(nx^n)]}{1/n + x^2} = +\infty.$$

If $x \leq -1$ the sequence is irregular, due to the oscillation of the term x^n in the
numerator. Summarizing

$$f(x) = \lim_{n \to +\infty} f_n(x) = \begin{cases} 0 & \text{if } |x| < 1 \text{ and } x \neq 0 \\ 2 & \text{if } x = 0 \\ +\infty & \text{if } x > 1 \\ 1 & \text{if } x = 1 \\ \text{does not exist} & \text{if } x \leq -1. \end{cases}$$

The domain of pointwise convergence is $D = (-1, 1]$.

Example 7.8 Consider the sequence of functions

$$\{f_n\} = \left\{ n\left(\sqrt{x + \frac{1}{n}} - \sqrt{x}\right) \right\}$$

in its context domain $E = [0, +\infty)$. If $x = 0$ it results in the numerical sequence $\sqrt{n}$, diverging to $+\infty$. If $x \neq 0$

$$\lim_{n \to +\infty} f_n(x) = \lim_{n \to +\infty} n \left(\sqrt{x + \frac{1}{n}} - \sqrt{x} \right) = \lim_{n \to +\infty} \left(\frac{1}{\sqrt{x + 1/n} + \sqrt{x}} \right) = \frac{1}{2\sqrt{x}}.$$

Summarizing

$$f(x) = \lim_{n \to +\infty} f_n(x) = \begin{cases} \dfrac{1}{2\sqrt{x}} & \text{if } x > 0 \\ +\infty & \text{if } x = 0. \end{cases}$$

Then, the domain of pointwise convergence is $D = (0, +\infty)$, with limit function $f(x) = \dfrac{1}{2\sqrt{x}}$.

Consider again the sequence $\{f_n(x)\} = \{x^n\} = \{x, x^2, x^3, \ldots, x^n, \ldots\}$ limiting to the interval $I = [0, 1]$, which is a subset of the domain of pointwise convergence. The functions are all continuous and differentiable in I whereas the limit function

$$f(x) = \lim_{n \to +\infty} f_n(x) = \begin{cases} 1 & \text{if } x = 1 \\ 0 & \text{if } x \in [0, 1) \end{cases}$$

exhibits a step discontinuity at $x = 1$. In other words, the limit function does not inherit the continuity from the functions of the sequence. It may also happen that the continuity is inherited, but not the differentiability. For example, the sequence

$$\{f_n\} = \left\{ \sqrt{x^2 + \frac{1}{n}} \right\}$$

is composed by functions that are differentiable in $\mathbb{R}$ and converges at every $x \in \mathbb{R}$ to the function $f(x) = |x|$, that is continuous in $\mathbb{R}$ but not differentiable at $x = 0$. Thus, one may ask what conditions must a sequence $\{f_n(x)\}$ satisfy in order that the limit function can inherit the properties of the functions f_n. To better understand the nature of the problem observe a further important phenomenon: consider the sequence $\{f_n(x)\} = \{e^{-nx}\}$ in its domain of pointwise convergence $D = [0, +\infty)$. The sequence converges in D to the limit function

$$f(x) = \lim_{n \to +\infty} f_n(x) = \begin{cases} 0 & \text{if } x \in (0, +\infty) \\ 1 & \text{if } x = 0 \end{cases}$$

that exhibits a step discontinuity at $x = 0$. Given $x_0 \in E$:

if $x_0 \in (0, +\infty)$

$$\lim_{x \to x_0} \left[\lim_{n \to +\infty} f_n(x) \right] = \lim_{x \to x_0} 0 = 0 \text{ and } \lim_{n \to +\infty} \left[\lim_{x \to x_0} f_n(x) \right] = \lim_{n \to +\infty} e^{-n x_0} = 0,$$

if $x_0 = 0$

$$\lim_{x \to 0} \left[\lim_{n \to +\infty} f_n(x) \right] = \lim_{x \to 0} 0 = 0 \text{ whereas } \lim_{n \to +\infty} \left[\lim_{x \to 0} f_n(x) \right] = \lim_{n \to +\infty} 1^n = 1,$$

so the equality (*inversion of limits*)

$$\lim_{x \to x_0} \left[\lim_{n \to +\infty} f_n(x) \right] = \lim_{n \to +\infty} \left[\lim_{x \to x_0} f_n(x) \right]$$

is not satisfied at the point of discontinuity.

The inversion of limits is a crucial issue for the inheritance of continuity. Observe that, if f is continuous at x_0, one has

$$\lim_{x \to x_0} \left[\lim_{n \to +\infty} f_n(x) \right] = \lim_{x \to x_0} f(x) = f(x_0)$$

and

$$\lim_{n \to +\infty} \left[\lim_{x \to x_0} f_n(x) \right] = \lim_{n \to +\infty} f_n(x_0) = f(x_0),$$

whence the equality

$$\lim_{x \to x_0} \left[\lim_{n \to +\infty} f_n(x) \right] = \lim_{n \to +\infty} \left[\lim_{x \to x_0} f_n(x) \right].$$

On the other hand, if the inversion of limits holds, one has

$$\lim_{x \to x_0} f(x) = \lim_{x \to x_0} \left[\lim_{n \to +\infty} f_n(x) \right] = \lim_{n \to +\infty} \left[\lim_{x \to x_0} f_n(x) \right] = \lim_{n \to +\infty} f_n(x_0) = f(x_0).$$

Then, the continuity of f is proven.

The above facts can be summarised in the following statement.

Let $\{f_n\}$ be a sequence of functions, with context domain $I = [a, b]$, convergent in I to the limit function f. If $x_0 \in I$ is a point of continuity to f_n, then f is continuous at x_0 if and only if the inversion of limits holds at x_0.

Consider again the sequence of continuous functions $\{f_n(x)\} = \{e^{nx}\}$, limiting its context domain to the interval $I = (0, 1)$. The sequence converges to the continuous limit function $f(x) = 0$ in I. At $x_0 = 0$

$$\lim_{x\to 0}\left[\lim_{n\to +\infty} f_n(x)\right] = \lim_{x\to 0} 0 = 0 \quad \text{whereas} \quad \lim_{n\to +\infty}\left[\lim_{x\to 0} f_n(x)\right] = \lim_{n\to +\infty} 1^n = 1,$$

thus the inversion of limits does not hold. This does not contradict the results stated above, since $x_0 = 0$ does not belong to I, but to its closure.

It is important to make it clear now what are the conditions allowing the inversion of limits. In order to give a satisfactory answer one needs to further investigate the concept of convergence of a sequence of functions.

7.2   Uniform Convergence

If one applies the formal definition of limit to the sequence of functions $\{f_n(x)\} = \{e^{-nx}\}$ in its domain of pointwise convergence $D = [0, +\infty)$ it must be stated that for each $\epsilon > 0$ exists $n_\epsilon(x)$ (in general depending on x) such that $|f_n(x) - f(x)| < \epsilon$. In the present case one finds

$$|f_n(x) - f(x)| = |e^{-nx} - 0| = e^{-nx} < \epsilon \quad \text{for } n > \frac{1}{x}\log\frac{1}{\epsilon} = n_\epsilon(x).$$

As it can be seen, n_ϵ inevitably depends on x and

$$\sup_{x\in D} n_\epsilon(x) = +\infty,$$

so there cannot exist n_ϵ such that, for $n > n_\epsilon$, $|f_n(x) - f(x)| < \epsilon$ for any x. For this sequence, as it was previously seen, the inversion of limits does not hold.

Example 7.9   Consider the sequence of functions

$$\{f_n\} = \left\{\frac{nx + 1}{n + 1}\right\}$$

in the context domain $E = [0, 1]$, that is also its domain of pointwise convergence. It results that

$$\lim_{n\to +\infty} f_n(x) = x.$$

Using the formal definition of limit, one can state that for each $\epsilon > 0$ exists $n_\epsilon(x)$ (in general depending on x) such that $|f_n(x) - f(x)| < \epsilon$ for $n > n_\epsilon(x)$. In this case

$$|f_n(x) - f(x)| = \left|\frac{nx + 1}{n + 1} - x\right| = \left|\frac{1 - x}{n + 1}\right| < \epsilon \quad \text{for } n > \frac{1 - x}{\epsilon} - 1 = n_\epsilon(x).$$

It can be seen that, although $n_\epsilon(x)$ depends on x, the boundedness of the domain of pointwise convergence $D = [0, 1]$ implies that

$$\sup_{x \in D} n_\epsilon(x) = 1/\epsilon = n_\epsilon$$

is finite (and, of course, independent of x). Thus, it can be stated that $|f_n(x) - f(x)| < \epsilon$ for $n > 1/\epsilon = n_\epsilon$ for any $x \in D$. For this sequence the inversion of limits holds at each x_0 that is an accumulation point of D:

$$\lim_{x \to x_0} \left[\lim_{n \to +\infty} \frac{nx+1}{n+1} \right] = \lim_{x \to x_0} x = x_0 \text{ and } \lim_{n \to +\infty} \left[\lim_{x \to x_0} \frac{nx+1}{n+1} \right] = \lim_{n \to +\infty} \frac{nx_0+1}{n+1} = x_0 \,.$$

The independence of n_ϵ of the variable x is the basic fact leading to the concept of uniform convergence, as it will be seen shortly, and it represents a crucial aspect that ensures the inversion of limits of a sequence. Formally, it can be stated as follows.

Let $\{f_n(x)\}$ be a sequence of functions with context domain E, $D \subseteq E$ a nonempty set, and let $f : D \to \mathbb{R}$ be a real function with domain D.
A sequence of functions $\{f_n(x)\}$ converges *uniformly* to f in D (in symbols $f_n \rightrightarrows f$) if

$$\lim_{n \to +\infty} \left[\sup_{x \in D} |f_n(x) - f(x)| \right] = 0 \,.$$

D is said to be the *domain of uniform convergence*.

Example 7.10 The sequence $\{f_n\} = \{e^{-nx}\}$ converges in $D = (0, +\infty)$ to the function $f(x) = 0$, but it does not converge uniformly. In fact

$$\lim_{n \to +\infty} \left[\sup_{x \in D} |f_n(x) - f(x)| \right] = \lim_{n \to +\infty} \left[\sup_{x \in D} |e^{-nx} - 0| \right] = \lim_{n \to +\infty} 1 = 1 \neq 0 \,.$$

The same sequence restricted to the subset of D, $D_1 = [a, +\infty)$ with $a > 0$, turns out to be uniformly convergent to the function $f(x) = 0$. In fact

$$\lim_{n \to +\infty} \left[\sup_{x \in D_1} |f_n(x) - f(x)| \right] = \lim_{n \to +\infty} \left[\sup_{x \in D_1} |e^{-nx} - 0| \right] = \lim_{n \to +\infty} e^{-na} = 0 \,.$$

Example 7.11 The sequence

$$\{f_n\} = \left\{ \frac{nx+1}{n+1} \right\}$$

converges pointwise in $\mathbb{R}$ to the function $f(x) = x$, but it does not converge uniformly. In fact

$$\lim_{n \to +\infty} \left[\sup_{x \in \mathbb{R}} |f_n(x) - f(x)| \right] = \lim_{n \to +\infty} \left[\sup_{x \in \mathbb{R}} \left| \frac{nx+1}{n+1} - x \right| \right]$$

$$= \lim_{n \to +\infty} \left[\sup_{x \in \mathbb{R}} \left| \frac{1-x}{n+1} \right| \right] = +\infty \neq 0 ;$$

but, if one restricts the sequence to a bounded domain, for example in the form $D = [-L, L]$, with $L > 0$, it results that

$$\lim_{n \to +\infty} \left[\sup_{x \in D} |f_n(x) - f(x)| \right] = \lim_{n \to +\infty} \left[\sup_{x \in D} \left| \frac{nx+1}{n+1} - x \right| \right]$$

$$= \lim_{n \to +\infty} \left[\sup_{x \in D} \left| \frac{1-x}{n+1} \right| \right] = \lim_{n \to +\infty} \left[\left| \frac{L+1}{n+1} \right| \right] = 0$$

and that is uniformly convergent in $[-L, L]$.

It is important to notice that

$$M_n = \sup_{x \in D} |f_n(x) - f(x)|$$

is a sequence of non-negative numbers (more precisely, $M_n \in (-\infty, +\infty)$). Then, if $f_n \rightrightarrows f$, that is if $\lim_{n \to +\infty} M_n = 0$, for each $\epsilon > 0$ there exists n_ϵ *independent* of x such that $|f_n(x) - f(x)| \leq M_n < \epsilon$ for $n > n_\epsilon$. Conversely, if for each $\epsilon > 0$ there exists n_ϵ *independent* of x such that $|f_n(x) - f(x)| < \epsilon/2$, then for $n > n_\epsilon$

$$M_n = \left[\sup_{x \in D} |f_n(x) - f(x)| \right] \leq \epsilon/2 < \epsilon$$

that is

$$\lim_{n \to +\infty} \left[\sup_{x \in D} |f_n(x) - f(x)| \right] = 0 .$$

This characterises the uniform convergence in the sense that can be stated as follows.

Let $\{f_n(x)\}$ be a sequence of functions with domain of uniform convergence D, and $f : D \to \mathbb{R}$ be a function with domain D. Then $f_n \rightrightarrows f$ if and only if for each $\epsilon > 0$ there exists n_ϵ *independent* of x such that $|f_n(x) - f(x)| < \epsilon$ for $n > n_\epsilon$.

A suggestive graphical interpretation of the concept of uniform convergence can be given: $f_n \rightrightarrows f$ means that for each $\epsilon > 0$ there exists n_ϵ *independent* of x such that for $n > n_\epsilon$ $f(x) - \epsilon \leq f_n(x) \leq f(x) + \epsilon$. Thus the curves representing the functions

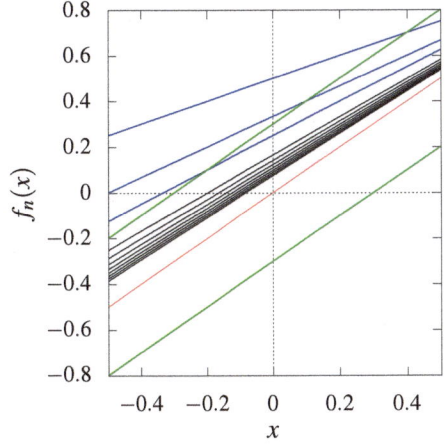

$$f_n(x) = \frac{nx+1}{n+1}, \ 5 \le n \le 12 \quad \text{———}$$

$$f_n(x), \ 1 \le n \le 3 \quad \text{———}$$

$$x \pm \varepsilon, \ \varepsilon = 0.3 \quad \text{———}$$

$$f(x) = x \quad \text{———}$$

Fig. 7.1 Uniform convergence of the sequence $\left\{ \dfrac{nx+1}{n+1} \right\}$, $x \in [-0.5, 0.5]$

f_n are definitively (*i.e.* for $n > n_\epsilon$) confined within the plane region bounded by the curves $f(x) - \epsilon$ and $f(x) + \epsilon$.

Consider the sequence

$$\{f_n\} = \left\{ \frac{nx+1}{n+1} \right\}, \quad x \in [-0.5, 0.5].$$

As it was pointed out in Example 7.11, it converges uniformly in any closed interval. If the bounded interval $[-0.5, 0.5]$ is considered, it is easily seen that, taking $\epsilon = 0.3$ it turns out to be $n_\epsilon = 4$. So, curves of f_n with $n \ge 5$ are confined within the strip delimited by the two straight lines $y = x - 0.3$ and $y = x + 0.3$, whereas curves of f_n with $n < 4$ will not. Selected curves from these two sets are depicted in Fig. 7.1.

The sequence $\{f_n\} = \{x^n\} = \{1, x, x^2, \dots, x^n, \dots\}$, $x \in (0, 1)$ is considered now. The sequence converges, but not uniformly, to the function $f(x) = 0$ as

$$\lim_{n \to +\infty} \left[\sup_{x \in (0,1)} \left| x^n - 0 \right| \right] = 1 \ne 0,$$

whose graphical representation of Fig. 7.2. The trend of curves suggests that there is no value of n above which the curves x^n can be expected to be confined within the strip bounded by the straight lines $y = -\epsilon$ and $y = +\epsilon$.

What has been illustrated in the previous considerations can now be summarised in the following statement.

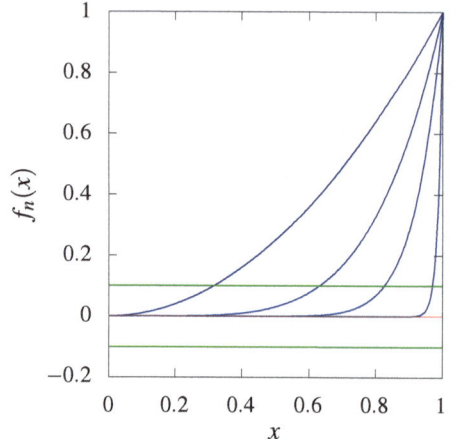

Fig. 7.2 Non uniform convergence of the sequence $\{x^n\}$, $x \in (0, 1)$

Let $\{f_n(x)\}$ be a sequence of functions with context domain E, $D \subseteq E$ its domain of uniform convergence and f its limit function. If x_0 is an accumulation point of D, and $\lim_{x \to x_0} f_n(x) = l_n$, then

$$\lim_{n \to +\infty} f_n(x) = l_n \quad \text{and it turns out to be} \quad \lim_{x \to x_0} f(x) = l.$$

In other words, the inversion of limits holds:

$$\lim_{x \to x_0} \left[\lim_{n \to +\infty} f_n(x) \right] = \lim_{n \to +\infty} \left[\lim_{x \to x_0} f_n(x) \right].$$

Indeed, $\{f_n(x)\}$ is a Cauchy sequence, that is for each $\epsilon > 0$ there exists n_ϵ (independent of x) such that for $n, m > n_\epsilon$, and for each $x \in D$, $|f_n(x) - f_m(x)| < \epsilon$. Taking the limit as $x \to x_0$ one obtains $|l_n - l_m| \leq \epsilon$, that is also $\{l_n\}$ is a Cauchy sequence, thus convergent, that is there exists

$$\lim_{n \to +\infty} l_n = l.$$

It is also known, from the hypotheses above, that for each n there exists $\delta_n(x_0)$ such that $|f_n(x) - l_n| < 0$ for $0 < |x - x_0| < \delta_n(x_0)$. Then, for $n, m > n_\epsilon$ and $0 < |x - x_0| < \delta_n(x_0)$ one has:

$$|f(x) - l| = |f(x) - f_n(x) + f_n(x) - l_n + l_n - l|$$

$$\leq |f(x) - f_n(x)| + |f_n(x) - l_n| + |l_n - l| < 3\epsilon \,.$$

In other words $\lim\limits_{x \to x_0} f(x) = l$. The next results immediately follows.

Let $\{f_n(x)\}$ be a sequence of functions with context domain E and $D \subseteq E$ its domain of uniform convergence. If the functions f_n are continuous at $x_0 \in D$, then the limit function f is continuous at x_0.

In fact, one has:

$$\lim_{x \to x_0} f(x) = \lim_{x \to x_0} \left[\lim_{n \to +\infty} f_n(x) \right] = \lim_{n \to +\infty} \left[\lim_{x \to x_0} f_n(x) \right]$$

$$= \lim_{n \to +\infty} f_n(x_0) = f(x_0) \,.$$

The sequence of continuous functions $f_n(x) = \{e^{-nx^2}\}$ converges pointwise in $D = (-\infty, +\infty)$ to the function

$$f(x) = \lim_{n \to +\infty} f_n(x) = \begin{cases} 0 & \text{if } x \neq 0 \\ 1 & \text{if } x = 0 \,, \end{cases}$$

which clearly does not inherit the property of continuity from the sequence. The previous results allows to state that convergence cannot be uniform in this case. It is worth noting that sequences of discontinuous functions may converge, uniformly or not uniformly, to continuous functions.

Example 7.12 The sequence $\{f_n\}$ of discontinuous functions defined by

$$f_n(x) = \begin{cases} e^{-nx^2} & \text{if } x \neq 0 \\ 0 & \text{if } x = 0 \end{cases}$$

converges everywhere in $\mathbb{R}$ to the continuous function $f(x) = 0$, but the convergence is not uniform. In fact, indicating $\mathbb{R}_0 := \mathbb{R} - \{0\}$, it results that

$$\lim_{n \to +\infty} \left[\sup_{x \in \mathbb{R}} |f_n(x) - f(x)| \right] = \lim_{n \to +\infty} \left[\sup_{x \in \mathbb{R}_0} \left| e^{-nx^2} - 0 \right| \right] = \lim_{n \to +\infty} 1 = 1 \neq 0 \,.$$

Example 7.13 The sequence $\{f_n\}$ of discontinuous functions defined by

$$f_n(x) = \begin{cases} 0 & \text{if } x \neq 0 \\ \dfrac{1}{n} & \text{if } x = 0 \end{cases}$$

converges everywhere in $\mathbb{R}$ to the continuous function $f(x) = 0$, and the convergence in this case is uniform. Indeed

$$\lim_{n \to +\infty} \left[\sup_{x \in \mathbb{R}} |f_n(x) - f(x)| \right] = \lim_{n \to +\infty} \left[\sup_{x=0} \left| \frac{1}{n} - 0 \right| \right] = \lim_{n \to +\infty} \frac{1}{n} = 0.$$

The continuity of the limit function does not guarantee the uniform convergence: actually, it may happen that sequences of continuous functions converge non uniformly (see Examples 7.10 and 7.11) to continuous functions. However, the following important condition of uniform convergence is stated.

> If $\{f_n(x)\}$ is a sequence of real continuous functions in the closed bounded interval $D = [a, b]$ such that:
>
> $-$ $f_n(x) \le f_{n+1}(x)$ $\forall x \in D$ or $(f_n(x) \ge f_{n+1}(x)$ $\forall x \in D)$,
> $-$ $\lim_{n \to +\infty} f_n(x) = f(x)$ $\forall x \in D$ with f continuous in D,
>
> then $\{f_n\}$ converges uniformly to f in D.

Example 7.14 The sequence of functions $\{f_n(x)\} = \{e^{-n^3(x-1/n)^2}\}$ considered in the closed and bounded domain $D = [0, 1]$ converges to the continuous function $f(x) = 0$, but the convergence is not uniform. Indeed

$$M_n = \sup_{x \in [0,1]} |f_n(x) - f(x)| = \sup_{x \in [0,1]} |f_n(x) - 0| \ge \left| f_n \left(\frac{1}{n} \right) \right| = 1.$$

Then $\lim_{n \to +\infty} M_n$ cannot be 0. The previous condition allows to say that the sequence is not monotone.

7.3 Taking Limits Under the Integral

Consider the sequence of continuous functions $\{f_n(x)\} = \{nxe^{-nx^2}\}$ in its context domain $E = [0, 1]$, that is also its domain of convergence to the (continuous) function $f(x) = 0$. Let's calculate

$$\lim_{n \to +\infty} \int_0^1 f_n(x)\,dx.$$

The temptation to take the limit under integral sign is strong, but it is totally unjustified. Indeed

$$\lim_{n \to +\infty} \int_0^1 f_n(x)\,dx = \lim_{n \to +\infty} \int_0^1 nxe^{-nx^2}\,dx = \int_0^1 \lim_{n \to +\infty} f_n(x)\,dx = 0,$$

whereas

$$\lim_{n\to+\infty}\left[-\frac{1}{2}e^{-nx^2}\right]_0^1 = \lim_{n\to+\infty}\left[-\frac{1}{2}e^{-n}+\frac{1}{2}\right] = \frac{1}{2} \neq 0,$$

It is natural, at this point, investigating which are the conditions allowing to exchange the order between the evaluation of the limit and the calculation of the integral. Once again, the uniform convergence plays a crucial role.

In fact, for each $\epsilon > 0$ there exists n_ϵ (independent of x) such that

$$sup_{x\in D}\,|f_n(x) - f(x)| < \epsilon.$$

Therefore

$$\left|\int_a^b f_n(x)\,dx - \int_a^b f(x)\,dx\right| = \left|\int_a^b [f_n(x) - f(x)]\,dx\right|$$

$$\leq \int_a^b |f_n(x) - f(x)|\,dx$$

$$\leq \sup_{x\in D}|f_n(x) - f(x)|\,(b-a) < \epsilon(b-a).$$

This can be summarised as follows.

If $\{f_n(x)\}$ is a sequence of continuous functions in $D = [a, b]$, where it is uniformly convergent to the function f, then it is allowed to take to the limit under integral sign as follows

$$\lim_{n\to+\infty}\int_a^b f_n(x)\,dx = \int_a^b \lim_{n\to+\infty} f_n(x)\,dx = \int_a^b f(x)\,dx.$$

It is worth noting that the previous result does not apply to integrals extending to unbounded domains, as it is shown below.

Example 7.15 Consider, in $\mathbb{R}$, the sequence $\{f_n\}$ defined by

$$f_n(x) = \begin{cases} \dfrac{x(n-x)}{n^3} & \text{if } 0 \leq x \leq n \\ 0 & \text{elsewhere}. \end{cases}$$

It is clear that

$$f(x) = \lim_{n\to+\infty} f_n(x) = 0, \quad \forall x \in \mathbb{R},$$

and that such convergence is uniform. In fact

$$M_n = \sup_{x \in \mathbb{R}} |f_n(x) - f(x)| = \sup_{x \in [0,n]} \left| \frac{x(n-x)}{n^3} - 0 \right| \ge \left| f_n\left(\frac{n}{2}\right) \right| = \frac{1}{4n},$$

and

$$\lim_{n \to +\infty} M_n = \lim_{n \to +\infty} \frac{1}{4n} = 0.$$

Nevertheless, it results that

$$\lim_{n \to +\infty} \int_{-\infty}^{+\infty} f_n(x)\,dx = \lim_{n \to +\infty} \int_0^n \frac{x(n-x)}{n^3}\,dx$$

$$= \lim_{n \to +\infty} \left[\frac{x^2}{2n^2} - \frac{x^3}{3n^3} \right]_0^n = \lim_{n \to +\infty} \left[+\frac{1}{2} - \frac{1}{3} \right] = \frac{1}{6} \ne 0,$$

whereas

$$\int_{-\infty}^{+\infty} \lim_{n \to +\infty} f_n(x)\,dx = 0.$$

Example 7.16 Consider, in $\mathbb{R}$, the sequence $\{f_n\} = \{nx(1-x^2)^n\}$ in the context domain $E = [0, 1]$. It can be easily seen that such domain is also the domain of pointwise convergence to the function $f(x) = 0$. It results that:

$$\lim_{n \to +\infty} \int_0^1 f_n(x)\,dx = \lim_{n \to +\infty} \int_0^1 nx(1-x^2)^n\,dx$$

$$= \lim_{n \to +\infty} -\frac{n}{2} \int_0^1 -2x(1-x^2)^n\,dx$$

$$= \lim_{n \to +\infty} -\frac{n}{2} \left[\frac{(1-x^2)^{n+1}}{n+1} \right]_0^1 = \lim_{n \to +\infty} -\frac{n}{2(n+1)} = \frac{1}{2} \ne 0,$$

whereas

$$\int_0^1 \lim_{n \to +\infty} f_n(x)\,dx = 0.$$

It follows that the convergence cannot be uniform.

It is worth noting that the uniform convergence is just a sufficient condition to allow taking the limit inside the integral sign, as it is shown below.

Example 7.17 Consider the sequence of continuous functions $\{f_n\} = \{nxe^{-nx}\}$ in the context domain $E = [0, 1]$. In E the sequence converges to the (continuous) function $f(x) = 0$, but the convergence is not uniform. In fact

$$\lim_{n \to +\infty} \sup_{x \in [0,1]} |f_n(x) - f(x)| = \lim_{n \to +\infty} \sup_{x \in [0,1]} |f_n(x)| \ge \lim_{n \to +\infty} \left| f_n\left(\frac{1}{n}\right) \right| = \frac{1}{e}$$

whereas

$$\int_0^1 \lim_{n \to +\infty} f_n(x)\, dx = 0.$$

Nevertheless, it results that:

$$\lim_{n \to +\infty} \int_0^1 f_n(x)\, dx = \lim_{n \to +\infty} \int_0^1 nxe^{-nx}\, dx = \lim_{n \to +\infty} \frac{1}{n} \int_0^n te^{-t}\, dt$$

$$= \lim_{n \to +\infty} \left[e^{-n} - \frac{e^{-n}}{n} + \frac{1}{n} \right] = 0.$$

7.4 Taking Limits Under the Derivative

Let $\{f_n\}$ be a sequence of real differentiable functions defined in the set E and there convergent to the function f. It has already been observed that, even in the case of uniform convergence, f does not necessarily inherit the property of differentiability. Actually, the aforementioned sequence $\{f_n\} = \{\sqrt{x^2 + 1/n}\}$ is a sequence of differentiable functions in $\mathbb{R}$, convergent for each $x \in \mathbb{R}$ to the function $f(x) = |x|$, that is continuous in $\mathbb{R}$ but not differentiable at $x = 0$, and the convergence is uniform:

$$\lim_{n \to +\infty} \sup_{x \in \mathbb{R}} |f_n(x) - f(x)| = \lim_{n \to +\infty} \sup_{x \in \mathbb{R}} \left| \sqrt{x^2 + \frac{1}{n}} - |x| \right|$$

$$= \lim_{n \to +\infty} \sup_{x \in \mathbb{R}} \frac{1/n}{\left| \sqrt{x^2 + \frac{1}{n}} - |x| \right|}$$

$$= \lim_{n \to +\infty} \left[e^{-n} - \frac{e^{-n}}{n} + \frac{1}{n} \right] = 0.$$

Furthermore, if the limit function f inherited the property of differentiability, this would not necessarily mean that

$$\lim_{n \to +\infty} f_n'(x) = f'(x)$$

(taking the limit under the derivative) as it is clearly shown below.

Example 7.18 Consider, in the domain $E = [0, 1]$, the sequence $\{f_n\} = \{x^n/n\}$. In such domain the sequence converges uniformly to the differentiable function $f(x) = 0$:

$$\lim_{n \to +\infty} \sup_{x \in [0,1]} |f_n(x) - f(x)| = \lim_{n \to +\infty} \sup_{x \in [0,1]} \left| \frac{x^n}{n} \right| = \lim_{n \to +\infty} \frac{1}{n} = 0.$$

The sequence of derivatives $\{f_n'(x) = \{x^{n-1}\}$ also converges in E to the (non continuous) function

$$g(x) = \begin{cases} 0 & \text{if } 0 \leq x \leq 1 \\ 1 & \text{if } x = 1 \end{cases}$$

and, at $x = 1$, it results:

$$0 = f'(1) \neq \lim_{n \to +\infty} f_n'(1) = \lim_{n \to +\infty} g(1) = \lim_{n \to +\infty} 1^{n-1} = 1 \,,$$

that is, taking the limit under the derivative is not allowed. It is worth noting that the convergence of the sequence $\{f_n'(x)\} = \{x^{n-1}\}$ to the function g is not uniform (the functions f_n' are continuous, whereas the function g is not).

To conclude, the uniform convergence of the sequence $\{f_n\}\} = \{x^n/n\}$ is not effective to the aim of inverting the order of taking the limit and taking the derivative. In this sense, on the contrary, it proves to be effective the uniform convergence of the sequence of derivatives. This circumstance is expressed by the following statement.

Let $\{f_n(x)\}$ be a sequence of real differentiable functions in $D = [a, b]$, where it is pointwise convergent to the function f. If the sequence $\{f_n'(x)\}$ converges uniformly in D to the function g then f is differentiable and it results that

$$\lim_{n \to +\infty} f_n'(x) = g(x) = f'(x) \,.$$

Stating the above result it is assumed that the functions f_n are differentiable with continuous derivative. Actually, it holds also in the weaker assumptions that the functions f_n be simply differentiable.

7.5 Suggested Readings

A concise introduction to sequences of functions is found in Apostol (1974, Chap. 9, pp. 218–223).

Reference

T. M. Apostol. *Mathematical Analysis: A Modern Approach to Advanced Calculus.* Pearson Education US, Hoboken, NJ, 2nd edition, 1974.

Chapter 8
Series of Functions

On the basis of the previous lesson, a comprehensive discussion is given here of the *series of functions*, an outstanding topic for the solution of differential problems in applied mathematics.

8.1 Introduction

Consider a sequence $\{f_n\}$ of real functions with context domain E. One can use $\{f_n\}$ to generate an associated sequence $\{S_n\}$ defined as follows: at every x

$$S_0(x) = f_0(x), \quad S_1(x) = f_0(x) + f_1(x), \quad S_2(x) = f_0(x) + f_1(x) + f_2(x), \quad \dots,$$

$$S_n(x) = f_0(x) + f_1(x) + f_2(x) + \cdots + f_n(x).$$

$S_n(x)$ is called the n-th *partial sum* of $\{f_n\}$ at x. The sequence $\{S_n\}$ is called a *series of functions* and it is traditionally (though inappropriately) denoted by

$$\sum_{k=0}^{+\infty} f_k(x).$$

Actually, this symbol should be more properly associated to the limit of the sequence $\{S_n\}$, when this limit exists; in other words, a natural writing would be as follows

$$\lim_{n \to \infty} S_n(x) = \sum_{k=0}^{+\infty} f_k(x) = S(x).$$

© The Author(s), under exclusive license to Springer Nature Switzerland AG 2024
G. Riccardi et al., *Multidimensional Differential and Integral Calculus*,
https://doi.org/10.1007/978-3-031-70326-3_8

The function $S(x)$ is said to be the *sum function* or, shortly, the *sum of the series* $\sum_{k=0}^{+\infty} f_k(x)$. This terminology stems from the observation that the symbol $\sum_{k=0}^{+\infty} f_k(x)$ represents the sum of infinitely many terms:

$$\sum_{k=0}^{+\infty} f_k(x) = f_0(x) + f_1(x) + \cdots + f_n(x) + \cdots .$$

Note that the terms of the sequence $\{S_n\}$, *i.e.* the functions $S_n(x) = f_0(x) + f_1(x) + f_2(x) + \cdots + f_n(x)$ (partial sums), being finite sums of functions, inherit all properties of boundedness, integrability and regularity of functions $f_n(x)$. However the convergence property of the sequence $\{f_n\}$ is not necessarily inherited: as it is well-known, the sequence $\{f_n\} = \{1/n\}$ converges to zero, whereas the associated series

$$\sum_{n=0}^{+\infty} \frac{1}{n} \quad \text{(harmonic series)}$$

diverges.

A series of functions $\sum_{k=0}^{+\infty} f_k(x)$ is actually the sequence of functions $\{S_n\}$, then such concepts as context domain, pointwise convergence and uniform convergence can be transferred to it. Moreover, at each $\bar{x}$ in the context domain, the series of functions $\sum_{k=0}^{+\infty} f_k(\bar{x})$ returns a series of numbers which all the tests of absolute convergence (root test, quotient test, etc.) can be applied to.

Example 8.1 Consider the series of functions

$$\sum_{k=0}^{+\infty} x^k = 1 + x + x^2 + \cdots + x^n + \cdots .$$

It was set, conventionally, $0^0 = 1$. The maximum context domain is $\mathbb{R}$. To find the domain of pointwise convergence, the root test is used. It results:

$$\lim_{k \to +\infty} \sqrt[k]{|x^k|} = |x|,$$

whence it is seen that the series converges absolutely when $|x| < 1$. When $|x| > 1$ the series cannot converge: it diverges when $x > 1$ and oscillates when $x < -1$. At $x = 1$ it results the series

$$\sum_{k=0}^{+\infty} 1^k = 1 + 1 + 1 + \cdots + 1 + \cdots ,$$

which diverges. At $x = -1$ it results the series

$$\sum_{k=0}^{+\infty} (-1)^k = 1 - 1 + 1 - 1 + 1 - \cdots,$$

which is undetermined (oscillating).

Consider now the issue of uniform convergence. Let $S(x)$ be the sum of the series and $R_n(x) = S(x) - S_n(x)$ its remainder; it results

$$\sup_{|x|<1} |R_n(x)| = \sup_{|x|<1} |S(x) - S_n(x)| = \sup_{|x|<1} |x^{n+1} + x^{n+2} + x^{n+3} + \cdots| = +\infty,$$

then

$$\lim_{n \to +\infty} \sup_{|x|<1} |S(x) - S_n(x)| = +\infty,$$

and it follows that there is not uniform convergence in $(-1, 1)$.

However, limiting the domain to $[-a, a]$, with $0 < a < 1$, the uniform convergence is achieved. In fact

$$\sup_{|x|\le a} |R_n(x)| = \sup_{|x|\le a} |x^{n+1} + x^{n+2} + x^{n+3} + \cdots| = \sup_{|x|\le a} \left| \frac{x^{n+1}}{1 - x} \right| = \frac{a^{n+1}}{1 - a}.$$

Then, since $0 < a < 1$,

$$\lim_{n \to +\infty} \sup_{|x|\le a} |S(x) - S_n(x)| = \lim_{n \to +\infty} \sup_{|x|\le a} |R_n(x)| = \lim_{n \to +\infty} \frac{a^{n+1}}{1 - a} = 0,$$

showing the uniform convergence in $[-a, a]$.

The example just discussed shows how difficult can be to establish the uniform convergence of a series directly applying the definition, due to the need of evaluating the least upper bound of the absolute value of the remainder $\sup_{x \in D} |R_n(x)|$. However, this was done quite easily in the example, since $R_n(x)$ had the form of a geometric sum, but it is rather complicated in most general cases. To circumvent such difficulties it can be useful to introduce the concept of *total convergence* of a series, introduced as follows.

Consider the series of functions $\sum_{k=0}^{+\infty} f_k(x)$, defined in the context domain E, and suppose there exists a sequence of real numbers $\{a_k\}$, with $a_k \ge 0$, such that at each $x \in D \subseteq E$ $|f_k(x)| \le a_k$.

If the series $\sum_{k=0}^{+\infty} a_k$ converges, one says that the series $\sum_{k=0}^{+\infty} f_k(x)$ is *totally convergent* in D, that is said to be the *domain of total convergence*.

Let $\sum_{k=0}^{+\infty} f_k(x)$ be a series of functions with domain of total convergence D, and let $S(x)$ its sum. Then, at each $x \in D \subseteq E$ there exists a sequence of real numbers $\{a_k\}$, with $a_k \geq 0$, such that $|f_k(x)| \leq a_k$ and the series converges, that is

$$\lim_{n \to +\infty} \left| \sum_{k=n+1}^{+\infty} a_k \right| = 0.$$

One can write

$$|S_n(x) - S(x)| = |R_n(x)| = \left| \sum_{k=n+1}^{+\infty} f_k(x) \right| \leq \sum_{k=n+1}^{+\infty} |f_k(x)| \leq \sum_{k=n+1}^{+\infty} |a_k|,$$

whence

$$\sup_{x \in D} |S_n(x) - S(x)| \leq \sum_{k=n+1}^{+\infty} |a_k|$$

and

$$0 \leq \lim_{n \to +\infty} \left[\sup_{x \in D} |S_n(x) - S(x)| \right] \leq \lim_{n \to +\infty} \sum_{k=n+1}^{+\infty} |a_k| = 0,$$

then

$$\lim_{n \to +\infty} \left[\sup_{x \in D} |S_n(x) - S(x)| \right] = 0,$$

that is $\sum_{k=0}^{+\infty} f_k(x)$ is *uniformly convergent*.

This can be formalised by the following statement.

> If the series of functions $\sum_{k=0}^{+\infty} f_k(x)$ is *totally convergent* then it is (absolutely) *uniformly convergent* in D.

However, it has to be observed that uniformly convergent series exist that are not totally convergent. As an example, consider the series

$$\sum_{k=0}^{+\infty} f_k(x) = \sum_{k=0}^{+\infty} (-1)^k \frac{x^k}{k+1} \qquad \text{in the domain } D = [0, 1).$$

According to the Leibniz[1] test the above series is simply (not absolutely) convergent in D. Furthermore, observe that

---

[1] https://mathshistory.st-andrews.ac.uk/Biographies/Leibniz/.

$$\sup_{x \in D} \left| (-1)^k \frac{x^k}{k+1} \right| = \sup_{x \in D} \left| \frac{x^k}{k+1} \right| = \frac{1}{k+1} = a_k.$$

In other words, $\{a_k\}$ is the "smallest" sequence dominating $|f_k(x)|$ $(|f_k(x)| \le a_k)$, and $\sum_{k=0}^{+\infty} a_k = \sum_{k=0}^{+\infty} \frac{1}{k+1}$ diverges, so that the series $\sum_{k=0}^{+\infty} (-1)^k \frac{x^k}{k+1}$ cannot be totally convergent. Nevertheless it is uniformly convergent in D. Indeed:

$$\sup_{x \in D} |S_n(x) - S(x)| = \sup_{x \in D} \left| \sum_{k=n+1}^{+\infty} (-1)^k \frac{x^k}{k+1} \right|$$

$$= \sup_{x \in D} \left| \frac{x^n}{n+1} - \frac{x^{n+1}}{n+2} + \frac{x^{n+2}}{n+3} - \cdots + \cdots \right|$$

$$\le \sup_{x \in D} \left| \frac{x^n}{n+1} \right| = \frac{1}{n+1},$$

whence

$$\lim_{n \to +\infty} \left[\sup_{x \in D} |S_n(x) - S(x)| \right],$$

that is the uniform convergence.

It was seen, in studying the sequences of functions, that the uniform convergence is a crucial condition to guarantee that the limit function inherits the regularity and integration properties of the functions in the sequence. In the language of series of functions the limit function represents the sum of the series and the results obtained in the theory of sequences take a particularly expressive form and, as it is seen in the following statements, of great significance for applications.

Consider the series of functions $\sum_{k=0}^{+\infty} f_k(x)$, with context domain E, and $D \subseteq E$ its domain of uniform convergence.

If the functions f_k are continuous at $x_0 \in D$ then the sum function S is continuous at x_0.

Consider now $\sum_{k=0}^{+\infty} f_k(x)$, a series of functions continuous in $D = [a, b]$ and there uniformly convergent to the sum function $S(x) = \sum_{k=0}^{+\infty} f_k(x)$ (continuous, according to the above result). The following relation holds, that allows *taking the series under the integral sign* (integration by series).

$$\int_a^b \sum_{k=0}^{+\infty} f_k(x)\,dx = \sum_{k=0}^{+\infty} \int_a^b f_k(x)\,dx\,.$$

Finally, let $\sum_{k=0}^{+\infty} f_k(x)$ be a series of differentiable functions with continuous derivative in $D = [a, b]$ and there pointwise convergent to the sum function S.

If the series $\sum_{k=0}^{+\infty} f_k'(x)$ is uniformly convergent in D to the function G, then S is differentiable and results

$$S'(x) = G(x) = \sum_{k=0}^{+\infty} f_k'(x)\,.$$

Example 8.2 Consider the series of functions

$$\sum_{k=0}^{+\infty} \frac{x^k}{k!} = 1 + x + \frac{x^2}{2} + \frac{x^3}{6} + \cdots$$

in the interval $[-a, a]$, with $a > 0$. It results that

$$\left|\frac{x^k}{k!}\right| \le \frac{a^k}{k!}$$

and the series $\sum_{k=0}^{+\infty} a^k/k!$ is convergent, as it is easily seen using the ratio test, so that $\sum_{k=0}^{+\infty} x^k/k!$ turns out to be totally convergent in $[-a, a]$. Let $S(x)$ be the sum of the series. Similarly

$$\left|\frac{kx^{k-1}}{k!}\right| \le \frac{a^{k-1}}{k-1!}$$

so also the series of derivatives $\sum_{k=1}^{+\infty} kx^{k-1}/k!$ is totally (and then uniformly) convergent in $[-a, a]$ to a function $G(x)$. By virtue of the third results above, it turns out to be $G(x) = S'(x)$.

Furthermore, observe that

$$S'(x) = G(x) = \sum_{k=1}^{+\infty} \frac{kx^{k-1}}{k!} = \sum_{k=1}^{+\infty} \frac{x^{k-1}}{k-1!} = 1 + x + \frac{x^2}{2} + \frac{x^3}{6} + \cdots = \sum_{k=0}^{+\infty} \frac{x^k}{k!} = S(x)\,.$$

Since it is also $S(0) = 0$, it can be stated that

$$\sum_{k=0}^{+\infty} \frac{x^k}{k!} = S(x) = e^x \quad \text{at each } x \in [-a, a].$$

Example 8.3 Consider the integral

$$\int_1^2 \frac{e^t}{t} \, dt \, .$$

As it is well-known, the antiderivative of $y = e^t / t$ cannot be expressed in closed form. Using the expansion $\sum_{k=0}^{+\infty} t^k / k! = e^t$, and observing that the series $\sum_{k=0}^{+\infty} t^{k-1} / k!$ is totally convergent in $[1, 2]$, one can write according to the second result stated above:

$$\int_1^2 \frac{e^t}{t} \, dt = \int_1^2 \frac{\sum_{k=0}^{+\infty} \frac{t^k}{k!}}{t} \, dt = \int_1^2 \sum_{k=0}^{+\infty} \frac{t^{k-1}}{k!} \, dt$$

$$= \sum_{k=0}^{+\infty} \int_1^2 \frac{t^{k-1}}{k!} \, dt = \left[\log t \right]_1^2 + \sum_{k=1}^{+\infty} \left[\frac{t^k}{kk!} \right]_1^2 .$$

8.2 Power Series

The series $\sum_{k=0}^{+\infty} x^k / k!$ considered above is a particular case of a series in the form

$$\sum_{k=0}^{+\infty} a_k (x - x_0)^k = a_0 + a_1(x - x_0) + a_2(x - x_0)^2 + \cdots \quad \text{called a } power \ series \, .$$

The numbers $a_k \in \mathbb{R}$ are the coefficients of the series, whereas x_0 is the pole. Using the root test one can observe that the series converges absolutely if

$$\lim_{k \to +\infty} \sqrt[k]{|a_k||x - x_0|^k} = |x - x_0| \lim_{k \to +\infty} \sqrt[k]{|a_k|} < 1$$

that is

$$|x - x_0| < \frac{1}{\lim_{k \to +\infty} \sqrt[k]{|a_k|}} =: R \, .$$

Using this expression for R (*radius of convergence*), one can define the *field of convergence* as the set $C = \{x \in \mathbb{R} : |x - x_0| < R\}$. It may happen that $\lim_{k \to +\infty} \sqrt[k]{|a_k|} = 0$. In this case one sets $R = +\infty$.

The root test can say nothing about (simple or absolute) convergence of the series at points $x = x_0 \pm R$. This case has to be investigated separately.

Example 8.4 Consider the series $\sum_{k=1} x^k/k$. It results that

$$R = \frac{1}{\lim\limits_{k \to +\infty} \sqrt[k]{|a_k|}} = \frac{1}{\lim\limits_{k \to +\infty} \sqrt[k]{1/k}} = 1 \, .$$

so that the field of absolute convergence is $C = \{x \in \mathbb{R} : |x| < 1\}$. Notice that the series converges simply (not absolutely) even at $x = -1$.

Sometimes it can be difficult to calculate the radius of convergence of the series $\sum_{k=0}^{+\infty} a_k(x - x_0)^k$ using the root test. In such a case, if the numbers a_k are definitively nonvanishing, it may be useful to apply the ratio test. It results that

$$\lim_{k \to +\infty} \left| \frac{a_{k+1}}{a_k} \right| \left| \frac{(x - x_0)^{k+1}}{(x - x_0)^k} \right| = \lim_{k \to +\infty} \left| \frac{a_{k+1}}{a_k} \right| |(x - x_0)|$$

and, set

$$R = \lim_{k \to +\infty} \left| \frac{a_k}{a_{k+1}} \right| \, ,$$

one can say that the series converges absolutely in the interval $C = \{x \in \mathbb{R} : |x - x_0| < R\}$. Even in this case, nothing can be said, until further investigation, about convergence at the endpoints $x = x_0 \pm R$.

Example 8.5 Consider the series $\sum_{k=0}^{+\infty} \frac{k^k}{e^k k!} x^k$. It is quite difficult to calculate the radius of convergence using the root test. However, since $a_k = k^k/(e^k k!) \neq 0$, one can set

$$R = \lim_{k \to +\infty} \frac{a_k}{a_{k+1}} = \lim_{k \to +\infty} \frac{k^k}{e^k k!} \frac{e^{k+1}(k+1)!}{(k+1)^{k+1}} = \lim_{k \to +\infty} \frac{e}{(1 + 1/k)^k} = 1 \, .$$

whence the field of absolute convergence is $C = \{x \in \mathbb{R} : |x| < 1\}$. Observe that the series diverges at $x = 1$. In fact, using the Stirling approximation for the factorial, $k! \simeq \sqrt{2\pi} \, e^{-k} k^{k+1/2}$ as $k \to +\infty$, it results that

$$\lim_{k \to +\infty} \frac{\sqrt{2\pi k} \, k^k e^{-k}}{k!} = 1 \, , \quad \text{whence} \quad \lim_{k \to +\infty} \sqrt{2\pi k} \, \frac{k^k}{e^k k!} = 1 \, , \quad \text{that is}$$

$$\left\{ \frac{k^k}{e^k k!} \right\} \sim \left\{ \frac{1}{\sqrt{2\pi k}} \right\} \, ,$$

and the series $\sum_{k=1}^{+\infty} 1/\sqrt{2\pi k}$ diverges. At $x = -1$ the series becomes $\sum_{k=0}^{+\infty}(-1)^k \dfrac{k^k}{e^k k!}$,

and it was noticed that

$$\frac{a_k}{a_{k+1}} = \frac{e}{(1+1/k)^k} > 1$$

so $a_k = k^k/(e^k k!)$ is a decreasing sequence, and also $\lim_{k\to+\infty} a_k = \lim_{k\to+\infty} 1/\sqrt{2\pi k} = 0$. The simple convergence follows from the Leibnitz test. To summarise, the following characteristic domains are defined for the series $\sum_{k=0}^{+\infty} \dfrac{k^k}{e^k k!} x^k$:

- field of absolute convergence $C = \{x \in \mathbb{R} : |x| < 1\}$,
- domain of convergence $D_c = \{x \in \mathbb{R} : -1 \le x < 1\}$,
- domain of absolute convergence $D_{ac} = \{x \in \mathbb{R} : |x| < 1\}$ coinciding with C in this case.

Example 8.6 Consider the series $\sum_{k=0}^{+\infty} \dfrac{x^k}{k^2}$. Notice first that at $|x| = 1$ the series converges absolutely. Furthermore

$$R = \frac{1}{\lim_{k\to+\infty} \sqrt[k]{|a_k|}} = \lim_{k\to+\infty} \sqrt[k]{k^2} = 1.$$

Thus, the following characteristic domains are defined for the series:

- field of absolute convergence $C = \{x \in \mathbb{R} : |x| < 1\}$,
- domain of convergence $D_c = \{x \in \mathbb{R} : |x| \le 1\}$,
- domain of absolute convergence $D_{ac} = \{x \in \mathbb{R} : |x| \le 1\}$, properly including C.

Example 8.7 Consider the series $\sum_{k=0}^{+\infty} a_k x^k$ with

$$a_k = \begin{cases} \dfrac{1}{2^k} & \text{if } k \text{ is even} \\[2mm] \dfrac{1}{3^k} & \text{if } k \text{ is odd}. \end{cases}$$

Observe that

$$\sqrt[k]{|a_k|} = \begin{cases} \dfrac{1}{2} & \text{if } k \text{ is even} \\[2mm] \dfrac{1}{3} & \text{if } k \text{ is odd}, \end{cases}$$

so that $\lim_{k\to+\infty}\sqrt[k]{|a_k|}$ does not exist! In this case the radius of convergence, and the relevant field of absolute convergence, can be determined as follows: observe that

$$\left|\sum_{k=0}^{n}a_kx^k\right|\leq\sum_{k=0}^{n}|a_kx^k|\leq\sum_{k=0}^{n}\frac{1}{2^kx^k}.$$

Taking the limit as $n\to+\infty$, it results that

$$\left|\sum_{k=0}^{+\infty}a_kx^k\right|\leq\sum_{k=0}^{+\infty}\frac{1}{2^kx^k},$$

and $\sum_{k=0}^{+\infty}1/(2^kx^k)$ converges as $|x|<2$ then, by comparison, the series $\sum_{k=0}^{+\infty}a_kx^k$ converges (absolutely) as $|x|<2$. Thus the field of absolute convergence is $C=\{x\in\mathbb{R}:|x|<2\}$, and the radius of convergence is $R=2$. Notice that at $|x|=2$ the series cannot converge, since $\lim_{k\to+\infty}a_k2^k$ does not exist. C coincides with the domain of absolute convergence.

Example 8.8 Consider the series $\sum_{k=0}^{+\infty}\frac{x^{2k}}{k^2}$. It can be thought of as a power series $\sum_{k=0}^{+\infty}a_kx^k$ with

$$a_k=\begin{cases}\dfrac{1}{k^2}&\text{if }k\text{ is even}\\[2mm]0&\text{if }k\text{ is odd}\end{cases}$$

and, as in the previous Example 8.7, $\lim_{k\to+\infty}\sqrt[k]{|a_k|}$ does not exist. In order to determine the field of absolute convergence, one might follow the same procedure as in the previous example, that is using a comparison technique. However, an alternative method. If one sets $t=x^2$ the series becomes $\sum_{k=0}^{+\infty}t^k/k^2$, that is an ordinary power series $\sum_{k=0}^{n}a_kt^k$ with $a_k=1/k^2$. Thus,

$$R=\frac{1}{\lim\limits_{k\to+\infty}\sqrt[k]{|a_k|}}=\lim_{k\to+\infty}\sqrt[k]{k^2}=1,$$

whence the field of absolute convergence is

$$C=\{t\in\mathbb{R}:|t|<1\}=\{x\in\mathbb{R}:|x^2|<1\}=\{x\in\mathbb{R}:|x|<1\}.$$

Notice that at $|x|=2$ the series converges absolutely also at $|x|=1$, so that domain of absolute convergence is $D_{ac}=\{x\in\mathbb{R}:|x|<1\}$, which properly includes C.

Some observation on the convergence quality of a power series are given in the following.

Consider the series $\sum_{k=0}^{+\infty} a_k (x - x_0)^k$, with R being its radius of convergence and $C = \{x \in \mathbb{R} : |x| < R\}$ its field of absolute convergence. Let T be a closed and bounded subset of C, defined by $T = \{x \in \mathbb{R} : |x - x_0| \le \rho\}$, with $0 < \rho < R$. It results that, evidently, $|a_k (x - x_0)^k| \le |a_k \rho^k|$ for each k, and the series $\sum_{k=0}^{+\infty} a_k \rho^k$ being convergent. In fact, using the root test, one gets

$$\lim_{k \to +\infty} \sqrt[k]{|a_k \rho^k|} = \rho \lim_{k \to +\infty} \sqrt[k]{|a_k|} = \frac{\rho}{R}.$$

It follows that T is domain of total (and then uniform) convergence. This result, due to Hadamard[2] can be summarised as:

The power series $\sum_{k=0}^{+\infty} a_k (x - x_0)^k$, with field of absolute convergence C, is totally convergent in any *compact* (closed and bounded) set T properly included in C.

An even more valuable result on the convergence quality of a power series is given by the following theorem, due to Abel.[3]

If the power series $\sum_{k=0}^{+\infty} a_k (x - x_0)^k$, converges pointwise at $x \in \mathbb{R}$, it is uniformly convergent at any $y \in [x_0, x]$.

This result does not add anything new to Hadamard's result as far as $x \in C$. On the other hand, it is a breakthrough when $x = x_0 + R$ or $x = x_0 - R$.

As an example, consider the series $\sum_{k=0}^{+\infty} x^k / k^2$. It is easily found that its radius of convergence is $R = 1$. Thus, any set $T = \{x \in \mathbb{R} : |x - x_0| \le \rho < 1\}$ is a domain of total (and then uniform) convergence. Indeed the series converges absolutely even at $|x| = 1$. So, by virtue of Abel's theorem, it converges absolutely uniformly in the closed intervals $[-1, 0]$ and $[0, 1]$, and then in the closed interval $[-1, 1] \supset C$.

Example 8.9 Consider the series $\displaystyle\sum_{k=0}^{+\infty} (-1)^k \frac{x^{k+1}}{k+1}$. It results that

$$R = \frac{1}{\lim\limits_{k \to +\infty} \sqrt[k]{|a_k|}} = \lim_{k \to +\infty} \sqrt[k]{k+1} = 1.$$

Furthermore, observe that the series converges simply at $x = 1$ and does not converge at $x = -1$. Thus, the characteristic sets for the convergence of the series are listed below:

[2] https://mathshistory.st-andrews.ac.uk/Biographies/Hadamard/.
[3] https://mathshistory.st-andrews.ac.uk/Biographies/Abel/.

- field of absolute convergence $C = \{x \in \mathbb{R} : |x| < 1\}$;
- domain of convergence $D_c = \{x \in \mathbb{R} : -1 < x \le 1\}$;
- domain of absolute convergence $D_{ac} = \{x \in \mathbb{R} : -1 < x < 1\} = C$.
- domain of total convergence $T = \{x \in \mathbb{R} : |x| \le \rho < 1\}$;
- domain of uniform convergence $D_{uc} = \{x \in \mathbb{R} : -\rho \le x \le 1\}$;
- domain of absolute uniform convergence $D_{auc} = \{x \in \mathbb{R} : |x| \le \rho < 1\} = T$

Given the power series $\sum_{k=0}^{+\infty} a_k (x - x_0)^k$ with radius of convergence R, observe that the series of derivatives $\sum_{k=1}^{+\infty} k a_k (x - x_0)^{k-1}$ is also a power series with the same radius of convergence. Indeed:

$$\frac{1}{\lim\limits_{k \to +\infty} \sqrt[k]{|k a_k|}} = \frac{1}{\lim\limits_{k \to +\infty} \sqrt[k]{k} \ \lim\limits_{k \to +\infty} \sqrt[k]{|a_k|}} = \frac{1}{\lim\limits_{k \to +\infty} \sqrt[k]{|a_k|}} = R \,.$$

Same radius of convergence means same field of total convergence but not necessarily same sets of convergence. For example, for the series $\sum_{k=1}^{+\infty} x^k / k$, it results:

$$D_{sc} = \{x \in \mathbb{R} : -1 \le x \le 1\} \supset D_{ac} = C = \{x \in \mathbb{R} : |x| < 1\}$$

whereas for the series of derivatives $\sum_{k=1}^{+\infty} x^{k-1}$, not convergent at $x = -1$, it results that:

$$D_{sc} = D_{ac} = C = \{x \in \mathbb{R} : |x| < 1\}.$$

Observe that if x belongs to the field of convergence C of the power series $\sum_{k=0}^{+\infty} a_k (x - x_0)^k$, since C is open, there exists a closed and bounded interval $[a, b]$ such that $x \in [a, b]$; in such interval the series is totally convergent as well as the series of derivatives. Thus, if $f(x) = \sum_{k=0}^{+\infty} a_k (x - x_0)^k$, it is also

$$f'(x) = \sum_{k=1}^{+\infty} k a_k (x - x_0)^{k-1} \quad \text{(differentiation by series)}.$$

This result can be seen reversed, in the sense that if $g(x) = \sum_{k=1}^{+\infty} k a_k (x - x_0)^{k-1}$, then

$$f(x) = \sum_{k=0}^{+\infty} \int a_k (x - x_0)^k$$

is the antiderivative of $g(x)$ (indefinite integration by series).

Example 8.10 Consider again the geometric series $\sum_{k=0}^{+\infty} x^k$. In its interval of absolute convergence $C = \{x \in \mathbb{R} : |x| < 1\}$, it results that

$$\sum_{k=0}^{+\infty} x^k = f(x) = \frac{1}{1 - x}.$$

According to previous observations, at every $x \in C$,

$$f'(x) = \frac{1}{(1-x)^2} = \sum_{k=1}^{+\infty} k x^{k-1}, \qquad f''(x) = \frac{2}{(1-x)^3} = \sum_{k=2}^{+\infty} k(k-1) x^{k-2}$$

and, with further iterations, in general

$$f^{(n)}(x) = \frac{n!}{(1-x)^{n+1}} = \sum_{k=n}^{+\infty} k(k-1) \ldots (k-n+1) x^{k-n}.$$

The above example highlights an important properties of power series. If $f(x) = \sum_{k=0}^{+\infty} \int a_k (x - x_0)^k$ (or, equivalently, f can be expanded as a power series) in the interval C, then f is of class C^∞ in C, that is f is differentiable infinitely many times in C; furthermore

$$f^{(n)}(x) = \sum_{k=n}^{+\infty} k(k-1) \ldots (k-n+1)(x - x_0)^{k-n}, \quad \text{and} \quad f^{(n)}(x_0) = n! a^n$$

whence

$$a_n = \frac{f^{(n)}(x_0)}{n!}.$$

In other words, if f can be expanded as a power series in C, such series is the Taylor series of f

$$f(x) = \sum_{k=0}^{+\infty} \int \frac{f^{(k)}(x_0)}{k!} (x - x_0)^k.$$

Example 8.11 Consider the function $f(x) = \dfrac{1}{1+x}$. When $|x| < 1$, f can be expressed as the sum of a geometric series

$$f(x) = \frac{1}{1+x} = \frac{1}{1-(-x)} = \sum_{k=0}^{+\infty} (-x)^k = \sum_{k=0}^{+\infty} (-1)^k x^k.$$

Performing an indefinite integration, it results that

$$F(x) = \int f(x)\, dx = \int \frac{1}{1+x}\, dx = \log(1+x) + c$$

$$= \sum_{k=0}^{+\infty} \int (-1)^k x^k\, dx = \sum_{k=0}^{+\infty} (-1)^k \frac{x^{k+1}}{k+1}.$$

At $x = 0$

$$F(0) = \log(1 + 0) + c = \sum_{k=0}^{+\infty}(-1)^k \frac{x^{k+1}}{k+1}\Bigg|_{x=0} = 0, \text{ whence } c = 0.$$

leading to the expansion

$$\log(1 + x) = \sum_{k=0}^{+\infty}(-1)^k \frac{x^{k+1}}{k+1}.$$

Notice that the above series converges also at $x = 1$ and, according to Abel's theorem, such convergence is uniform in $[a, 1]$, with $-1 < a < 1$.

Example 8.12 Consider the function $f(x) = \dfrac{1}{1 + x^2}$. When $|x| < 1$, f can be expressed as the sum of a geometric series

$$f(x) = \frac{1}{1 + x^2} = \frac{1}{1 - (-x^2)} = \sum_{k=0}^{+\infty}(-x^2)^k = \sum_{k=0}^{+\infty}(-1)^k x^{2k}.$$

Performing an indefinite integration it results that

$$F(x) = \int f(x)\,dx = \int \frac{1}{1 + x^2}\,dx = \arctan x + c$$

$$= \sum_{k=0}^{+\infty} \int (-1)^k x^{2k}\,dx = \sum_{k=0}^{+\infty}(-1)^k \frac{x^{2k+1}}{2k+1}.$$

At $x = 0$

$$F(0) = \arctan 0 + c = \sum_{k=0}^{+\infty}(-1)^k \frac{x^{2k+1}}{2k+1}\Bigg|_{x=0} = 0, \text{ whence } c = 0.$$

leading to the expansion

$$\arctan x = \sum_{k=0}^{+\infty}(-1)^k \frac{x^{2k+1}}{2k+1}.$$

Notice that the above series converges also at $x = 1$ and, according to Abel's theorem, such convergence is uniform in $[-1, 1]$. Since at $x = 1$ is

$$\frac{\pi}{4} = \arctan 1 = \sum_{k=0}^{+\infty} \frac{(-1)^k}{2k+1},$$

and one finally obtains the representation

$$\pi = 4 \sum_{k=0}^{+\infty} \frac{(-1)^k}{2k+1}.$$

It was observed that if f can be expanded in a power series in C, that is $f(x) = \sum_{k=0}^{+\infty} a_k(x - x_0)^k$, then f is of class C^∞ in C, that is f is differentiable infinitely many times in C, and the coefficients a_k can be expressed indeed as $a_k = f^{(k)}(x_0)/k!$. However, the inverse of this statement is not true. As an example consider the function (Cauchy's basin)

$$f(x) = \begin{cases} e^{-\frac{1}{x^2}} & \text{if } x \neq 0 \\ 0 & \text{if } x = 0. \end{cases}$$

It is easily seen that $f^{(k)}(0) = 0$ for every k, so the corresponding Taylor series with pole $x_0 = 0$ turns out to be $\sum_{k=0}^{+\infty}$, that is identically zero in $\mathbb{R}$ whereas the function f is not.

At this point it is natural to ask what conditions a function of class C^∞ has to meet to be expanded in a power series. The answer comes from the following result

Let $x_0 \in (a, b)$ and let $f \in C^\infty(a, b)$. If there exist $L, M \in \mathbb{R}$ such that at every $x \in (a, b)$ and every $k \in \mathbb{N}$ it results $\left| f^{(k)}(x) \leq LM^k \right|$ then

$$f(x) = \sum_{k=0}^{+\infty} \frac{f^{(k)}(x_0)}{k!} (x - x_0)^k \text{ in } (a, b).$$

Indeed

$$\lim_{n \to +\infty} \left| f(x) - \sum_{k=0}^{n} \frac{f^{(k)}(x_0)}{k!} (x - x_0)^k \right| = \lim_{n \to +\infty} R_n(x, x_0)$$

where $R_n(x, x_0)$ is the remainder of the Taylor series. Using Lagrange's representation formula for this, it results that

$$R_n(x, x_0) = \frac{f^{(n+1)}(\xi)}{(n+1)!} (x - x_0)^{n+1}, \quad \text{with } \xi \in (x_0, x).$$

Thus

$$\lim_{n \to +\infty} \left| f(x) - \sum_{k=0}^{n} \frac{f^{(k)}(x_0)}{k!} (x - x_0)^k \right| = \lim_{n \to +\infty} \left| \frac{f^{(n+1)}(\xi)}{(n+1)!} (x - x_0)^{n+1} \right|$$

$$\leq \lim_{n \to +\infty} \left| \frac{LM^{n+1}}{(n+1)!} (b - a)^{n+1} \right|$$

whence the possibility of expansion of f in a power series follows.

8.3   Remarkable Developments

The condition $\left| f^{(k)}(x) \right| \leq LM^k$ is met, in particular, if $\left| f^{(k)}(x) \right| \leq L$, that is when the function f has derivatives of any order equally bounded. It is so, for example, for the circular functions $f(x) = \sin x$ and $g(x) = \cos x$, for which $\left| f^{(k)} \right| \leq 1$ and $\left| g^{(k)} \right| \leq 1$ at every $x \in \mathbb{R}$. So this functions can be expanded as power series in $\mathbb{R}$. To determine such expansions observe that, for $f(x) = \sin x$

$$f^{(k)}(0) = \begin{cases} 0 & \text{if } k \text{ is even} \\ 1 & \text{if } k = 1, 5, 9, \ldots \\ -1 & \text{if } k = 3, 7, 11, \ldots \end{cases}$$

Recalling that $a_k = f^{(k)}(0)/k!$, it results that

$$\sin x = \sum_{k=0}^{+\infty} (-1)^k \frac{x^{2k+1}}{(2k+1)!}$$

and, differentiating by series,

$$\cos x = \sum_{k=0}^{+\infty} (-1)^k \frac{x^{2k}}{(2k)!} .$$

The exponential function $f(x) = e^x$ has equally bounded derivatives in every compact set $[-a, a]$, with a a nonnegative real number. In fact, in $[-a, a]$, $\left| f^{(k)}(x) \right| = e^x \leq e^a$; since a is arbitrary, it follows that the function can be expanded in a power series in $\mathbb{R}$.

Using the same reasoning, it can be easily seen that the functions $f(x) = \sinh x$ and $g(x) = \cosh x$ can also be expanded in power series in $\mathbb{R}$:

Table 8.1 Maclaurin series of main elementary functions

Function	Domain of pointwise convergence
$e^x = \displaystyle\sum_{k=0}^{+\infty} \dfrac{x^k}{k!}$	$\mathbb{R}$
$\log(1+x) = \displaystyle\sum_{k=0}^{+\infty} (-1)^k \dfrac{x^{k+1}}{k+1}$	$(-1, 1]$
$\sin x = \displaystyle\sum_{k=0}^{+\infty} (-1)^k \dfrac{x^{2k+1}}{(2k+1)!}$	$\mathbb{R}$
$\cos x = \displaystyle\sum_{k=0}^{+\infty} (-1)^k \dfrac{x^{2k}}{(2k)!}$	$\mathbb{R}$
$\sinh x = \displaystyle\sum_{k=0}^{+\infty} \dfrac{x^{2k+1}}{(2k+1)!}$	$\mathbb{R}$
$\cosh x = \displaystyle\sum_{k=0}^{+\infty} \dfrac{x^{2k}}{(2k)!}$	$\mathbb{R}$
$\arctan x = \displaystyle\sum_{k=0}^{+\infty} (-1)^k \dfrac{x^{2k+1}}{2k+1}$	$[-1, 1]$

$$f^{(k)}(0) = \begin{cases} 0 & \text{if } k \text{ is even} \\ 1 & \text{if } k \text{ is odd} \end{cases}$$

whence

$$\sinh x = \sum_{k=0}^{+\infty} \frac{x^{2k+1}}{(2k+1)!}$$

and, differentiating by series,

$$\cosh x = \sum_{k=0}^{+\infty} \frac{x^{2k}}{(2k)!}.$$

The series expansions of Maclaurin[4] (pole at $x_0 = 0$) for the main elementary functions are summarised in Table 8.1.

8.4 The Binomial Series

Consider, for $\alpha \neq 0$, the series $\sum_{k=0}^{+\infty} \binom{\alpha}{k} x^k$, called the binomial series. The radius of convergence is

---

[4] https://mathshistory.st-andrews.ac.uk/Biographies/Maclaurin/.

$$R = \lim_{k \to +\infty} \left| \frac{a_k}{a_{k+1}} \right| = \lim_{k \to +\infty} \left| \frac{\binom{\alpha}{k}}{\binom{\alpha}{k+1}} \right|$$

$$= \lim_{k \to +\infty} \left| \frac{\frac{\alpha(\alpha-1)...(\alpha-k+1)}{k!}}{\frac{\alpha(\alpha-1)...(\alpha-k)}{(k+1)!}} \right| = \lim_{k \to +\infty} \left| \frac{k+1}{\alpha-k} \right| = 1.$$

Thus the field of absolute convergence is $C = \{x \in \mathbb{R} : |x| < 1\}$. In this field we set

$$S(x) = \sum_{k=0}^{+\infty} \binom{\alpha}{k} x^k$$

and, after some algebra, one gets:

$$(1+x)S'(x) = \sum_{k=0}^{+\infty} \binom{\alpha}{k} \alpha x^k = \alpha S(x).$$

Integrating both members of the equality $(1+x)S'(x) = \alpha S(x)$, and considering that $|x| < 1$, it results that

$$\log |S(x)| = \log c |1 + x|^\alpha = \log c (1+x)^\alpha .$$

with c a positive constant that can be determined by observing it has to be $S(0) = 1$. So $\log c = \log 1$, thus $c = 1$. In conclusion $|S(x)| = (1+x)^\alpha$ and, being $S(0) = 1$, it must be $S(x) = (1+x)^\alpha$. Then there exists in C the expansion

$$(1+x)^\alpha = \sum_{k=0}^{+\infty} \binom{\alpha}{k} x^k \qquad (\alpha \neq 0).$$

Remark 8.1 If $\alpha = n$ is a natural number, then $\binom{n}{k} = 0$ for $k > n$ and the expansion is simply the Newton[5] binomial

$$(1+x)^n = \sum_{k=0}^{n} \binom{n}{k} x^k.$$

Remark 8.2 If $\alpha = 1/2$, then the expansion becomes

$$(1+x)^{\frac{1}{2}} = \sqrt{1+x} = \sum_{k=0}^{+\infty} \binom{1/2}{k} x^k$$

---

[5] https://mathshistory.st-andrews.ac.uk/Biographies/Newton/.

$$= 1 + \frac{1}{2}x - \frac{1}{2!2^2}x^2 + \frac{3}{3!2^3}x^3 - \frac{3 \cdot 5}{4!2^4}x^4 + \frac{3 \cdot 5 \cdot 7}{5!2^5}x^5 - \cdots.$$

or, more synthetically,

$$\sqrt{1+x} = 1 + \frac{1}{2}x - \sum_{k=2}^{+\infty}(-1)^{k-1}\frac{(2k-3)!!}{k!2^k}x^k$$

$$= 1 + \frac{1}{2}x - \sum_{k=2}^{+\infty}(-1)^{k-1}\frac{(2k-3)!!}{(2k)!!}x^k.$$

Remark 8.3 If $\alpha = -1/2$, then the expansion becomes

$$(1+x)^{-\frac{1}{2}} = \frac{1}{\sqrt{1+x}} = \sum_{k=0}^{+\infty}\binom{1/2}{k}x^k$$

$$= 1 - \frac{1}{2}x + \frac{3}{2!2^2}x^2 - \frac{3 \cdot 5}{3!2^3}x^3 + \frac{3 \cdot 5 \cdot 7}{4!2^4}x^4 - \cdots.$$

or, more synthetically,

$$\frac{1}{\sqrt{1+x}} = 1 + \sum_{k=1}^{+\infty}(-1)^k\frac{(2k-1)!!}{k!2^k}x^k = 1 + \sum_{k=1}^{+\infty}(-1)^k\frac{(2k-1)!!}{(2k)!!}x^k.$$

If one replaces x with $-x^2$ in the above expansion, the following is obtained:

$$\frac{1}{\sqrt{1-x^2}} = 1 + \sum_{k=1}^{+\infty}(-1)^k\frac{(2k-1)!!}{k!2^k}(-x^2)^k = 1 + \sum_{k=1}^{+\infty}\frac{(2k-1)!!}{(2k)!!}x^{2k}.$$

and, integrating by series, one finally gets

$$\arcsin x = \sum_{k=0}^{+\infty}\frac{(2k-1)!!}{(2k)!!}\frac{x^{2k+1}}{2k+1} \qquad (|x| < 1).$$

Supplement.

The behaviour of the binomial series at the endpoints $x = -1$ and $x = 1$ is now investigated.

As a first step, it is necessary to recall the convergence test due to Raabe[6]: let $\sum_{k=0}^{+\infty} a_k$ be a series with *nonnegative* terms, and let

$$\lim_{k \to +\infty} k \left(\frac{a_k}{a_{k+1}} - 1 \right) = l \in [0, +\infty].$$

Then

(a) if $l > 1$ the series converges;
(b) if $l < 1$ the series diverges;
(c) if $l = 1$ nothing can be said about the character of the series.

The test is now used for the series $\sum_{k=0}^{+\infty} \left| \binom{\alpha}{k} x^k \right|$. At $|x| = 1$ it becomes $\sum_{k=0}^{+\infty} \left| \binom{\alpha}{k} \right|$. It results that

$$\lim_{k \to +\infty} k \left(\frac{a_k}{a_{k+1}} - 1 \right) = \lim_{k \to +\infty} k \left(\frac{\binom{\alpha}{k}}{\binom{\alpha}{k+1}} - 1 \right) = \lim_{k \to +\infty} k \left(\left| \frac{k+1}{\alpha - k} \right| - 1 \right)$$

$$= \lim_{k \to +\infty} k \left(\frac{k+1}{\alpha - k} - 1 \right) = \lim_{k \to +\infty} k \left(\frac{1+\alpha}{k - \alpha} \right) = 1 + \alpha.$$

Thus, there is *absolute convergence* when $1 + \alpha > 1$, that is when $\alpha > 0$.

For $\alpha < 0$ the series *does not converge absolutely* at $|x| = 1$, but it can be shown that the binomial series *converges simply* also at $x = 1$.

To summarise:

$$(1 + x)^\alpha = \sum_{k=0}^{+\infty} \binom{\alpha}{k} x^k = \begin{cases} \text{converges absolutely } \forall x \in [-1, 1], & \text{if } \alpha > 0 \\ \text{converges } \forall x \in (-1, 1], & \text{if } -1 < \alpha < 0 \\ \text{converges } \forall x \in (-1, 1), & \text{if } \alpha \leq -1. \end{cases}$$

8.5 Trigonometric Series and Fourier Series

Sound is the acoustic (pressure) result of the oscillation of an elastic body. The simplest, non-trivial, oscillation is the harmonic one (pure tone): it is well described by the following equation

$$y = A \sin(\omega t - \varphi),$$

---

[6] https://mathshistory.st-andrews.ac.uk/Biographies/Raabe/.

where A is the *amplitude* of pressure oscillation, related to the intensity, ω is the *angular frequency* (2π times the frequency, or pitch) of sound, and $varphi$ the *phase*.

It is useful to express the above equation in a different form:

$$y = A \cos\varphi \sin(\omega t) - A \sin\varphi \cos(\omega t)$$

$$= a \cos(\omega t) + b \sin(\omega t) = a \cos\left(\frac{2\pi}{T} t\right) + b \sin\left(\frac{2\pi}{T} t\right),$$

where $T = 2\pi/\omega$ is the oscillation period, $a = -A \sin\varphi$, $b = A \cos\varphi$, so that $A = \sqrt{a^2 + b^2}$.

A pure tone of frequency $\omega/(2\pi)$, or period T, is then well described by the equation

$$y = a \cos(\frac{2\pi}{T} t) + b \sin(\frac{2\pi}{T} t).$$

Not all sounds are pure tones. However any sound can be represented as a superposition of a number (possibly infinite) of pure sounds with frequencies multiple of a fundamental frequency. In other words if $y = f(t)$ is a sound, and $\omega/(2\pi)$ its fundamental frequency, then f can be expressed as a series of pure tones of frequency kv, $k = 1, 2, \ldots$ (harmonics, determining the *timbre* of the sound)

$$y_k = a_k \cos(k \frac{2\pi}{T} t) + b_k \sin(k \frac{2\pi}{T} t),$$

so that

$$f(t) = \sum_{k=0}^{+\infty} \left[a_k \cos(k \frac{2\pi}{T} t) + b_k \sin(k \frac{2\pi}{T} t) \right]. \tag{8.1}$$

In order to simplify some formulas, as it will be clear later, the right-hand side of Eq. (8.1) is usually written with the constant term (corresponding to $k = 0$) isolated and denoted as $a_0/2$. Thus Eq. (8.1) becomes

$$f(t) = \frac{a_0}{2} + \sum_{k=1}^{+\infty} \left[a_k \cos(k \frac{2\pi}{T} t) + b_k \sin(k \frac{2\pi}{T} t) \right]. \tag{8.2}$$

The right-hand side of Eq. (8.2) is called a *trigonometric series* with coefficients a_k, b_k, $k = 1, 2, \ldots$.

Equality (8.2) requires the above series to be convergent. This implies that the sequences of coefficients are infinitesimal of proper order.

One may ask what conditions a T-periodic function f has to meet in order to satisfy Eq. (8.1), or as it is commonly said, to be represented as a trigonometric series. To answer this question one needs to better understand the structure of a trigonometric series.

In the upcoming developments it will be useful to consider that the following (easy to check) relations hold for $h, k \geq 1$.

$$\int_{-L}^{L} \cos(k \frac{2\pi}{T} t) \cos(h \frac{2\pi}{T} t) \, dt = \begin{cases} 0 & \text{if } h \neq k \\ L & \text{if } h = k \end{cases}$$

$$\int_{-L}^{L} \sin(k \frac{2\pi}{T} t) \sin(h \frac{2\pi}{T} t) \, dt = \begin{cases} 0 & \text{if } h \neq k \\ L & \text{if } h = k \end{cases} \tag{8.3}$$

$$\int_{-L}^{L} \sin(k \frac{2\pi}{T} t) \cos(h \frac{2\pi}{T} t) \, dt = 0.$$

Suppose now that the series (8.2) is convergent to the function $f(t)$ in $[-L, L]$ and that this convergence is uniform. This happens, for example, if the series

$$\sum_{k=1}^{+\infty} \left(|a_k|^2 + |b_k|^2 \right)$$

is convergent; in this case the function f is continuous (and bounded) in $[-L, L]$, so that one can integrate by series to obtain

$$\int_{-L}^{L} f(t) \, dt = \int_{-L}^{L} \left\{ \frac{a_0}{2} + \sum_{k=1}^{+\infty} \left[a_k \cos(k \frac{2\pi}{T} t) + b_k \sin(k \frac{2\pi}{T} tt) \right] \right\} dt$$

$$= \int_{-L}^{L} \frac{a_0}{2} \, dt + \sum_{k=1}^{+\infty} \left[\int_{-L}^{L} a_k \cos(k \frac{2\pi}{T} t) dt + \int_{-L}^{L} b_k \sin(k \frac{2\pi}{T} t) dt \right]$$

$$= L \, a_0.$$

With similar procedure, using properly the relations (8.3), one finds for $k = 1, 2, \ldots$

$$\int_{-L}^{L} f(t) \cos\left(k \frac{2\pi}{T} t \right) dt = L a_k, \qquad \int_{-L}^{L} f(t) \sin\left(k \frac{2\pi}{T} t \right) dt = L b_k.$$

Under the same assumptions, taking into account again the relations (8.3) one obtains

$$\int_{-L}^{L} [f(t)]^2 \, dt = \int_{-L}^{L} \left\{ \frac{a_0}{2} + \sum_{k=1}^{+\infty} \left[a_k \cos(k \frac{2\pi}{T} t) + b_k \sin(k \frac{2\pi}{T} t) \right] \right\}^2 dt$$

$$= L \left[\frac{a_0^2}{2} + \sum_{k=1}^{+\infty} \left(a_k^2 + b_k^2 \right) \right].$$

The above results can be summarised as follows.

Let f be a $2L$-periodic function, uniformly expandable in a trigonometric series, that is

$$f(t) = \frac{a_0}{2} + \sum_{k=1}^{+\infty} \left[a_k \cos(k \frac{2\pi}{T} t) + b_k \sin(k \frac{2\pi}{T} t) \right],$$

and the series converges uniformly. The following relations hold.

$$a_0 = \frac{1}{L} \int_{-L}^{L} f(t) \, dt,$$

$$a_k = \frac{1}{L} \int_{-L}^{L} f(t) \cos(k \frac{2\pi}{T} t) \, dt, \qquad b_k = \frac{1}{L} \int_{-L}^{L} f(t) \sin(k \frac{2\pi}{T} t) \, dt,$$

$$k = 1, 2, \ldots$$

$$\int_{-L}^{L} [f(t)]^2 \, dt = L \left[\frac{a_0^2}{2} + \sum_{k=1}^{+\infty} (a_k^2 + b_k^2) \right].$$

The latter equation is known as the Parseval's[7] identity.

The perspective is now reversed. Suppose that the T-periodic ($T = 2L$) function f is such that the following integrals are meaningful (in case, in a generalised sense):

$$a_0 = \frac{1}{L} \int_{-L}^{L} f(t) \, dt,$$

$$a_k = \frac{1}{L} \int_{-L}^{L} f(t) \cos(k \frac{2\pi}{T} t) \, dt, \qquad b_k = \frac{1}{L} \int_{-L}^{L} f(t) \sin(k \frac{2\pi}{T} t) \, dt,$$

$$k = 1, 2, \ldots$$

Thus the trigonometric series (Fourier series)

$$\frac{a_0}{2} + \sum_{k=1}^{+\infty} \left[a_k \cos(k \frac{2\pi}{T} t) + b_k \sin(k \frac{2\pi}{T} t) \right]$$

can be associated to the function f, that is

---

[7] https://mathshistory.st-andrews.ac.uk/Biographies/Parseval/.

$$f(t) \approx \frac{a_0}{2} + \sum_{k=1}^{+\infty} \left[a_k \cos(k \frac{2\pi}{T} t) + b_k \sin(k \frac{2\pi}{T} t) \right].$$

Example 8.13 Consider the function $f(t) = t$, with $x \in [-\pi, 0) \bigcup (0, \pi)$, and its 2π-periodic extension. f is an *odd* function, $f(-t) = -f(t)$, and so are the functions $f(t) \cos(kt)$, $k = 0, 1, 2, \ldots$; it follows that

$$a_0 = \frac{1}{\pi} \int_{-\pi}^{\pi} f(t) \, dt = 0; \qquad a_k = \frac{1}{\pi} \int_{-\pi}^{\pi} f(t) \cos(kt) \, dt = 0, \quad k = 1, 2, \ldots;$$

On the other hand, the functions $F(t) = f(t) \sin(kt)$ turn out to be *even*, $f(-t) = f(t)$, so that

$$b_k = \frac{1}{\pi} \int_{-\pi}^{\pi} f(t) \sin(kt) \, dt = \frac{2}{\pi} \int_{0}^{\pi} f(t) \sin(kt) \, dt = (-1)^{k+1} \frac{2}{k}, \quad k = 1, 2, \ldots.$$

In conclusion the Fourier series associated with f is a series of only sins:

$$f(t) \approx \sum_{k=1}^{+\infty} (-1)^{k+1} \frac{2}{k} \sin(kt).$$

The above example introduces to a general result that can be formalised as follows.

The Fourier series of a $2L$-periodic *odd* function consists of only sins ($a_k = 0$, $k = 0, 1, 2, \ldots$):

$$f(t) \approx \sum_{k=1}^{+\infty} b_k \sin(k \frac{2\pi}{T} t).$$

The Fourier series of a $2L$-periodic *even* function consists of only cosins ($b_k = 0$, $k = 1, 2, \ldots$):

$$f(t) \approx \frac{a_0}{2} + \sum_{k=1}^{+\infty} a_k \cos(k \frac{2\pi}{T} t).$$

One may ask what conditions f has to satisfy in order to have

$$f(t) = \frac{a_0}{2} + \sum_{k=1}^{+\infty} a_k \cos(k \frac{2\pi}{T} t).$$

that is, for the series to be pointwise convergent. The following trivial example shows that this is not always the case.

Let f have the form

$$f(t) = \begin{cases} 0 \text{ if } t \in [-\pi, 0) \bigcup (0, \pi) \\ 1 \text{ if } t = 0 \end{cases}$$

and extend f as a 2π-periodic function. It is clear that the Fourier coefficients of f are identically zero as well as the trigonometric series associated with f. Since $f(0) = 1 \neq 0$, then 0 in the series does not converge to the function. In fact it was also shown that the condition of continuity is not enough to ensure the convergence of the Fourier series associated with f to the function itself.

At present no necessary and sufficient conditions are for a periodic function f to be the sum of its Fourier series. However, the following are useful sufficient conditions for the pointwise convergence of a piecewise smooth function.

If f is T-periodic and piecewise smooth, then the sum S_f of the Fourier series associated with f at x_0 is:

$$S_f(x_0) = \frac{f(x_0^+) + f(x_0^-)}{2},$$

where $f(x_0^+) = \lim_{x \to x_0^+} f(x)$ and $f(x_0^-) = \lim_{x \to x_0^-} f(x)$.

It is clearly seen that $S_f(x_0) = f(x_0)$ at every x_0 where f is continuous.

For the pointwise convergence of bounded piecewise monotone functions the following condition holds.

If f is T-periodic, bounded and piecewise monotone, then the sum S_f of the Fourier series associated with f at x_0 is:

$$S_f(x_0) = \frac{f(x_0^+) + f(x_0^-)}{2},$$

where $f(x_0^+) = \lim_{x \to x_0^+} f(x)$ and $f(x_0^-) = \lim_{x \to x_0^-} f(x)$.

It is clearly seen that $S_f(x_0) = f(x_0)$ at every x_0 where f is continuous.

If one looks at the *quality* of convergence, the following results hold for uniform and total convergence, respectively.

If f is T-periodic and piecewise smooth, then the Fourier series associated with f converges *uniformly* to f in every interval where f is continuous.

If f is T-periodic, piecewise smooth and continuous in $\mathbb{R}$, then the Fourier series associated with f converges *totally* to f.

Example 8.14 Consider the function

$$f(t) = \frac{\pi - t}{2}, \quad \text{with } t \in [0, 2\pi),$$

and its 2π-periodic extension. It is better to redefine the function as

$$f(t) = \begin{cases} \dfrac{\pi - t}{2} & \text{if } t \in [0, 2\pi) \\ -\dfrac{t + \pi}{2} & \text{if } t \in [-\pi, 0) \end{cases}$$

whence it is evident that f is an *odd* function, so that $a_k = 0, k = 0, 1, 2, \ldots$, whereas

$$b_k = \frac{1}{\pi} \int_{-\pi}^{\pi} f(t) \sin(kt)\, dt = \frac{1}{\pi} \int_{0}^{2\pi} \frac{t - \pi}{2} \sin(kt)\, dt = \frac{1}{k}, \quad k = 1, 2, \ldots.$$

Thus

$$f(t) \approx \sum_{k=1}^{+\infty} \frac{1}{k} \sin(kt), \quad k = 1, 2, \ldots.$$

More precisely, since f is piecewise smooth,

$$\sum_{k=1}^{+\infty} \frac{1}{k} \sin(kt) = \begin{cases} f(t) & \text{if } t \in (0, 2\pi) \\ 0 & \text{if } t = 0 \end{cases} \quad (\text{mod } 2\pi).$$

In particular, if the series is convergent at $t = 1$, it results that

$$\sum_{k=1}^{+\infty} \frac{1}{k} \sin k = f(1) = \frac{\pi - 1}{2}.$$

Moreover, Parseval's identity provides

$$\pi \sum_{k=1}^{+\infty} \frac{1}{k^2} = \int_{0}^{2\pi} \left[\frac{\pi - t}{2} \right]^2 dt = \frac{\pi^3}{6}, \quad \text{whence} \quad \sum_{k=1}^{+\infty} \frac{1}{k^2} = \frac{\pi^2}{6}.$$

The equality Parseval's identity has been shown previously under the restrictive assumption that f was uniformly expandable in a Fourier series. Actually this relationship exists in much more general terms.

Suppose the $2L$-periodic function f is square summable in $[-L, L]$, that is the integral

$$\int_{-L}^{L} [f(t)]^2 dt$$

exists with a finite value. Under this assumption it does make sense to define the Fourier coefficients

$$a_k = \frac{1}{L} \int_{-L}^{L} f(t) \cos(k \frac{2\pi}{T} t) \, dt, \quad k = 0, 1, 2, \ldots$$

$$b_k = \frac{1}{L} \int_{-L}^{L} f(t) \sin(k \frac{2\pi}{T} t) \, dt, \quad k = 1, 2, \ldots$$

and the following relation (Parseval's identity) holds

$$\int_{-L}^{L} [f(t)]^2 \, dt = L \left[\frac{a_0^2}{2} + \sum_{k=1}^{+\infty} \left(a_k^2 + b_k^2\right) \right].$$

An immediate consequence of Parseval's identity is the convergence of the series

$$\sum_{k=1}^{+\infty} \left(a_k^2 + b_k^2\right)$$

and then of the two series

$$\sum_{k=1}^{+\infty} a_k^2 \quad \text{and} \quad \sum_{k=1}^{+\infty} b_k^2.$$

Thus it is evident that (Riemann-Lebesgue[8] Lemma)

---

[8] https://mathshistory.st-andrews.ac.uk/Biographies/Lebesgue/.

If f is a $2L$-periodic function square summable in $[-L, L]$, then

$$\lim_{k \to +\infty} a_k = \lim_{k \to +\infty} \frac{1}{L} \int_{-L}^{L} f(t) \cos(k \frac{2\pi}{T} t) \, dt = 0,$$

$$\lim_{k \to +\infty} b_k = \lim_{k \to +\infty} \frac{1}{L} \int_{-L}^{L} f(t) \sin(k \frac{2\pi}{T} t) \, dt = 0.$$

One may ask now if every trigonometric series with coefficients that behave as above can be associated to a $2L$-periodic function square summable in $[-L, L]$. For example, consider the series

$$\sum_{k=1}^{+\infty} \frac{1}{\sqrt{k}} \sin(kt),$$

and check whether it can be the Fourier series associated with a 2π-periodic function square summable in $[-\pi, \pi]$. The Parseval's identity allows to give a negative answer. In fact, in that case it should be

$$\int_{-\pi}^{\pi} [f(t)]^2 dt = \pi \sum_{k=1}^{+\infty} b_k^2 = \pi \sum_{k=1}^{+\infty} \frac{1}{k}.$$

However the series in the far right-hand side above diverges, whereas for every square-integrable function f the integral in the left-hand side is finite. This leads to the following general remark.

Not every trigonometric series, albeit with coefficients that meet the conditions

$$\lim_{k \to +\infty} a_k = \lim_{k \to +\infty} \frac{1}{L} \int_{-L}^{L} f(t) \cos(k \frac{2\pi}{T} t) \, dt = 0,$$

$$\lim_{k \to +\infty} b_k = \lim_{k \to +\infty} \frac{1}{L} \int_{-L}^{L} f(t) \sin(k \frac{2\pi}{T} t) \, dt = 0,$$

can be the Fourier series associated with a $2L$-periodic function square-integrable in $[-\pi, \pi]$.

8.6 Differentiability and Integrability of Fourier Series

Taking the derivative (term by term) of the series

$$\frac{a_0}{2} + \sum_{k=1}^{+\infty} \left[a_k \cos(k \frac{2\pi}{T} t) + b_k \sin(k \frac{2\pi}{T} t) \right],$$

one gets

$$\sum_{k=1}^{+\infty} \left[-k \frac{2\pi}{T} a_k \sin(k \frac{2\pi}{T} t) + k \frac{2\pi}{T} b_k \cos(k \frac{2\pi}{T} t), \right]$$

and, upon defining,

$$\alpha_k = k \frac{2\pi}{T} b_k, \quad \text{and} \quad \beta_k = -k \frac{2\pi}{T} a_k,$$

the series

$$\sum_{k=1}^{+\infty} \left[\alpha_k \cos(k \frac{2\pi}{T} t) + \beta_k \sin(k \frac{2\pi}{T} t) \right],$$

that is still a trigonometric series. It is evident that the appearance of the k factor in new the coefficients α_k, β_k, could prejudice the convergence of the derivative series. This aspect emphasises the particular attitude of Fourier series to describe non-smooth functions (unlike power series that describe C^∞ functions).

One wonder what should be the level of smoothness of a function f expandable in a Fourier series so that it can be differentiated by series. If this were possible it should occur

$$f(t) = \frac{a_0}{2} + \sum_{k=1}^{+\infty} \left[a_k \cos(k \frac{2\pi}{T} t) + b_k \sin(k \frac{2\pi}{T} t) \right]$$

and

$$f'(t) = \sum_{k=1}^{+\infty} \left[\alpha_k \cos(k \frac{2\pi}{T} t) + \beta_k \sin(k \frac{2\pi}{T} t) \right], \text{ with } \alpha_k = k \frac{2\pi}{T} b_k, \ \beta_k = -k \frac{2\pi}{T} a_k.$$

It is important to note the following circumstance: if one calculate directly, for example, α_k coefficient one finds:

$$\alpha_k = \frac{1}{L} \int_{-L}^{L} f'(t) \cos(k \frac{2\pi}{T} t)$$

$$= \frac{1}{L} \left[f(t) \cos(k \frac{2\pi}{T} t) \right]_{-L}^{L} + \frac{k\pi}{L} \int_{-L}^{L} f(t) \cos(k \frac{2\pi}{T} t)$$

$$= \frac{\cos(k\pi)}{L} [f(L) - f(-L)] + k \frac{2\pi}{T} b_k.$$

Thus,

$$\alpha_k = k \frac{2\pi}{T} b_k$$

if and only if the matching condition $f(L) = f(-L)$ is satisfied (for example, if f is continuous in $\mathbb{R}$). Analogous considerations hold for β_k. Combining the latter observations with the conditions for the uniform and total convergence stated earlier, one obtains the following result.

Let f be $2L$-periodic and C^1 in $[-L, L]$, with piecewise smooth derivative f'. If the matching condition $f(-L) = f(L)$ holds, then the Fourier series of f is differentiable by series in every interval $[a, b] \subset (-L, L)$. If also the matching condition $f'(-L) = f'(L)$ holds, the series is differentiable in $[-L, L]$.

Suppose now that f is a $2L$-periodic function, continuous in $\mathbb{R}$ (that is, the matching condition $f(-L) = f(L)$ holds) and differentiable with derivative f' square-integrable in $[-L, L)$. Under such assumptions, the Parseval identity holds

$$\int_{-L}^{L} [f'(t)]^2 dt = L \left[\sum_{k=1}^{+\infty} [(a'_k)^2 + (b'_k)^2] \right].$$

so that coefficients

$$b'_k = -k \frac{2\pi}{T} a_k \text{ and } a'_k = k \frac{2\pi}{T} b_k \text{ go to 0 as } k \to +\infty.$$

Thus one gets

$$a_k = o(\frac{1}{k}) \text{ and } b_k = o(\frac{1}{k}) \text{ as } k \to +\infty.$$

If f were a $2L$-periodic function C^1 in $\mathbb{R}$ (so that the matching conditions $f(-L) = f(L)$ and $f'(-L) = f'(L)$ are both satisfied) and twice differentiable with derivative f'' square summable in $[-L, L)$, the Parseval identity would apply to f''

$$\int_{-L}^{L} [f''(t)]^2 dt = L \left[\sum_{k=1}^{+\infty} [(a''_k)^2 + (b''_k)^2] \right].$$

so that coefficients

$$b_k'' = -k^2 \left(\frac{2\pi}{T}\right)^2 b_k \text{ and } a_k'' = -k^2 \left(\frac{2\pi}{T}\right)^2 a_k \text{ go to 0 as } k \to +\infty.$$

In this case one gets

$$a_k = o(\frac{1}{k^2}) \text{ and } b_k = o(\frac{1}{k^2}) \text{ as } k \to +\infty.$$

When the above reasoning is iterated, the following general result if found.

If f is a $2L$-periodic function C^n in $\mathbb{R}$ $n+1$ differentiable with derivative $f^{(n+1)}$ square-integrable in $[-L, L)$, then

$$a_k = o(\frac{1}{k^{n+1}}) \text{ and } b_k = o(\frac{1}{k^{n+1}}) \text{ as } k \to +\infty.$$

Finally, the following result shows how easily the Fourier series can be integrated.

Let f be a $2L$-periodic piecewise continuous function and

$$\frac{a_0}{2} + \sum_{k=1}^{+\infty} \left[a_k \cos(k\frac{2\pi}{T}t) + b_k \sin(k\frac{2\pi}{T}t) \right],$$

the Fourier series associated with f (warning: no assumption on convergence). In every interval $(x_1, x_2) \subseteq [-L, L]$ the following relation (*integration by series*)

$$\int_{x_1}^{x_2} f(t)\, dt = \int_{x_1}^{x_2} \frac{a_0}{2}\, dt + \sum_{k=1}^{+\infty} \int_{x_1}^{x_2} \left[a_k \cos(k\frac{2\pi}{T}t) + b_k \sin(k\frac{2\pi}{T}t) \right] dt$$

holds with uniform convergence.

8.7   Complex Form of Fourier Series

Let

$$f(t) \approx \frac{a_0}{2} + \sum_{k=1}^{+\infty} \left[a_k \cos(k \frac{2\pi}{T} t) + b_k \sin(k \frac{2\pi}{T} t) \right].$$

Using the Euler's formulas one can write

$$\frac{a_0}{2} + \sum_{k=1}^{+\infty} \left[a_k \cos(k \frac{2\pi}{T} t) + b_k \sin(k \frac{2\pi}{T} t) \right]$$

$$= \frac{a_0}{2} + \sum_{k=1}^{+\infty} \left[a_k \frac{e^{(ik \frac{\pi}{L} t} + e^{-ik \frac{\pi}{L} t}}{2} + b_k \frac{e^{ik \frac{\pi}{L} t} - e^{-ik \frac{\pi}{L} t}}{2} \right]$$

$$= \frac{a_0}{2} + \sum_{k=1}^{+\infty} \left[\frac{a_k - ib_k}{2} e^{ik \frac{\pi}{L} t} + \frac{a_k + ib_k}{2} e^{-ik \frac{\pi}{L} t} \right]$$

$$= \sum_{k=-\infty}^{+\infty} c_k e^{ik \frac{\pi}{L} t},$$

where it has been set

$$c_0 = \frac{a_0}{2}, \quad c_k = \frac{a_k - ib_k}{2}, \quad c_{-k} = \frac{a_k + ib_k}{2}, \quad k = 1, 2, 3, \ldots$$

whence

$$a_0 = 2c_0, \quad a_k = c_k + c_{-k}, \quad b_k = i(c_k - c_{-k}), \quad k = 1, 2, 3, \ldots$$

or, more explicitly

$$c_0 = \frac{a_0}{2} = \frac{1}{2L} \int_{-L}^{L} f(t) \, dt,$$

$$c_k = \frac{a_k - ib_k}{2} = \frac{1}{2L} \int_{-L}^{L} f(t) \left[\cos(k \frac{2\pi}{T} t) - i \sin(k \frac{2\pi}{T} t) \right] dt$$

$$= \frac{1}{2L} \int_{-L}^{L} f(t) \, e^{-ik \frac{\pi}{L} t}, \quad k = 1, 2, 3, \ldots,$$

$$c_{-k} = \frac{a_k + ib_k}{2} = \frac{1}{2L} \int_{-L}^{L} f(t) \left[\cos(k \frac{2\pi}{T} t) + i \sin(k \frac{2\pi}{T} t) \right] dt$$

$$= \frac{1}{2L} \int_{-L}^{L} f(t) \, e^{-ik \frac{\pi}{L} t}, \quad k = 1, 2, 3, \ldots.$$

In conclusion, one can state the following result.

The Fourier series of a function f can be expressed in the complex form:

$$f(t) \approx \sum_{k=-\infty}^{+\infty} c_k e^{ik\frac{\pi}{L}t}, \quad \text{with } c_k = \frac{1}{2L} \int_{-L}^{L} f(t) e^{-ik\frac{\pi}{L}t}, \quad k \in \mathbb{N}.$$

If f is square summable in $[-L, L]$, then the Parseval's identity

$$\int_{-L}^{L} [f(t)]^2 \, dt = 2L \sum_{k=-\infty}^{+\infty} c_k^2.$$

holds.

8.8 Gibbs Phenomenon

Consider the 2π-periodic function defined by

$$f(t) = \begin{cases} 1 \text{ if } t \in [0, \pi) \\ -1 \text{ if } t \in [-\pi, 0) \end{cases} \tag{8.4}$$

For the Fourier coefficients one finds:

$$a_o = a_k = 0,$$

$$b_k = \frac{2}{\pi} \int_0^{\pi} f(t) \sin(kt) \, dt = \frac{2}{\pi} \left[\frac{-\cos(kt)}{k} \right]_0^{\pi} = \begin{cases} \dfrac{4}{k\pi} \text{ if } k \text{ is odd} \\ 0 \text{ if } k \text{ is even} \end{cases}$$

Thus

$$f(t) \approx \sum_{k=1}^{+\infty} \frac{4}{\pi} \frac{\sin[(2k-1)t]}{2k-1}, \quad \text{and, at } t \neq k\pi \ f(t) = \sum_{k=1}^{+\infty} \frac{4}{\pi} \frac{\sin[(2k-1)t]}{2k-1}.$$

Consider now the function $S_n(t)$, that is the partial sum of the series:

$$S_n(t) = \frac{4}{\pi} \sum_{k=1}^{n} \frac{\sin[(2k-1)t]}{2k-1}. \tag{8.5}$$

This can be written in the integral form

$$S_n(t) = \frac{4}{\pi} \sum_{k=1}^{n} \int_0^t \cos[(2k-1)x],$$

so that its derivative with respect to t is easily obtained as

$$S_n'(t) = \frac{4}{\pi} \sum_{k=1}^{n} \cos[(2k-1)t].$$

The latter can be manipulated using Werner's formulas as follows:

$$S_n'(t) = \frac{4}{\pi} \sum_{k=1}^{n} \frac{\cos[(2k-1)t]\sin t}{\sin t} = \frac{4}{\pi} \sum_{k=1}^{n} \frac{\sin(2kt) - \sin(2kt - 2t)}{2\sin t}$$

$$= \frac{4}{\pi} \frac{\sin(2t) - 0 + \sin(4t) - \sin(2t) + \sin(6t) - \sin(4t) + \cdots}{2\sin t}$$

$$= \frac{4}{\pi} \frac{\sin(2nt)}{2\sin t} = \frac{2}{\pi} \frac{\sin(2nt)}{\sin t}.$$

Thus, one finally gets

$$S_n(t) = \frac{4}{\pi} \sum_{k=1}^{n} \int_0^t \cos[(2k-1)x] = \frac{4}{\pi} \int_0^t \frac{\sin(2nx)}{2\sin x}\,dx, \quad S_n'(t) = \frac{2}{\pi} \frac{\sin(2nt)}{\sin t}.$$

So, there are local extrema whenever $2nt = h\pi$, that is at points $t = h\pi/(2n)$. In particular, for $h = 1$ there is a local maximum at $t = \pi/(2n)$. For large n this maximum is close to the discontinuity point $t = 0$, and its height can be estimated as follows:

$$S_n\left(\frac{\pi}{2n}\right) = \frac{4}{\pi} \int_0^{\frac{\pi}{2n}} \frac{\sin(2nx)}{2\sin x}\,dx = \frac{2}{\pi} \int_0^{\pi} \frac{\sin t}{2n\sin(\frac{t}{2n})}\,dt$$

$$\text{for large } n \Rightarrow \quad \cong \frac{2}{\pi} \int_0^{\pi} \frac{\sin t}{t}\,dt \cong 1.1789.$$

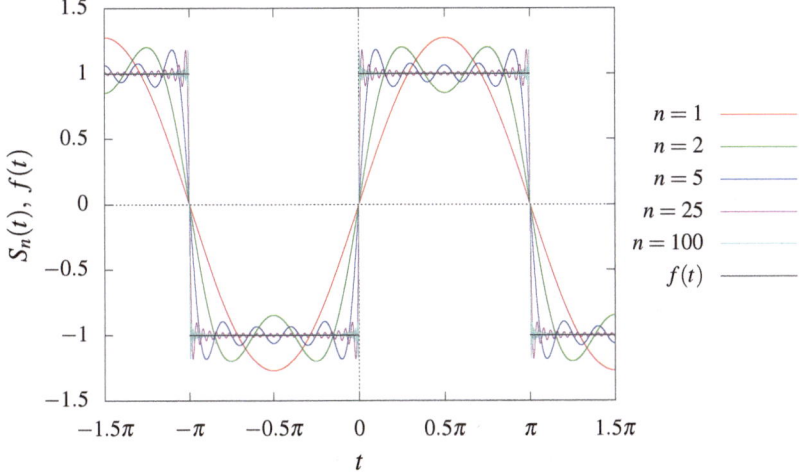

Fig. 8.1 Approximation of the function defined by Eq. (8.4) using partial sums of Eq. (8.5) with increasing order

This means that for large n, in the vicinity of the point of discontinuity, the partial sums S_n exhibit a swing of about 18% higher than the function value at the jump and this does not tend to disappear as n increases. This phenomenon, pathological in applications, is known as the Gibbs[9] phenomenon.

A typical example of Gibbs phenomenon is illustrated in the next page.

Figure 8.1 illustrates the approximation of the *square wave* represented by Eq. (8.4) through several partial sums of its Fourier series. At increasing n, the maxima approach the jump, showing an asymptotical trend toward a value of about 1.18, as it is highlighted in the zoom view of the jump neighbourhood of Fig. 8.2.

8.9 Suggested Readings

A comprehensive analysis of series of functions is given in Apostol (1974, Chap. 9, pp. 223–252). An in-depth analysis of the properties of Fourier series is found in Lang (1987, Appendix), Apostol (1974, Chap. 9; Chap. 11, pp. 306–313).

---

[9] https://mathshistory.st-andrews.ac.uk/Biographies/Gibbs/.

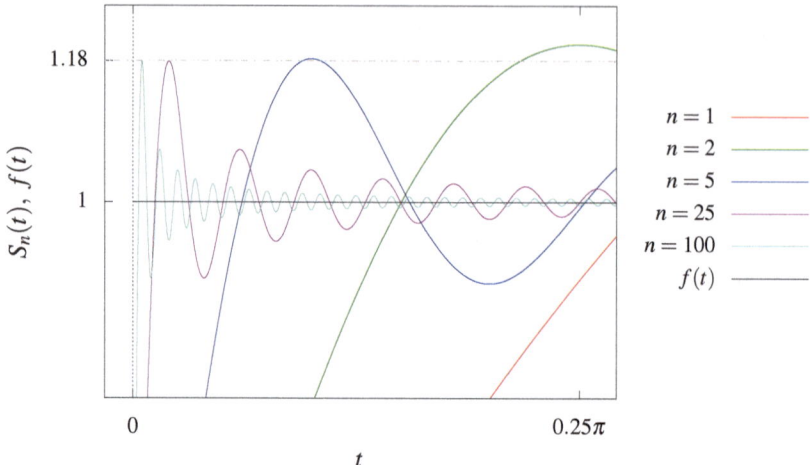

Fig. 8.2 Gibbs phenomenon illustrated by the zoom view of Fig. 8.1, showing the behaviour of partial sums Eq. (8.5) close to the jump

References

T. M. Apostol. *Mathematical Analysis: A Modern Approach to Advanced Calculus*. Pearson Education US, Hoboken, NJ, 2nd edition, 1974.

S. Lang. *Calculus of Several Variables*. Springer, New York, 3rd edition, 1987.

Chapter 9
Taylor Series for Functions of Several Variables

In the previous lessons, the concept of differential of a function $f : \mathbb{R}^n \to \mathbb{R}$ at a point x_0 of its existence field has been introduced and discussed. Considered an arbitrary point x lying inside a (sufficiently small) neighbourhood of x_0, the differential of f in x_0 is the part of the corresponding increment of this function ($f(x) - f(x_0)$) which is linear in the increment of the independent variable, *i.e.* in $x - x_0$ (named as Δx before). Another way to reread the differential lies in considering the approximation of a differentiable f in a (sufficiently small) neighbourhood of x_0. In order to do this, it is sufficient to rewrite the result (6.12) in the following form:

$$f(x) \simeq f(x_0) + \sum_{k=1}^{n} \partial_k f(x_0)(x_k - x_{0k}),$$

where contributions that vanish as $x \to x_0$ faster than the principal infinitesimal $|x - x_0|$ have been neglected. In order to write relations as the above one in a more coincise way, the rule of neglecting the symbol of the sum when two indices are identical will be adopted hereafter. In this way the above approximation takes the following short form:

$$f(x) \simeq f(x_0) + \partial_k f(x_0)(x_k - x_{0k}). \tag{9.1}$$

9.1 Taylor Series for a Function $f : \mathbb{R}^n \to \mathbb{R}$

The approximation (9.1) is now generalised, leading to better approximations of f in a neighbourhood of x_0. The price to be paid lies in requiring stronger constraints on the regularity of the function f. In particular, if it is required that the error on the approximation is of order $m + 1$ (in the approximation (9.1) $m = 1$), *i.e.* the

G. Riccardi et al., *Multidimensional Differential and Integral Calculus*,
https://doi.org/10.1007/978-3-031-70326-3_9

neglected terms vanish at least as $|x - x_0|^{m+1}$, f must be $(m+1)$-times partially differentiable, with continuous $(m+1)$th derivatives.

The Taylor approximation of the function f of the n real variables $(x_1, x_2, \ldots, x_n)$ grouped in the vector x is now deduced, using the well known 1-dimensional Taylor series. Once an arbitrary direction is fixed (in the following, a unit vector d will specify this direction), the increment Δx is rewritten by means of the real variable σ as $\Delta x = \sigma d$, so that $x = x_0 + \sigma d$ and $\sigma = 0$ corresponds to the point x_0. The values of the function f along the straight line $x = x_0 + \sigma d$ define the following function of the variable σ:

$$g(\sigma) := f(x_0 + \sigma d). \tag{9.2}$$

Due to the hypothesis about the smoothness of f, the function g possesses $(m+1)$th continuous derivatives, so that it can be expanded in the following Taylor series of order m about the point $\sigma = \sigma_0$ (σ_0 will be set to 0 below, but now it is convenient to leave it arbitrary):

$$g(\sigma) = \sum_{p=0}^{m} \frac{g^{(p)}(\sigma_0)}{p!} (\sigma - \sigma_0)^p + R_m(\sigma_0, \sigma), \tag{9.3}$$

$g^{(p)}$ being the pth derivative of g and R_m the remainder of order m, so that the expansion (9.4) is exact. It can be written in the forms[1]:

$$R_m(\sigma_0, \sigma) = \begin{cases} \dfrac{g^{(m+1)}(\eta')}{(m+1)!} (\sigma - \sigma_0)^{m+1} & \text{with } \eta' \in (\sigma_0, \sigma) & \text{(Lagrange)} \\[3mm] \dfrac{1}{m!} \displaystyle\int_{\sigma_0}^{\sigma} d\xi \, (\sigma - \xi)^m g^{(m+1)}(\xi) & & \text{(integral)} \\[3mm] \dfrac{g^{(m+1)}(\eta'')}{m!} (\sigma - \sigma_0)(\sigma - \eta'')^m & \text{with } \eta'' \in (\sigma_0, \sigma) & \text{(Cauchy)} \\[3mm] o[(\sigma - \sigma_0)^m] & & \text{(Peano)} \end{cases} \tag{9.4}$$

The forms (9.4) are now briefly justified, for the Reader convenience. The Lagrange remainder is deduced considering the function:

$$w(\sigma) = g(\sigma) - \sum_{k=0}^{m} \frac{g^{(k)}(\sigma_0)}{k!} (\sigma - \sigma_0)^k - a (\sigma - \sigma_0)^{m+1},$$

a being a constant to be calculated. It is defined enforcing that $w^{(m)}(\sigma) \equiv 0$, which leads to the definition of a:

---

[1] It is implicitly assumed that $\sigma > \sigma_0$, otherwise $\eta' \in (\sigma, \sigma_0)$ and $\eta'' \in (\sigma, \sigma_0)$ have to be used.

$$a \equiv \frac{1}{(m+1)!} \frac{g^{(m)}(\sigma) - g^{(m)}(\sigma_0)}{\sigma - \sigma_0} = \frac{g^{(m+1)}(\eta')}{(m+1)!} \quad \text{with } \eta' \in (\sigma_0, \sigma),$$

due to a famous theorem of Lagrange. The integral form is obtained starting from the fundamental theorem of the integral calculus:

$$g(\sigma) = g(\sigma_0) + \int_{\sigma_0}^{\sigma} d\xi \, g'(\xi)$$

and integrating m times by parts:

$$g(\sigma) = g(\sigma_0) + \int_{\sigma_0}^{\sigma} d(\xi - \sigma) \, g'(\xi)$$

$$= g(\sigma_0) + \frac{g'(\sigma_0)}{1!}(\sigma - \sigma_0) + \frac{1}{1!} \int_{\sigma_0}^{\sigma} d\xi \, (\sigma - \xi) \, g''(\xi)$$

$$= g(\sigma_0) + \frac{g'(\sigma_0)}{1!}(\sigma - \sigma_0) - \frac{1}{2!} \int_{\sigma_0}^{\sigma} d(\xi - \sigma)^2 \, g''(\xi)$$

$$\vdots$$

$$= g(\sigma_0) + \frac{g'(\sigma_0)}{1!}(\sigma - \sigma_0) + \ldots + \frac{g^{(m)}(\sigma_0)}{m!}(\sigma - \sigma_0)^m$$
$$+ \frac{1}{m!} \int_{\sigma_0}^{\sigma} d\xi \, (\sigma - \xi)^m \, g^{(m+1)}(\xi) .$$

The Cauchy remainder is obtained applying the mean theorem to the above integral form:

$$R_m(\sigma_0, \sigma) = \frac{1}{m!} \int_{\sigma_0}^{\sigma} d\xi \, (\sigma - \xi)^m g^{(m+1)}(\xi)$$

$$= \frac{g^{(m+1)}(\eta'')}{m!} (\sigma - \sigma_0)(\sigma - \eta'')^m \quad \text{con } \eta'' \in (\sigma_0, \sigma).$$

Finally, the Peano remainder summarises all the other forms, without specifying the coefficient that multiplies $(\sigma - \sigma_0)^{m+1}$.

Going back to the series (9.3), σ_0 is set to 0 and the first term $\sigma g'(0)$ is rewritten by means of the partial derivatives of f:

$$\sigma g'(0) = \sigma d_i \, \partial_i f(x_0) = (x_i - x_{0i}) \, \partial_i f(x_0) = (x - x_0) \cdot \nabla f(x_0), \qquad (9.5)$$

using the definitions of σ and d for writing $\sigma d_i = \Delta x_i = x_i - x_{0i}$. The result (9.5) confirms what has been discussed before (9.1) assuming f differentiable in x_0, only. But now it is possible to write $\sigma^2 g''(0)/2$ in terms of the second derivatives of f:

$$\frac{\sigma^2}{2} g''(0) = \frac{\sigma^2}{2} d_i d_j \, \partial^2_{ij} f(\boldsymbol{x}_0) = \frac{1}{2} (x_i - x_{0i})(x_j - x_{0j}) \partial^2_{ij} f(\boldsymbol{x}_0). \qquad (9.6)$$

The above term can be rewritten in the shorter form that uses vectorial quantities. In order to do this, the following $n \times n$ matrix of the second derivatives of f is introduced:

$$\boldsymbol{H}(\boldsymbol{x}) = \begin{pmatrix} \partial^2_{11} f(\boldsymbol{x}) & \partial^2_{12} f(\boldsymbol{x}) & \cdots & \partial^2_{1n} f(\boldsymbol{x}) \\ \partial^2_{21} f(\boldsymbol{x}) & \partial^2_{22} f(\boldsymbol{x}) & \cdots & \partial^2_{2n} f(\boldsymbol{x}) \\ \vdots & \vdots & & \vdots \\ \partial^2_{n1} f(\boldsymbol{x}) & \partial^2_{n2} f(\boldsymbol{x}) & \cdots & \partial^2_{nn} f(\boldsymbol{x}) \end{pmatrix}. \qquad (9.7)$$

It is named as *Hessian matrix* of f. Note that it depends on the point $\boldsymbol{x}$ in which these derivatives are evaluated and is a first example of function from $\mathbb{R}^n$ to $\mathbb{R}^n \times \mathbb{R}^n$. Moreover, it is symmetric due to the Schwarz theorem discussed in Sect. 6.4. The third term (9.6) in the series (9.3) takes the following vector form:

$$\frac{\sigma^2}{2} g''(0) = \frac{1}{2} (\boldsymbol{x} - \boldsymbol{x}_0)^T \cdot \boldsymbol{H}(\boldsymbol{x}_0) \cdot (\boldsymbol{x} - \boldsymbol{x}_0)$$

and the approximation for $m = 2$ becomes:

$$f(\boldsymbol{x}) = f(\boldsymbol{x}_0) + (\boldsymbol{x} - \boldsymbol{x}_0) \cdot \boldsymbol{\nabla} f(\boldsymbol{x}_0) + \frac{1}{2} (\boldsymbol{x} - \boldsymbol{x}_0)^T \cdot \boldsymbol{H}(\boldsymbol{x}_0) \cdot (\boldsymbol{x} - \boldsymbol{x}_0) + o(|\boldsymbol{x} - \boldsymbol{x}_0|^2), \qquad (9.8)$$

the remainder being at least of the third order with respect to the principal infinitesimal $|\boldsymbol{x} - \boldsymbol{x}_0|$, or shortly $o(|\boldsymbol{x} - \boldsymbol{x}_0|^2)$. This truncated Taylor series will be used in the next lesson for locating maxima, minima and saddle points of the surface $z = f(x, y)$.

The general term $\sigma^p g^{(p)}(0)/p!$ in the series (9.3) is written as:

$$\frac{\sigma^p}{p!} g^{(p)}(0) = \frac{\sigma^p}{p!} d_{i_1} d_{i_2} \cdot \ldots \cdot d_{i_p} \, \partial^p_{i_1 i_2 \cdots i_p} f(\boldsymbol{x}_0)$$

$$= \frac{1}{p!} (x_{i_1} - x_{0i_1})(x_{i_2} - x_{0i_2}) \cdot \ldots \cdot (x_{i_p} - x_{0i_p}) \partial^p_{i_1 i_2 \cdots i_p} f(\boldsymbol{x}_0), \qquad (9.9)$$

where the indices run from 1 to n and, as before, the symbols of the sums on repeated indices are omitted. From the relation (9.9) the Taylor expansion of order $m \geq 1$ follows:

$$f(\boldsymbol{x}) = \sum_{p=0}^{m} \frac{1}{p!} (x_{i_1} - x_{0i_1})(x_{i_2} - x_{0i_2}) \cdot \ldots \cdot (x_{i_p} - x_{0i_p}) \partial^p_{i_1 i_2 \cdots i_p} f(\boldsymbol{x}_0) + R_m(\boldsymbol{x}_0, \boldsymbol{x}). \qquad (9.10)$$

In order to explain the use of the formula (9.10), it is now applied in the case $n = 2$, *i.e.* to functions of the two real variables $x_1 = x$ and $x_2 = y$. A partial derivative of order p is made deriving r times in x and s times in y with $0 \le r, s \le p$ and $r + s = p$. How many pth derivatives exist? They are as much as the number of combinations of p things taken r (or s) at a time:

$$\binom{p}{r} = \frac{p!}{r! \, s!} .$$

Moreover, these derivatives are equal, being continuous all the partial derivatives up to the order $m + 1$. Gathering the pth derivatives made r times in x and s times in y in the general form (9.10) of the Taylor series, the series expansion of a function f of the two variables x and y

$$f(x) = \sum_{p=0}^{m} \sum_{r+s=p} \frac{1}{r! s!} \frac{\partial^r}{\partial x^r} \frac{\partial^s}{\partial y^s} f(x_0, y_0) \, (x - x_0)^r (y - y_0)^s + o(|x - x_0|^m)$$

(9.11)

follows. The remainder is written in the Peano's form, for the sake of simplicity. Note that the terms in the series are symmetrical with respect to the two indices (r, s) and this symmetry will be always preserved. Few sample cases are discussed below.

Example 9.1 The function $f(x, y) = \exp(xy)$ is now expanded in Taylor series about an arbitrary point (x_0, y_0). Its Taylor series is easily written, noting that:

$$\frac{\partial^r}{\partial x^r} f(x, y) = y^r \exp(xy) .$$

The sth derivative in y is evaluated using Leibniz's rule:

$$\frac{\partial^s}{\partial y^s} \left[\frac{\partial^r}{\partial x^r} f(x, y) \right] = \sum_{l=0}^{s} \binom{s}{l} \frac{d^l}{dy^l} y^r \frac{\partial^{s-l}}{\partial y^{s-l}} f(x, y)$$

$$= f(x, y) \sum_{l=0}^{\min(r,s)} \frac{1}{l!} \frac{s!}{(s-l)!} \frac{r!}{(r-l)!} y^{r-l} x^{s-l}$$

$$= f(x, y) \, r! s! \sum_{l=0}^{\min(r,s)} \frac{1}{l!} \frac{x^{s-l} y^{r-l}}{(s-l)!(r-l)!} .$$

The Taylor series follows as:

$$f(x, y) = f(x_0, y_0) \sum_{p=0}^{m} \sum_{r+s=p} \left[\sum_{l=0}^{\min(r,s)} \frac{1}{l!} \frac{x_0^{s-l} y_0^{r-l}}{(s-l)!(r-l)!} \right] (x - x_0)^r (y - y_0)^s$$

$$+ o(|x - x_0|^m) .$$

(9.12)

The partial sum (9.12) is drawn in Fig. 9.1 with blue lines for x running in a square neighbourhood of the point $x_0 = (-1, -0.5)$. The corresponding values of the function f are also drawn with red lines.

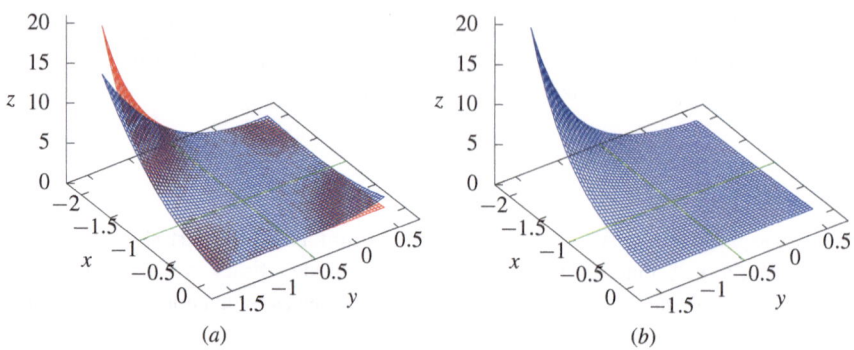

Fig. 9.1 Graphical representation of the function $f(x, y) = \exp(xy)$ drawing the surface $z = f(x, y)$ (red lines) in a neighbourhood of the point $x_0 = (-1, -0.5)$ (located at the intersection between the green lines) and partial sums of the Taylor series (9.12) (blue lines) with $m = 4$ and 14. The maximum remainder is 6 for $m = 4$ and becomes $5 \cdot 10^{-3}$ for $m = 14$. It decreases up to $5 \cdot 10^{-14}$ for $m = 40$

Example 9.2 Consider the function: $f(x, y) = (x^2 - y^2)^\alpha$, with $\alpha \in \mathbb{R}$. The existence field $\mathscr{E}$ of this function is the region between the two bisecting lines $y = +x$ and $y = -x$. As a sample case, take the point x_0 inside the part of $\mathscr{E}$ lying on the half-plane $y > 0$. The function is written as the product: $f(x, y) = (x - y)^\alpha (x + y)^\alpha$, being $x - y$ and $x + y$ positive in that region of $\mathbb{R}^2$. If the point x_0 would belong to the part of $\mathscr{E}$ lying on the other half-plane, the function would be rewritten as $f(x, y) \equiv (-x + y)^\alpha (-x - y)^\alpha$, being $-x + y$ and $-x - y$ positive.

Using the Leibniz rule, the rth partial derivative in x of the above function is evaluated as:

$$\frac{\partial^r}{\partial x^r} f(x, y) = \sum_{l'+l''=r} \binom{r}{l'} \alpha(\alpha - 1) \cdot \ldots \cdot (\alpha - l' + 1)\,(x - y)^{\alpha - l'} \times$$

$$\times\, \alpha(\alpha - 1) \cdot \ldots \cdot (\alpha - l'' + 1)\,(x + y)^{\alpha - l''}$$

$$= r! \sum_{l'+l''=r} \binom{\alpha}{l'} \binom{\alpha}{l''} (x - y)^{\alpha - l'} (x + y)^{\alpha - l''}$$

and the same rule leads to the sth partial derivative in y of the derivative $\partial^r f/\partial x^r$:

$$\frac{1}{r! s!} \frac{\partial^s}{\partial y^s} \left[\frac{\partial^r}{\partial x^r} f(x, y) \right] = \sum_{l'+l''=r} \binom{\alpha}{l'} \binom{\alpha}{l''} \times$$

$$\times \sum_{m'+m''=s} (-1)^{m'} \binom{\alpha - l'}{m'} \binom{\alpha - l''}{m''} (x - y)^{\alpha - l' - m'} (x + y)^{\alpha - l'' - m''}$$

$$=: C_{rs}(x).$$

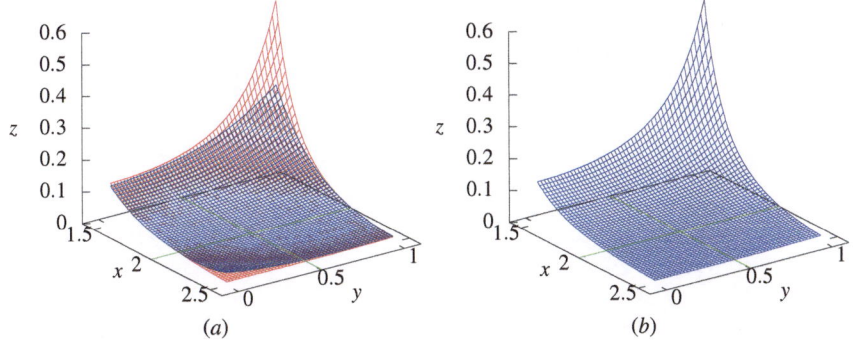

Fig. 9.2 Function $z = f(x, y) = (x^2 - y^2)^{-2.5}$ (red lines) in a neighbourhood of the point $x_0 = (2, 0.5)$ (located at the intersection between the green lines) and partial sums of the Taylor series (9.12) with $m = 4$ and 20. The maximum remainder is 0.3 for $m = 4$ and becomes $2 \cdot 10^{-3}$ for $m = 20$. It decreases up to 10^{-6} for $m = 40$

This derivative is employed for writing the partial sum of order m of the Taylor series:

$$f(x) = \sum_{p=0}^{m} \sum_{r+s=p} C_{rs}(x_0)\,(x - x_0)^r (y - y_0)^s + o(|x - x_0|^m). \tag{9.13}$$

The partial sums (9.13) for $\alpha = -2.5$ are shown in Fig. 9.2 (blue lines) for x belonging to a square neighbourhood of x_0. In the same figure, these sums are compared with the values of the function $f(x, y) = (x^2 - y^2)^{-2.5}$.

Example 9.3 Consider the function $f(x, y) = 1/(1 + xy)$, defined on $\mathbb{R}^2$ deprived of the hyperbola $y = -1/x$. The rth partial derivative in x of this function is:

$$\frac{\partial^r}{\partial x^r} f(x, y) = (-1)^r r! \frac{y^r}{(1 + xy)^{r+1}}$$

and the sth partial derivative in y of the above derivative in x is evaluated by means of the Leibniz rule:

$$\frac{\partial^s}{\partial y^s}\left[\frac{\partial^r}{\partial x^r} f(x, y)\right] = (-1)^r r! \sum_{l=0}^{s} \binom{s}{l} \frac{d^l}{dy^l} y^r \frac{\partial^{s-l}}{\partial y^{s-l}} \frac{y^r}{(1 + xy)^{r+1}}$$

$$= \frac{(-1)^p r! s!}{(1 + xy)^{p+1}} \sum_{l=0}^{\min(r,s)} (-1)^l \binom{r}{l} \binom{p-l}{r} x^{s-l} y^{r-l} (1 + xy)^l,$$

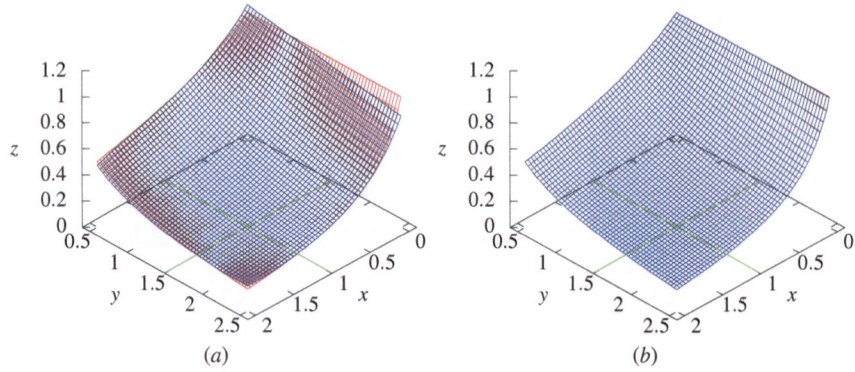

Fig. 9.3 Function $z = f(x, y) = 1/(1 + xy)$ (red lines) in a neighbourhood of the point $x_0 = (1, 1.5)$ (located at the intersection of the two green lines) and partial sums of the Taylor series (9.14) (blue lines) with $m = 5$ and 15. The maximum remainder is 0.15 for $m = 5$ and becomes $7 \cdot 10^{-3}$ for $m = 15$. It decreases up to $4 \cdot 10^{-6}$ for $m = 40$

$r + s$ being equal to p. The mth partial sum of the Taylor series follows:

$$f(x) = \sum_{p=0}^{m} \frac{(-1)^p}{(1 + x_0 y_0)^{p+1}} \sum_{r=0}^{p} \left[\sum_{l=0}^{\min(r,s)} (-1)^l \binom{r}{l} \binom{p-l}{r} x_0^{p-r-l} y_0^{r-l} (1 + x_0 y_0)^l \right] \times$$

$$\times (x - x_0)^r (y - y_0)^{p-r} + o(|x - x_0|^m).$$

$$(9.14)$$

The partial sums (9.14) for $m = 5$ and 15 are shown in Fig. 9.3 (blue lines) with x belonging to a square neighbourhood of the point $x_0 = (1, 1.5)$. In the same figure, the sums are compared with the values of the function $f(x, y)$ (red lines).

9.2   Suggested Readings

Chapter VI of Lang (1987) is completely devoted to analysis and applications of the Taylor series expansion for functions of several variables. Examples are also given in Adams and Essex (2010, Chap. 12, pp. 735–739) and Apostol (1974, Chap. 12, pp. 361–366).

References

R. A. Adams and C. Essex. *Calculus: Several Variables*. Pearson Education Canada, Toronto, ON, 7th edition, 2010.

T. M. Apostol. *Mathematical Analysis: A Modern Approach to Advanced Calculus*. Pearson Education US, Hoboken, NJ, 2nd edition, 1974.

S. Lang. *Calculus of Several Variables*. Springer, New York, 3rd edition, 1987.

References

Chapter 10
Applications of the Taylor Series

In the present lesson a few applications of the Taylor series for functions $f : \mathbb{R}^n \to \mathbb{R}$ will be discussed. The analysis is limited to functions of two or three variables. The first application deals with the search for the extreme and saddle points of a surface given by means of implicit (*e.g.*, $f(x, y, z) = $ constant) or explicit (*e.g.*, $z = z(x, y)$) relations.

10.1 Calculation of the Extreme and Saddle Points

In this section the calculation of the extreme and of the saddle points of a surface Σ in $\mathbb{R}^3$ is addressed. Firstly, this issue is investigated using the *explicit representation*, *e.g.* $z = z(x, y)$. This function gives the third coordinate z in correspondence to any point (x, y) belonging to an open set $A \subseteq \mathbb{R}^2$, such that $(x, y, z(x, y))$ lies on Σ. Consider now a *stationary point* of the function $z = z(x, y)$, *i.e.* a point $(\xi, \eta) \in A$ which solves the system:

$$\begin{cases} \partial_x z(\xi, \eta) = 0 \\ \partial_y z(\xi, \eta) = 0 . \end{cases} \tag{10.1}$$

At that point, $z(\xi, \eta) =: \zeta$ and the partial sum of the Taylor series (9.8) up to second order terms takes the form:

$$z(x, y) = \zeta + \frac{1}{2}(x - \xi, y - \eta) \cdot \boldsymbol{H}(\xi, \eta) \cdot \begin{pmatrix} x - \xi \\ y - \eta \end{pmatrix} + o[(x - \xi)^2 + (y - \eta)^2], \tag{10.2}$$

due to the fact that the first derivatives vanish ((ξ, η) solves the system (10.1)).

If the Hessian matrix[1] H is not zero at the stationary point (ξ, η), the expansion (10.2) enables to study the function $z = z(x, y)$ in a small neighbourhood $B \subseteq A$ of its stationary point (ξ, η), in a similar way to the use of the second derivative for the functions of one variable. Indeed, assume that H is such that:

$$\forall \boldsymbol{u} \in \mathbb{R}^2, \, \boldsymbol{u} \neq \boldsymbol{0}: \quad \boldsymbol{u}^\mathsf{T} \cdot \boldsymbol{H} \cdot \boldsymbol{u} < 0, \tag{10.3}$$

i.e. H is *definite negative*. It follows that (ξ, η) is a (relative) *maximum*. Indeed, the expansion (10.2) shows that the value of z at a point $(x, y) \in B - (\xi, \eta)$ is smaller than the one (ζ) reached at (ξ, η). In a symmetric way, if H is *definite positive*:

$$\forall \boldsymbol{u} \in \mathbb{R}^2, \, \boldsymbol{u} \neq \boldsymbol{0}: \quad \boldsymbol{u}^\mathsf{T} \cdot \boldsymbol{H} \cdot \boldsymbol{u} > 0, \tag{10.4}$$

the point (ξ, η) is a (relative) *minimum*. Indeed, the value of z at a point $(x, y) \in B - (\xi, \eta)$ is larger than $\zeta = z(\xi, \eta)$. If the matrix H does not satisfy the condition (10.3), nor the one (10.4), (ξ, η) is a *saddle point*. In this case, there are at the same time directions along which z decreases as well as directions along which it grows. Finally, if $H(\xi, \eta) = \boldsymbol{0}$ the expansion (10.2) does not give any information about the behaviour of f in B and a more refined partial sum is needed.

Assume now that Σ is described by means of an[2] *implicit representation*: Σ is the set of points $(x, y, z) \in \mathbb{R}^3$ such that $f(x, y, z) = c$, f being a two-times partially differentiable function having continuous second derivatives and c a (given) constant. Note that, in order to assure the existence of the normal vector, ∇f must be non-vanishing on Σ. In particular, here it is assumed that its component along z does not vanish, i.e. $\partial_z f(\xi, \eta, \zeta) \neq 0$.

Fixed the coordinate direction along which the component of ∇f does not vanish, a *stationary point* is defined as a solution (ξ, η, ζ) of the system:

$$\begin{cases} \partial_x f(\xi, \eta, \zeta) = 0 \\ \partial_y f(\xi, \eta, \zeta) = 0 \\ f(\xi, \eta, \zeta) = c, \end{cases} \tag{10.5}$$

in which the first two equations imply that $\nabla f(\xi, \eta, \zeta)$ is directed exactly along z, whereas the third assures that the point lies on Σ.

In lesson 9 it has been shown that, if the function $f : \mathbb{R}^3 \to \mathbb{R}$ possesses continuous second partial derivatives, the surface $f(\boldsymbol{x}) = c$ can be approximated in a neighbourhood of an arbitrary point $\boldsymbol{\xi} = (\xi, \eta, \zeta) \in \Sigma$ by means of the first order ($m = 1$) partial sum of the Taylor series of f:

$$c = f(\boldsymbol{x}) \simeq f(\boldsymbol{\xi}) + (\boldsymbol{x} - \boldsymbol{\xi}) \cdot \nabla f(\boldsymbol{\xi}) = c + (\boldsymbol{x} - \boldsymbol{\xi}) \cdot \nabla f(\boldsymbol{\xi}),$$

---

[1] Hereafter, the Hessian matrix is evaluated at the stationary point (ξ, η), only. For this reason its dependence on that point will be omitted, for the sake of shortness.

[2] This representation is not unique.

and then with the plane:

$$(x - \xi) \cdot \nabla f(\xi) = 0. \tag{10.6}$$

As discussed in Sect. 6.1, the gradient $\nabla f(\xi)$ is normal to the surface Σ at its point ξ and then the plane (10.6), containing all the points x such that the vector $x - \xi$ is orthogonal to the above gradient, is *tangent* to Σ at ξ. Now the point ξ is chosen among the solutions of the system (10.5) and, due to the fact that $\nabla f(\xi, \eta, \zeta)$ is directed along z, the tangent plane (10.6) is simply given by $z = \zeta$. As a consequence, in the coordinate plane $z = 0$ a sufficiently small neighbourhood B of the point (ξ, η) exists, at any point (x, y) of which it is possible to find a unique value of z (depending on (x, y)) such that the point $(x, y, z(x, y))$ lies on Σ. Moreover, at the point (ξ, η) the coordinate z takes the value ζ.

The function $z = z(x, y)$ gives a *local explicit representation* of Σ in the neighbourhood B of (ξ, η), where the implicit representation takes the form $f(x, y, z(x, y)) = c$. Using the chain rule for differentiating composite functions and the first two equations in the system (10.5), the system:

$$\begin{cases} 0 = \partial_x f(\xi) + \partial_z f(\xi) \, \partial_x z(\xi, \eta) & = \partial_z f(\xi) \, \partial_x z(\xi, \eta) \\ 0 = \partial_y f(\xi) + \partial_z f(\xi) \, \partial_y z(\xi, \eta) & = \partial_z f(\xi) \, \partial_y z(\xi, \eta) \end{cases}$$

follows. Hence, also locally, the condition $\partial_z f(\xi) \neq 0$ leads to the conditions (10.1).

It is worth noting that it is possible to evaluate the Hessian matrix of $z(x, y)$ by means of the corresponding matrix of the function f, without calculating the local explicit representation $z = z(x, y)$ (solving with respect to z the equation $f(x, y, z) = c$), that often is a very difficult task. Applying the chain rule to evaluate the second derivatives of the function $f[x, y, z(x, y)] \equiv c$ the following relations are found:

$$0 \equiv \partial_{xx}^2 f + 2\partial_{xz}^2 f \, \partial_x z + \partial_{zz}^2 f \, (\partial_x z)^2 + \partial_z f \, \partial_{xx}^2 z$$

$$0 \equiv \partial_{xy}^2 f + \partial_{xz}^2 f \, \partial_y z + \partial_{yz}^2 f \, \partial_x z + \partial_{zz}^2 f \, \partial_x z \partial_y z + \partial_z f \, \partial_{xy}^2 z$$

$$0 \equiv \partial_{yy}^2 f + 2\partial_{yz}^2 f \, \partial_y z + \partial_{zz}^2 f \, (\partial_y z)^2 + \partial_z f \, \partial_{yy}^2 z.$$

These relations are evaluated at the point (ξ, η) accounting for the conditions (10.1), so that:

$$\boldsymbol{H}(\xi, \eta) = \begin{pmatrix} \partial_{xx}^2 z(\xi, \eta) & \partial_{xy}^2 z(\xi, \eta) \\ \partial_{yx}^2 z(\xi, \eta) & \partial_{yy}^2 z(\xi, \eta) \end{pmatrix} = -\frac{1}{\partial_z f(\xi)} \begin{pmatrix} \partial_{xx}^2 f(\xi) & \partial_{xy}^2 f(\xi) \\ \partial_{yx}^2 f(\xi) & \partial_{yy}^2 f(\xi) \end{pmatrix}. \tag{10.7}$$

The relation (10.7), when $\partial_z f(\xi) \neq 0$, gives the Hessian matrix for an implicit representation of the surface Σ. In the following several examples are discussed in which the knowledge of the Hessian matrix is sufficient to define the behaviour of Σ in a neighbourhood of a stationary point.

Example 10.1 Consider the function $f(x) = |x| = c > 0$, that gives an implicit representation of a sphere of center on the origin and radius c. The conditions (10.5) specify in the following ones:

$$\begin{cases} x/c = 0 \\ y/c = 0 \\ |x| = c. \end{cases}$$

From the first two it follows that $x = y = 0$, whereas from the third condition $z = \pm c$. This surface possesses two stationary points: $\boldsymbol{\xi}_1 = (0, 0, +c)$ and $\boldsymbol{\xi}_2 = (0, 0, -c)$. Once the following derivatives:

$$\partial_z f = \frac{z}{|x|}, \quad \partial_{xx}^2 f = \frac{y^2 + z^2}{|x|^3}, \quad \partial_{xy}^2 f = \partial_{yx}^2 f = -\frac{xy}{|x|^3}, \quad \partial_{yy}^2 f = \frac{x^2 + z^2}{|x|^3},$$

have been evaluated, the Hessian matrices at the points $\boldsymbol{\xi}_{1,2}$ follow from the relation (10.7):

$$H_{1,2} = \mp \frac{1}{c} \begin{pmatrix} 1 & 0 \\ 0 & 1 \end{pmatrix}.$$

H_1 is negative definite, whereas H_2 is positive definite. It follows that $\boldsymbol{\xi}_1$ is a local maximum and $\boldsymbol{\xi}_2$ a local minimum. It is easy to show that they are also global maximum and minimum.

Example 10.2 Consider the part with $z > 0$ of the surface: $f(x, y, z) = x^2 + y^2 + z^4 + 2x - 6y = 1$. The only stationary point is $(-1, 3, 11^{1/4})$. The Hessian matrix is constant:

$$H \equiv -\frac{1}{2 \, 11^{3/4}} \begin{pmatrix} 1 & 0 \\ 0 & 1 \end{pmatrix},$$

and it is definite negative, so that the stationary point is a maximum.

Example 10.3 Consider the part with $z > 0$ of the surface: $f(x, y, z) = x^3 + yz^2 + 2x^2 - 2y^2 = 4$. The only stationary point is: $(0, 2^{1/2}, 2^{5/4})$. The Hessian matrix evaluated at that point is written as:

$$H = 2^{-3/4} \begin{pmatrix} -1 & 0 \\ 0 & 1 \end{pmatrix},$$

which is not definite. As a consequence, the stationary point is a saddle.

Example 10.4 Consider the part with $z > 0$ of the surface $f(x, y, z) = z^2 \cos x \sin y = 1$, where it is assumed $z > 0$. Its stationary points verify the conditions:

$$\begin{cases} -z^2 \sin x \sin y = 0 \\ z^2 \cos x \cos y = 0. \end{cases}$$

Due to the fact that $\cos x$ and $\sin y$ cannot vanish, the stationary points verify the conditions:

$$\sin x = 0, \quad \cos y = 0.$$

As a consequence, the stationary points are given by $(k\pi, \pi/2 + m\pi, 1)$ with $k + m$ even. In these points, the Hessian matrix is evaluated as:

$$H = \frac{1}{2}\begin{pmatrix} 1 & 0 \\ 0 & 1 \end{pmatrix},$$

so that the stationary points are minima.

10.2   Approximation of the Solution of a Partial Differential Problem

The Taylor series is often used for writing numerical schemes able to approximate the solutions of partial differential problems. As a sample case, assume that the function $f : \mathbb{R}^2 \to \mathbb{R}$ is two times partially differentiable with continuous second derivatives and satisfies the *Poisson problem*:

$$\begin{cases} \nabla^2 f = p & x \in B \\ f = q & x \in \partial B, \end{cases} \tag{10.8}$$

in which B is the rectangle $(a, b) \times (c, d) \subset \mathbb{R}^2$ and p, q are given continuous functions. In particular, p is taken as two times partially differentiable with continuous second derivatives, whereas q is assumed as four times differentiable with continuous fourth derivatives. It can be shown that the solution of the problem (10.8) exists and is unique. Moreover, f results to be four times partially differentiable with continuous fourth derivatives.

The solution f of the problem (10.8) is approximated on a grid of $n_x \times n_y$ points $\{(x_i, y_j); i = 1, \ldots, n_x, j = 1, \ldots, n_y\}$ with: $x_i = a + (i - 1)h_x$, $y_j = c + (j - 1)h_y$ and $h_x = (b - a)/(n_x - 1)$, $h_y = (d - c)/(n_y - 1)$. Both the steps of the above grid (h_x and h_y) are assumed of the same order of magnitude: $O(h_x) = O(h_y) = O(h)$. Moreover, the points (x_1, y_j), (x_{n_x}, y_j) lie on the segments parallel to y of ∂B, whereas (x_i, y_1), (x_i, y_{n_y}) belong to the segments along x. On these points the solution f is given in terms of the function b, from the second equation in the problem (10.8) that consequently is called *boundary condition*. The problem consists in approximating the solution f on the grid points that lie inside the domain B.

It is now presented a way to approximate the partial differential equation in the first row of the problem (10.8) at the internal points (x_i, y_j) (with $2 \le i \le n_x - 1$ and $2 \le j \le n_y - 1$) using the values $f(x_i, y_j) =: f_{i,j}$ on the grid points, only. The second partial derivative along x can be approximated by means of the following Taylor expansions centered on (x_i, y_j):

$$f_{i+1,j} = f_{i,j} + h_x \partial_x f_{i,j} + \frac{1}{2}h_x^2 \partial_{xx}^2 f_{i,j} + \frac{1}{6}h_x^3 \partial_{xxx}^3 f_{i,j} + O(h_x^4)$$

$$f_{i-1,j} = f_{i,j} - h_x \partial_x f_{i,j} + \frac{1}{2}h_x^2 \partial_{xx}^2 f_{i,j} - \frac{1}{6}h_x^3 \partial_{xxx}^3 f_{i,j} + O(h_x^4).$$

The sum of the above expansions gives an approximation of the derivative:

$$\partial_{xx}^2 f_{i,j} = \frac{f_{i+1,j} - 2f_{i,j} + f_{i-1,j}}{h_x^2} + O(h_x^2). \tag{10.9}$$

An approximation of the second derivative in y is deduced in an analogous way, so that the approximation of the partial differential equation in the problem (10.8) with the *finite difference equation*:

$$\nabla^2 f(x_i, y_j) \simeq \frac{f_{i+1,j} - 2f_{i,j} + f_{i-1,j}}{h_x^2} + \frac{f_{i,j+1} - 2f_{i,j} + f_{i,j-1}}{h_y^2} = p_{i,j} \tag{10.10}$$

follows. An iterative procedure is now built for solving the Eq. (10.10), without using linear system solvers.

First of all, Eq. (10.10) is solved in $f_{i,j}$. Once the quantities $\beta = h_x/h_y$, $m_x := 1/[2(1 + \beta^2)]$, $m_y := \beta^2/[2(1 + \beta^2)]$ and $m := \beta/[2(1 + \beta^2)]$ have been defined, $f_{i,j}$ is written as:

$$f_{i,j} = m_x(f_{i+1,j} + f_{i-1,j}) + m_y(f_{i,j+1} + f_{i,j-1}) - mh_x h_y p_{i,j}.$$

Then, at the right hand side of the above equation $f_{i,j}$ is added and subtracted:

$$f_{i,j} = f_{i,j} + \left[-f_{i,j} + m_x(f_{i+1,j} + f_{i-1,j}) + m_y(f_{i,j+1} + f_{i,j-1}) - mh_x h_y p_{i,j} \right].$$

The iterative procedure is based on a suitable rereading of the above equation: the value of $f_{i,j}$ at the lth iteration ($l = 1, 2, 3, \ldots$), indicated by $f_{i,j}^{(l)}$, at the left hand side is the sum of the value of f at the same point at the previous iteration ($l - 1$, *i.e.* $f_{i,j}^{(l-1)}$), plus the "correction" term in the square brackets. In this latter, the values of f are taken at the iteration l, if they are available, or at the $l - 1$th, if they are not. Finally, this "correction" term is multiplied times a relaxation parameter $\omega \in (0, 2)$, so that the iterative procedure is written as:

$$f_{i,j}^{(l)} = (1 - \omega)f_{i,j}^{(l-1)}$$
$$+ \omega \left[m_x(f_{i+1,j}^{(l-1)} + f_{i-1,j}^{(l)}) + m_y(f_{i,j+1}^{(l-1)} + f_{i,j-1}^{(l)}) - mh_x h_y p_{i,j} \right] \tag{10.11}$$

for $2 \leq i \leq n_x - 1$ and $2 \leq j \leq n_y - 1$. In the above equation, $f_{i,j}^{(0)}$ for $2 \leq i \leq n_x - 1, 2 \leq j \leq n_y - 1$ are assumed zero, whereas the values of f on ∂B are given by the boundary conditions of the problem (10.8), *i.e.* by the function q.

The procedure (10.11) converges quite rapidly to the approximate solution of the problem (10.8). In order to measure this convergence, the "distance" between an iteration and the previous one:

$$\delta^{(l)} = \sqrt{h_x h_y \sum_{i=2}^{n_x-1} \sum_{j=2}^{n_y-1} \left[f_{i,j}^{(l)} - f_{i,j}^{(l-1)} \right]^2}. \tag{10.12}$$

is evaluated during the calculation. Note that the "distance" (10.12), accounts only for two successive iterations, without considering the analytical solution of the problem (10.8) which remains, in general, unknown. As sample cases, in Fig. 10.1 three approximations (black lines) $f_{i,j}^{(1)}$ (*a*), $f_{i,j}^{(41)}$ (*b*) and $f_{i,j}^{(81)}$ (*c*) are superimposed to the analytical solution of the problem (10.8) (red).

Exercise 10.1 Evaluate an approximation of the solution of the problem (10.8) in $B = (0, 1) \times (0, 2)$ assuming:

$$p(x, y) = \frac{4(y^2 - x^2)(y^2 + x^2 + 3)}{(x^2 + y^2 + 1)^3}$$

and the boundary conditions:

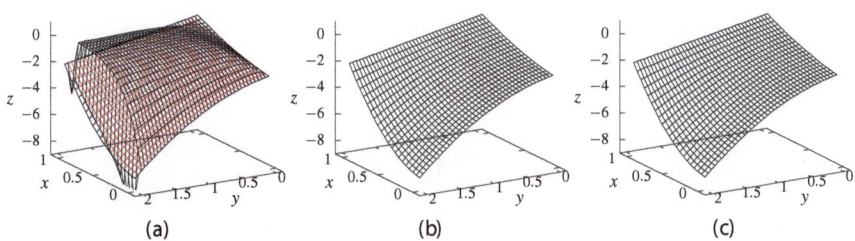

(a)   (b)   (c)

Fig. 10.1 Approximations of the solution $f(x, y) = -2 + x - y + x^2 - y^2 + x^2 y^2$ (red lines) of the problem (10.8) in the domain $B = (0, 1) \times (0, 2)$, with $n_x = 21$, $n_y = 41$ and $\omega = 1.6$. The approximate values (black lines) on the grid are obtained for $l = 1$ **a**, 41 **b** and 81 **c**. The distance (10.12) becomes $\sim 10^{-15}$ for $l > 500$

$$y = 0, \quad 0 < x < 1: \quad q(x, 0) = \frac{x^2}{x^2 + 1}$$

$$y = 2, \quad 0 < x < 1: \quad q(x, 2) = \frac{x^2 - 4}{x^2 + 5}$$

$$x = 0, \quad 0 < y < 2: \quad q(0, y) = -\frac{y^2}{y^2 + 1}$$

$$x = 1, \quad 0 < y < 2: \quad q(1, y) = \frac{1 - y^2}{2 + y^2}.$$

The above set of data corresponds to the analytical solution:

$$f(x, y) = \frac{x^2 - y^2}{x^2 + y^2 + 1}.$$

Chapter 11
Integration of Functions of Two Variables

In this lesson, the integration of functions of several variables will be addressed. The first step consists in a one-dimensional integration of a function of two variables.

11.1 Derivative of a Definite Integral with Variable Limits

Given a function $f : \mathbb{R}^2 \to \mathbb{R}$ partially differentiable with continuous first derivatives, the new function defined by means of the integral:

$$\int_{\alpha(x)}^{\beta(x)} dy \, f(x, y) =: F(x) \tag{11.1}$$

is here considered. The limits of the integral (11.1) are two functions $\alpha, \beta : \mathbb{R} \to \mathbb{R}$ that are assumed differentiable. The variable x plays the role of a parameter inside the integral, which is calculated with respect to an independent variable y. Integrals like the one in Eq. (11.1) often appear in integrating functions of two or more variables.

An interesting issue concerns the calculation of the derivative of the new function F (11.1), which involves the functions f, α and β as well as their derivatives. The incremental ratio corresponding to an arbitrary increment Δx of the variable x is written as:

$$\frac{F(x + \Delta x) - F(x)}{\Delta x} =$$

$$= \frac{1}{\Delta x} \left[\int_{\alpha(x+\Delta x)}^{\beta(x+\Delta x)} dy \, f(x + \Delta x, y) - \int_{\alpha(x)}^{\beta(x)} dy \, f(x, y) \right]$$

$$\equiv \int_{\alpha(x+\Delta x)}^{\beta(x+\Delta x)} dy \, \frac{f(x + \Delta x, y) - f(x, y)}{\Delta x}$$

$$+ \frac{1}{\Delta x} \left[\int_{\alpha(x+\Delta x)}^{\beta(x+\Delta x)} dy \, f(x, y) - \int_{\alpha(x)}^{\beta(x)} dy \, f(x, y) \right]$$

$$= \int_{\alpha(x)}^{\beta(x)} dy \, \frac{f(x + \Delta x, y) - f(x, y)}{\Delta x}$$

$$+ \left[- \int_{\alpha(x)}^{\alpha(x+\Delta x)} dy \, \frac{f(x + \Delta x, y) - f(x, y)}{\Delta x} + \int_{\beta(x)}^{\beta(x+\Delta x)} dy \, \frac{f(x + \Delta x, y) - f(x, y)}{\Delta x} \right]$$

$$+ \frac{1}{\Delta x} \left[- \int_{\alpha(x)}^{\alpha(x+\Delta x)} dy \, f(x, y) + \int_{\beta(x)}^{\beta(x+\Delta x)} dy \, f(x, y) \right],$$

where the well-known property of the 1D integral about the partition of the range of integration:

$$\int_a^b dx \, g(x) = \int_a^c dx \, g(x) + \int_c^b dx \, g(x),$$

a, b and c being arbitrary (finite) numbers, has been applied. The last four integrals are handled by means of the mean value theorem, which implies that the four numbers: $\eta_{1,2} \in (\alpha(x), \alpha(x + \Delta x))$ and $\eta_{3,4} \in (\beta(x), \beta(x + \Delta x))$ exist such that:

$$\frac{F(x + \Delta x) - F(x)}{\Delta x} = \int_{\alpha(x)}^{\beta(x)} dy \, \frac{f(x + \Delta x, y) - f(x, y)}{\Delta x}$$

$$- \left[\alpha(x + \Delta x) - \alpha(x) \right] \frac{f(x + \Delta x, \eta_1) - f(x, \eta_1)}{\Delta x}$$

$$+ \left[\beta(x + \Delta x) - \beta(x) \right] \frac{f(x + \Delta x, \eta_3) - f(x, \eta_3)}{\Delta x}$$

$$- \frac{\alpha(x + \Delta x) - \alpha(x)}{\Delta x} f(x, \eta_2) + \frac{\beta(x + \Delta x) - \beta(x)}{\Delta x} f(x, \eta_4).$$

The limit as $\Delta x \to 0$ of both sides in the previous relation is evaluated accounting for that $\eta_{1,2} \to \alpha(x)$ and $\eta_{3,4} \to \beta(x)$, so that it follows:

$$F'(x) = \int_{\alpha(x)}^{\beta(x)} dy \, \partial_x f(x, y) - \alpha'(x) f[x, \alpha(x)] + \beta'(x) f[x, \beta(x)]. \qquad (11.2)$$

The rule (11.2) is now applied in a few sample cases.

Example 11.1 The derivative of the integral:

$$F(x) = \int_x^{x^2} \frac{dy}{1 + x^2 y^2} \tag{11.3}$$

is now evaluated applying the rule (11.2) as well as by a direct computation. In the first way the derivative is written as:

$$F'(x) = -\int_x^{x^2} dy \frac{2xy^2}{(1 + x^2 y^2)^2} - \frac{1}{1 + x^2 x^2} \cdot 1 + \frac{1}{1 + x^2 x^4} \cdot 2x ,$$

where the integral is evaluated by means of the change of variable $\eta = xy$:

$$-\int_x^{x^2} dy \frac{2xy^2}{(1 + x^2 y^2)^2} = \frac{1}{x^2} \int_{x^2}^{x^3} \eta \, d\frac{1}{1 + \eta^2}$$

$$= \frac{x}{1 + x^6} - \frac{1}{1 + x^4} - \frac{\arctan(x^3) - \arctan(x^2)}{x^2} .$$

Accounting for the above result, the derivative F' becomes:

$$F'(x) = -\frac{\arctan(x^3) - \arctan(x^2)}{x^2} + \frac{3x}{1 + x^6} - \frac{2}{1 + x^4} . \tag{11.4}$$

In the direct way the integral (11.3) is calculated (with the same change of variable as before):

$$F(x) = \frac{1}{x} \int_{x^2}^{x^3} \frac{d\eta}{1 + \eta^2} = \frac{\arctan(x^3) - \arctan(x^2)}{x}$$

calculating the derivative of F the result (11.4) is obtained.

Example 11.2 Calculate the derivative of the integral:

$$F(x) = \int_0^{2 \arctan(x)} \frac{dy}{1 + x \cos y} , \tag{11.5}$$

with $x \in (-1, 1)$. The direct calculation is performed by means of the change of variable $t = \tan(y/2)$ that implies:

$$\int \frac{dy}{1 + x \cos y} = \frac{2}{1 + x} \int \frac{dt}{1 + \frac{1 - x}{1 + x} t^2}$$

and again changing variable $\eta = \sqrt{(1 - x)/(1 + x)} \, t$:

$$\int \frac{dy}{1 + x \cos y} = \frac{2}{\sqrt{1 - x^2}} \arctan\left(\sqrt{\frac{1-x}{1+x}} \, \tan\frac{y}{2}\right).$$

It follows for F:

$$F(x) = \frac{2}{\sqrt{1-x^2}} \arctan\left(x\sqrt{\frac{1-x}{1+x}}\right), \tag{11.6}$$

the derivative of which is calculated as:

$$F'(x) = \frac{2x}{(1-x^2)^{3/2}} \arctan\left(x\sqrt{\frac{1-x}{1+x}}\right) + 2\frac{1 - x - x^2}{(1-x^2)(1 + x + x^2 - x^3)}. \tag{11.7}$$

Now the rule (11.2) is applied, leading to the following form of the derivative F':

$$F'(x) = -\int_0^{2\arctan(x)} dy \frac{\cos y}{(1 + x \cos y)^2} + \frac{1}{1 + x \cos(2\arctan(x))} \frac{2}{1 + x^2}$$

$$= -\frac{F(x)}{x} + \frac{1}{x}\int_0^{2\arctan(x)} \frac{dy}{(1 + x \cos y)^2} + \frac{2}{1 + x + x^2 - x^3}. \tag{11.8}$$

Using the previous changes of variable ($t = \tan(y/2)$, then $\eta = \sqrt{(1 - x)/(1 + x)}\, t$), the integral in Eq. (11.8) becomes:

$$\frac{1}{x}\int_0^{2\arctan(x)} \frac{dy}{(1 + x \cos y)^2} =$$

$$= \frac{2}{(1-x^2)^{3/2}}\left[-2x\int_0^{x\sqrt{(1-x)/(1+x)}} \frac{d\eta}{(1+\eta^2)^2} + (1+x)\int_0^{x\sqrt{(1-x)/(1+x)}} \frac{d\eta}{1+\eta^2}\right],$$

where the first integral on the right hand side is evaluated by parts:

$$\int \frac{d\eta}{(1+\eta^2)^2} = -\frac{1}{2}\int \frac{1}{\eta} d \frac{1}{1+\eta^2} = \frac{\eta}{2(1+\eta^2)} + \frac{1}{2}\arctan\eta + \text{const.}$$

from which it follows:

$$\frac{1}{x}\int_0^{2\arctan(x)} \frac{dy}{(1 + x \cos y)^2} = \frac{2}{(1-x^2)^{3/2}}\arctan\left(x\sqrt{\frac{1-x}{1+x}}\right)$$

$$-\frac{2x^2}{(1-x^2)(1 + x + x^2 - x^3)}. \tag{11.9}$$

Inserting the integrals (11.6), (11.9) inside the form (11.8) of the derivative, the derivative (11.7) follows again.

Example 11.3 Consider the function defined by the following integral:

$$F(x) = \int_{1/x}^{x} dy \sqrt{1 + xy^2}, \tag{11.10}$$

with $x > 0$. A direct calculation of the integral (11.10) by means of the change of variable $\sqrt{x}y = \sinh t$ leads to:

$$\int dy \sqrt{1 + xy^2} = \frac{1}{2\sqrt{x}} \left[\sqrt{x}y\sqrt{1 + xy^2} + \log\left(\sqrt{x}y + \sqrt{1 + xy^2}\right) \right] + \text{const.}$$

which is now evaluated in the limits x and $1/x$:

$$F(x) = \frac{1}{2\sqrt{x}} \left[x^{3/2}\sqrt{1 + x^3} - \frac{\sqrt{1 + x}}{x} + \log\left(x^{3/2} + \sqrt{1 + x^3}\right) + \frac{1}{2} \log(x) - \log\left(1 + \sqrt{1 + x}\right) \right]. \tag{11.11}$$

The derivative of the function F (11.11) is:

$$F'(x) = \frac{5}{4}\sqrt{1 + x^3} + \frac{3}{4}\frac{\sqrt{1 + x}}{x^{5/2}} - \frac{1}{4x^{3/2}} \log\left(x^{3/2} + \sqrt{1 + x^3}\right)$$
$$- \frac{1}{8x^{3/2}} \log x + \frac{1}{4x^{3/2}} \log\left(1 + \sqrt{1 + x}\right). \tag{11.12}$$

The application of the rule (11.2) directly gives the derivative:

$$F'(x) = \int_{1/x}^{x} dy \frac{y^2}{2\sqrt{1 + xy^2}} - \sqrt{1 + x\frac{1}{x^2}}\left(-\frac{1}{x^2}\right) + \sqrt{1 + x x^2}$$
$$= \frac{F(x)}{2x} - \frac{1}{2x} \int_{1/x}^{x} \frac{dy}{\sqrt{1 + xy^2}} + \frac{\sqrt{1 + x}}{x^{5/2}} + \sqrt{1 + x^3},$$

where the function F (11.11) is inserted and the integral is evaluated by means of the same change of variable $\sqrt{x}y = \sinh t$ used before. In this way, the result (11.12) follows again.

Exercise 11.1 Evaluate the derivative of the following functions:

$$(a) \quad F_1(x) := \int_{1/x^2}^{x^2} dy \, \exp(x^2 y), \qquad (b) \quad F_2(x) := \int_{x^2}^{x^4} dy \sqrt{y + x^2},$$

with the rule (11.2) as well as calculating the integral and then its first derivative.

Solution 11.1

$$(a) \quad F_1'(x) = 2e^{x^4}\left(2x - \frac{1}{x^3}\right) + \frac{2}{x^3}$$

$$(b) \quad F_2'(x) = 2x^2\left[(1 + 2x^2)\sqrt{x^2 + 1} - 2\sqrt{2}\right].$$

11.2   Double Integrals

It is now considered a continuous function of two variables $f : \mathbb{R}^2 \to \mathbb{R}$ in a bounded domain A of the (x, y)-plane. Such a function can be visualised in a geometric fashion thinking at the surface Σ in $\mathbb{R}^3$ obtained assuming x and y as parameters and $f(x, y)$ as the z-coordinate of a point. The integral of f on A is now introduced, explaining at the same time its geometrical meaning.

On the domain A a tagged partition with norm δ is performed: a coordinate grid (*i.e.* a grid defined by segments parallel to the coordinate axes) on a region of the (x, y)-plane containing A is built. Its cells c_i ($i = 1, \ldots, n$) have diameters not longer than δ and have non-empty intersections with A. For any cell c_i, its intersection with A called a_i, having area $|a_i|$. By definition, the set A is the nonempty union of all interactions:

$$A = \bigcup_{i=1}^{n} a_i \tag{11.13}$$

and the formula (11.13) is just called as "*coordinate partition*" of A. This partition being tagged, for any set a_i an arbitrary interior point $\boldsymbol{\xi}_i = (\xi_i, \eta_i)$ is chosen and the following sum is defined:

$$S(\delta) = \sum_{i=1}^{n} |a_i|\, f(\boldsymbol{\xi}_i). \tag{11.14}$$

Although the sum S (11.14) depends on all the details of the tagged partition (in particular, on the choice of the points $\{\boldsymbol{\xi}_i,\ i = 1, \ldots, n\}$), the more interesting dependence is on the norm δ and, for this reason, it will be the only one to be specified. Moreover, the sum (11.14) gives an approximation even better for decreasing δ (which implies increasing n, but the viceversa is not true) of the volume included between the surface Σ and the coordinate plane (x, y).

For δ going to 0, the sum (11.14) converges to a number that will be called *integral of f on A* and indicated by the symbol:

$$\lim_{\delta \to 0} S(\delta) =: \int_A dA(\boldsymbol{x})\, f(\boldsymbol{x}). \tag{11.15}$$

In the above notation, the symbol $dA(x)$ indicates the *area element* in which the position is named x. This symbol is often replaced by the equivalent one $dxdy$. Moreover, the symbol of integral is often used in a different way: in order to specify that the integral is calculated on a two-dimensional domain, this symbol is doubled

$$\iint_A dx \, dy \, f(x, y),$$

but this notation will not been adopted here. On the basis of the previous remark on S (11.14), the integral (11.15) measures the volume (with sign, *i.e.* it is negative if Σ lies mainly on the half-space with negative z) included between the surface Σ and the coordinate plane (x, y).

For the new integral (11.15) the same rules of the one-dimensional integral hold:

(1) is a *functional*, *i.e.* it works on a function and gives a (finite) number;
(2) is linear, so that the integral of a linear combination (with constant coefficients) of functions is the linear combination (with the same coefficients) of the integrals;
(3) depends in a *"continuous way"* on the domain of integration (A) and on the integrand function (f);
(4) verifies the additional property with respect to the domain of integration, *i.e.* if A is decomposed in the union of a finite number (N) of sub-domains B_i ($i = 1$, ..., N) having intersections no more than one-dimensional, the integral on A is the sum of the integrals on the sub-domains B_i, for $i = 1, \ldots, N$. In symbols:

$$A = \bigcup_{i=1}^N B_i \quad \text{such that:} \quad B_k \cap B_h = 0 \quad \text{if } k \neq h$$

$$\Downarrow$$

$$\int_A dA(x) \, f(x) = \sum_{i=1}^N \int_{B_i} dA(x) \, f(x).$$

The meaning of the statement (3) is, at least at the moment, quite unclear, even if it can be explained by means of the following qualitative explanation. The integral changes little if the domain of integration is slightly modified, *e.g.* adding or subtracting a region of small area, with respect to the one of A. Moreover, the same holds also for little changes of the integrand function f, even if the measure of these changes is a rather complicated task.

How the integral (11.15) on a two-dimensional domain can be effectively evaluated using the well-known 1d integral? First of all, a way for covering the entire A changing in a simple way the coordinates x and y of the point $x \in A$ has to be defined. To this regard, consider the simple case of the domain A shown in Fig. 11.1a: two numbers $a < b$ exist, as well as two continuous functions $f_{1,2} : (a, b) \to \mathbb{R}$ (with $f_1(x) \leq f_2(x)$, for any $a < x < b$) such that if $x = (x, y) \in A$, then $a < x < b$ and $f_1(x) < y < f_2(x)$. This integration domain is named as *"normal with respect to the axis x"* and it can be covered moving x from a to b and, in correspondence to the fixed value of x, y from $f_1(x)$ to $f_2(x)$. Accounting for this way to cover A, the two-dimensional integral (11.15) is decomposed in two successive one-dimensional ones:

$$\int_A dA(\boldsymbol{x})\, f(\boldsymbol{x}) = \int_a^b dx \int_{f_1(x)}^{f_2(x)} dy\, f(x, y), \qquad (11.16)$$

the inner integral being calculated on the interval $(f_1(x), f_2(x))$ in which moves y and the outer one on the interval (a, b) in which lies x. The integration domain shown in Fig. 11.1b is handled in an analogous way, with y in place of x and viceversa. For this domain two numbers $c < d$ exist as well as two continuous functions $g_{1,2}$: $(c, d) \to \mathbb{R}$, with $g_1(y) \le g_2(y)$ for any $y \in (c, d)$. A is said to be "*normal with respect to the y axis*" and in correspondence the integral (11.15) is decomposed in the following way:

$$\int_A dA(\boldsymbol{x})\, f(\boldsymbol{x}) = \int_c^d dy \int_{g_1(y)}^{g_2(y)} dx\, f(x, y). \qquad (11.17)$$

A more general integration domain A is decomposed in the (finite) union of sub-domains that are normal with respect to one of the two coordinate axes (as sample case, see Fig. 11.2) and the integral (11.15) is written by means of the above property (4) as the sum of integrals made on normal domains, which are evaluated by means of one of the decomposition rules (11.16), (11.17). Sample cases of calculation of two-dimensional integrals on normal domains are discussed below, whereas calculations on more complicated domains (as the one in Fig. 11.2) will be addressed in the next lesson.

Example 11.4 Consider the function $f(x, y) = 1/(3x + y)$ and integrate it on the domain (normal with respect to the x axis):

$$A = \{(x, y), 1 < x < 2, x - 3 < y < 4(2 - x)(x - 1)\}.$$

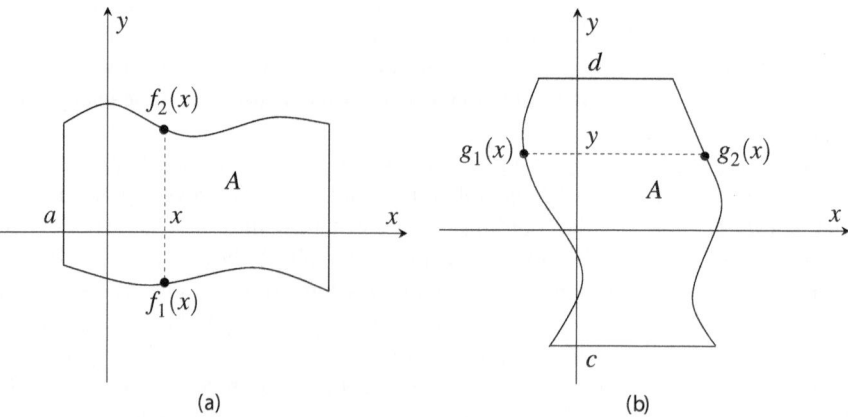

Fig. 11.1 Normal domains with respect to the coordinate axis x **a** and y **b**

Fig. 11.2 The integration domain A is decomposed in the union of sub-domains that are normal with respect to the axis x ($A_{1,3}$) and to the axis y ($A_{2,4,5,6}$)

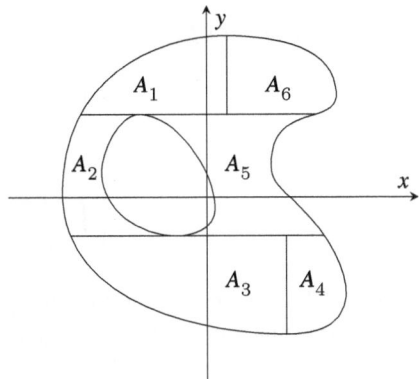

$$\int_A dA(\boldsymbol{x})\, f(\boldsymbol{x}) = \int_1^2 dx \int_{x-3}^{4(2-x)(x-1)} \frac{dy}{3x+y}$$

$$= \int_1^2 dx \left[\log(-4x^2 + 15x - 8) - \log(4x - 3) \right]$$

$$= \log 3 + 2\log 2 - \frac{5}{4}\log 5 - 1 + \frac{15 - \sqrt{97}}{8}\log\frac{\sqrt{97}-7}{\sqrt{97}+1}$$

$$+ \frac{15 + \sqrt{97}}{8}\log\frac{\sqrt{97}+7}{\sqrt{97}-1}.$$

Example 11.5 Consider the function $f(x, y) = \log(x + y)$ and integrate it on the domain (normal with respect to the x axis):

$$A = \{(x, y), 0 < x < 1, x - x^2 < y < x\}.$$

$$\int_A dA(\boldsymbol{x})\, f(\boldsymbol{x}) = \int_0^1 dx \int_{x-x^2}^x dy \,\log(x + y)$$

$$= \int_0^1 dx \left[y\log(x + y)\big|_{y=x-x^2}^{y=x} - \int_{x-x^2}^x dy \,\frac{y}{x+y} \right]$$

$$= \frac{5}{18} - \frac{\log 2}{3}.$$

Example 11.6 Consider the function $f(x, y) = x\sqrt{x^2 + y}$ and integrate it on the domain (normal with respect to the y axis):

$$A = \{(x, y), 0 < y < 1, \sqrt{y}/2 < x < \sqrt{y}\}.$$

$$\int_A dA(\boldsymbol{x})\, f(\boldsymbol{x}) = \int_0^1 dy \int_{\sqrt{y}}^{\sqrt{y}/2} dx\, x\sqrt{x^2 + y}$$

$$= \frac{1}{3}\left[\left(\frac{5}{4}\right)^{3/2} \int_0^1 dy\, y^{3/2} - 2^{3/2}\int_0^1 dy\, y^{3/2}\right]$$

$$= \frac{1}{3}\left(\frac{\sqrt{5}}{4} - \frac{4\sqrt{2}}{5}\right).$$

Example 11.7 Consider the function $f(x, y) = 1/(x + y)$ and integrate it on the domain (normal with respect to the y axis):

$$A = \{(x, y), 0 < y < 1, (1 - y)/2 < x < \sqrt{1 - y}\}$$

$$\int_A dA(\boldsymbol{x})\, f(\boldsymbol{x}) = \int_0^1 dy \int_{(1-y)/2}^{\sqrt{1-y}} \frac{dx}{x + y}$$

$$= \int_0^1 dy\, \log\left(\sqrt{1 - y} + y\right) - \int_0^1 dy\, \log(y + 1) + \log 2$$

$$= \sqrt{5}\, \log \frac{\sqrt{5} + 1}{\sqrt{5} - 1} - 1 - \log 2.$$

11.3   Suggested Readings

For an overview of topics in this lesson, see: Lax and Terrell (2017, Chap. 6, pp. 205–244), Adams and Essex (2010, Chap. 14, pp. 790–807), Apostol (1969, Chap. 11, pp. 353–404), Apostol (1974, Chap. 14, pp. 388–404).

References

R. A. Adams and C. Essex. *Calculus: Several Variables*. Pearson Education Canada, Toronto, ON, 7th edition, 2010.

T. M. Apostol. *Calculus, Volume II*. Wiley, Hoboken, NJ, 2nd edition, 1969.

T. M. Apostol. *Mathematical Analysis: A Modern Approach to Advanced Calculus*. Pearson Education US, Hoboken, NJ, 2nd edition, 1974.

P. D. Lax and M. S. Terrell. *Multivariable Calculus with Applications*. Springer, Cham, CH, 2017.

Chapter 12
Samples of Two-Dimensional Integration and Change of Variables

In this lesson more complicated samples of two-dimensional integral are evaluated. The main difficulty in handling these integrals lies in the shape of the domain of integration, which needs to be decomposed in the union of normal domains. It is shown that these calculations are strongly simplified by introducing suitable changes of the integration variables.

12.1 A Quite Complicated Integral

Consider the function:

$$f(x, y) = \frac{1}{x + y + 2R} \quad \text{with} \quad R > 0 \tag{12.1}$$

and integrate it on the hollow disk A having center on the origin, inner radius r and outer one R ($R > r$, so that $R/r =: \chi > 1$), as shown in Fig. 12.1. The function (12.1) is defined and continuous on $\mathbb{R}^2$ deprived of the straight line $y = -x - 2R$, which does not intersect the outer boundary of A. As a consequence, it is continuous on the hollow disk and its integral on that domain can be evaluated.

In this case, the domain of integration is not normal with respect to both coordinate axes, so that it must be decomposed in the union of normal domains. As it is obvious from Fig. 12.1, this partition is not unique, even if the value of the integral does not depend on this choice. In the present calculation, A is decomposed in the union of the domains $A = A_1 \cup A_2 \cup A_3 \cup A_4$ normal with respect to the y axis (see Fig. 12.1):

G. Riccardi et al., *Multidimensional Differential and Integral Calculus*, https://doi.org/10.1007/978-3-031-70326-3_12

Fig. 12.1 Decomposition of
the hollow disk with center
on the origin, inner radius r
and outer one R in the union
of four domains normal with
respect to the y axis

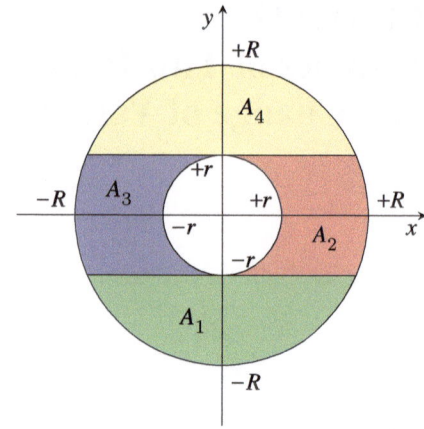

$$A_1 = \{(x, y), -R < y < -r, -\sqrt{R^2 - y^2} < x < +\sqrt{R^2 - y^2}\}$$

$$A_2 = \{(x, y), -r < y < +r, +\sqrt{r^2 - y^2} < x < +\sqrt{R^2 - y^2}\}$$

$$A_3 = \{(x, y), -r < y < +r, -\sqrt{R^2 - y^2} < x < -\sqrt{r^2 - y^2}\}$$

$$A_4 = \{(x, y), +r < y < +R, -\sqrt{R^2 - y^2} < x < +\sqrt{R^2 - y^2}\}.$$

In correspondence to the above decomposition of the domain A, the integral is
split in the sum of the following four integrals:

$$\int_A dA(\boldsymbol{x})\, f(\boldsymbol{x}) =$$

$$= \int_{-R}^{-r} dy \int_{-\sqrt{R^2-y^2}}^{+\sqrt{R^2-y^2}} \frac{dx}{x+y+2R} + \int_{-r}^{+r} dy \int_{+\sqrt{r^2-y^2}}^{+\sqrt{R^2-y^2}} \frac{dx}{x+y+2R}$$

$$+ \int_{-r}^{+r} dy \int_{-\sqrt{R^2-y^2}}^{-\sqrt{r^2-y^2}} \frac{dx}{x+y+2R} + \int_{+r}^{+R} dy \int_{-\sqrt{R^2-y^2}}^{+\sqrt{R^2-y^2}} \frac{dx}{x+y+2R}$$

$$= \int_{-R}^{-r} dy \left[\log(+\sqrt{R^2 - y^2} + y + 2R) - \log(-\sqrt{R^2 - y^2} + y + 2R) \right] \qquad (12.2)$$

$$+ \int_{-r}^{+r} dy \left[\log(+\sqrt{R^2 - y^2} + y + 2R) - \log(+\sqrt{r^2 - y^2} + y + 2R) \right]$$

$$+ \int_{-r}^{+r} dy \left[\log(-\sqrt{r^2 - y^2} + y + 2R) - \log(-\sqrt{R^2 - y^2} + y + 2R) \right]$$

$$+ \int_{+r}^{+R} dy \left[\log(+\sqrt{R^2 - y^2} + y + 2R) - \log(-\sqrt{R^2 - y^2} + y + 2R) \right].$$

The integrals (12.2) can be evaluated by means of two successive changes of variables. First of all, the integrals containing the square root of $R^2 - y^2$ are separated by the ones containing the square root of $r^2 - y^2$. In the first ones, the variable is changed in $y = R \sin \Theta$ and the angle $\Theta^\star := \arcsin(1/\chi) \in (0, \pi/2)$ is introduced. In the second integrals the variable is changed in $y = r \sin \theta$. In these new variables Θ and θ the integrals (12.2) are written as:

$$\int_A dA(x)\, f(x) =$$

$$= R \int_{-\pi/2}^{-\Theta^\star} d\sin\Theta \left[\log(+\cos\Theta + \sin\Theta + 2) - \log(-\cos\Theta + \sin\Theta + 2)\right]$$

$$+ R \int_{-\Theta^\star}^{+\Theta^\star} d\sin\Theta \left[\log(+\cos\Theta + \sin\Theta + 2) - \log(-\cos\Theta + \sin\Theta + 2)\right]$$

$$+ R \int_{+\Theta^\star}^{+\pi/2} d\sin\Theta \left[\log(+\cos\Theta + \sin\Theta + 2) - \log(-\cos\Theta + \sin\Theta + 2)\right]$$

$$- r \int_{-\pi/2}^{+\pi/2} d\sin\theta \left[\log(+\cos\theta + \sin\theta + 2\chi) - \log(-\cos\theta + \sin\theta + 2\chi)\right]$$

$$= R \int_{-\pi/2}^{+\pi/2} d\sin\Theta \left[\log(+\cos\Theta + \sin\Theta + 2) - \log(-\cos\Theta + \sin\Theta + 2)\right]$$

$$- r \int_{-\pi/2}^{+\pi/2} d\sin\theta \left[\log(+\cos\theta + \sin\theta + 2\chi) - \log(-\cos\theta + \sin\theta + 2\chi)\right].$$

Integrations by parts lead to the following form of the integrals:

$$\int_A dA(x)\, f(x) = -R \int_{-\pi/2}^{+\pi/2} d\Theta\, \sin\Theta \left(\frac{-\sin\Theta + \cos\Theta}{\cos\Theta + \sin\Theta + 2} - \frac{\sin\Theta + \cos\Theta}{-\cos\Theta + \sin\Theta + 2}\right)$$

$$+ r \int_{-\pi/2}^{+\pi/2} d\theta\, \sin\theta \left(\frac{-\sin\theta + \cos\theta}{\cos\theta + \sin\theta + 2\chi} - \frac{\sin\theta + \cos\theta}{-\cos\theta + \sin\theta + 2\chi}\right).$$

$$(12.3)$$

A second change of variables is performed in the integrals (12.3): the new variables $T = \tan(\Theta/2)$ and $t = \tan(\theta/2)$ are used in the first and second integrals, respectively. In this way, the integrals (12.3) reduce to integral of ratio in the first and second integrals, respectively. In this way, the integrals (12.3) reduce to the following integral of rational functions:

$$\int_A dA(x)\, f(x) = -4R \int_{-1}^{+1} dT\, \frac{T}{(1+T^2)^2} \left(\frac{-T^2 - 2T + 1}{T^2 + 2T + 3} + \frac{T^2 - 2T - 1}{3T^2 + 2T + 1} \right)$$

$$+ 4r \int_{-1}^{+1} dt\, \frac{t}{(1+t^2)^2} \left[\frac{-t^2 - 2t + 1}{(2\chi - 1)t^2 + 2t + (2\chi + 1)} \right.$$

$$\left. + \frac{t^2 - 2t - 1}{(2\chi + 1)t^2 + 2t + (2\chi - 1)} \right].$$

$$(12.4)$$

In order to evaluate the integrals (12.4), the rational functions must be decomposed in simple fractions. The first fraction is decomposed searching the coefficients A, $\ldots$, F in the following formula:

$$\frac{T}{(1+T^2)^2}\, \frac{-T^2 - 2T + 1}{T^2 + 2T + 3} \equiv \frac{AT + B}{T^2 + 1} + \frac{CT + D}{(T^2 + 1)^2} + \frac{ET + F}{T^2 + 2T + 3},$$

due to the fact that the quadratic polynomial $T^2 + 2T + 3$ does not have real roots. Using the polynomial identity principle the coefficients result to be $A = 1/2$, $B = -1$, $C = 0$, $D = 1$, $E = -1/2$, $F = 0$ so that the above decomposition becomes:

$$\frac{T}{(1+T^2)^2}\, \frac{-T^2 - 2T + 1}{T^2 + 2T + 3} \equiv \frac{1}{2}\frac{T - 2}{T^2 + 1} + \frac{1}{(T^2 + 1)^2} - \frac{1}{2}\frac{T}{T^2 + 2T + 3}. \quad (12.5)$$

In an analogous way the other rational functions in the integral (12.4) are written as:

$$\frac{T}{(1+T^2)^2}\, \frac{T^2 - 2T - 1}{3T^2 + 2T + 1} \equiv -\frac{1}{2}\frac{T}{T^2 + 1} - \frac{1}{(T^2 + 1)^2} + \frac{1}{2}\frac{3T + 2}{3T^2 + 2T + 1}$$

$$\frac{t}{(1+t^2)^2}\, \frac{-t^2 - 2t + 1}{(2\chi - 1)t^2 + 2t + (2\chi + 1)} \equiv \frac{1}{2}\frac{\chi t - (\chi + 1)}{t^2 + 1} + \frac{1}{(t^2 + 1)^2}$$

$$+ \frac{1}{2}\frac{-\chi(2\chi - 1)t + (2\chi + 1)(\chi - 1)}{(2\chi - 1)t^2 + 2t + (2\chi + 1)}$$

$$\frac{t}{(1+t^2)^2}\, \frac{t^2 - 2t - 1}{(2\chi + 1)t^2 + 2t + (2\chi - 1)} \equiv -\frac{1}{2}\frac{\chi t + (\chi - 1)}{t^2 + 1} - \frac{1}{(t^2 + 1)^2}$$

$$+ \frac{1}{2}\frac{\chi(2\chi + 1)t + (2\chi - 1)(\chi + 1)}{(2\chi + 1)t^2 + 2t + (2\chi - 1)}.$$

$$(12.6)$$

The integrals of the rational functions at the right hand sides of the decompositions (12.5) and (12.6) are evaluated without difficulties:

$$\int_{-1}^{+1} \frac{dt}{t^2+1} = \frac{\pi}{2}, \qquad \int_{-1}^{+1} dt \, \frac{t}{t^2+1} = 0,$$

$$\int_{-1}^{+1} \frac{dt}{t^2+2t+3} = \frac{\arctan(\sqrt{2})}{\sqrt{2}},$$

$$\int_{-1}^{+1} dt \, \frac{t}{t^2+2t+3} = \frac{1}{2}\left[\log 3 - \sqrt{2}\arctan(\sqrt{2})\right], \tag{12.7}$$

$$\int_{-1}^{+1} \frac{dt}{3t^2+2t+1} = \frac{\arctan(2\sqrt{2})+\arctan(\sqrt{2})}{\sqrt{2}},$$

$$\int_{-1}^{+1} dt \, \frac{t}{3t^2+2t+1} = \frac{1}{6}\left\{\log 3 - \sqrt{2}[\arctan(2\sqrt{2})+\arctan(\sqrt{2})]\right\},$$

$$\int_{-1}^{+1} \frac{dt}{(2\chi-1)t^2+2t+(2\chi+1)} = \frac{g(\chi)+g_-(\chi)}{\sqrt{4\chi^2-2}},$$

$$\int_{-1}^{+1} dt \, \frac{t}{(2\chi-1)t^2+2t+(2\chi+1)} = \frac{1}{2(2\chi-1)}\left\{\log\frac{2\chi+1}{2\chi-1} - \frac{2[g(\chi)+g_-(\chi)]}{\sqrt{4\chi^2-2}}\right\},$$

$$\int_{-1}^{+1} \frac{dt}{(2\chi+1)t^2+2t+(2\chi-1)} = \frac{g_+(\chi)+g(\chi)}{\sqrt{4\chi^2-2}}, \tag{12.8}$$

$$\int_{-1}^{+1} dt \, \frac{t}{(2\chi+1)t^2+2t+(2\chi-1)} = \frac{1}{2(2\chi+1)}\left\{\log\frac{2\chi+1}{2\chi-1} - \frac{2[g_+(\chi)+g(\chi)]}{\sqrt{4\chi^2-2}}\right\},$$

having used the following definitions:

$$g(\chi) := \arctan\left(\frac{2\chi}{\sqrt{4\chi^2-2}}\right), \qquad g_\pm(\chi) := \arctan\left(\frac{2\chi\pm2}{\sqrt{4\chi^2-2}}\right),$$

for the sake of shortness. Inserting the integrals (12.7) and (12.8) inside the form (12.4) of the integral on the hollow disk, it becomes:

$$\frac{1}{r}\int_A dA(\boldsymbol{x})\, f(\boldsymbol{x}) = \sqrt{4\chi^2-2}\left[g_-(\chi)+2g(\chi)+g_+(\chi)\right]+ \tag{12.9}$$
$$-\sqrt{2}\chi\left[2\arctan\sqrt{2}+\arctan(2\sqrt{2})\right].$$

On this result, several considerations have to be carried out.

First of all, it is shown that $g_- + g \in (0, \pi/2)$. Indeed, both the functions g_- and g belong to the above interval and their sum does not exceed $\pi/2$, due to the fact that its tangent is positive:

$$\tan\left[g_-(\chi)+g(\chi)\right] = \frac{\tan[g(\chi)]+\tan[g_-(\chi)]}{1-\tan[g(\chi)]\tan[g_-(\chi)]} = \sqrt{4\chi^2-2}.$$

From the above relation, it follows that:

$$g_-(\chi) + g(\chi) = \arctan\left[\sqrt{4\chi^2 - 2}\right]. \tag{12.10}$$

A similar discussion implies that $g + g_+ \in (\pi/2, \pi)$, both functions belonging to this interval and the tangent of their sum being negative:

$$\tan\left[g(\chi) + g_+(\chi)\right] = \frac{\tan[g(\chi)] + \tan[g_+(\chi)]}{1 - \tan[g(\chi)]\tan[g_+(\chi)]} = -\sqrt{4\chi^2 - 2}.$$

As before, the above relation leads to the shorter form of the sum $g + g_+$:

$$g(\chi) + g_+(\chi) = -\arctan\left[\sqrt{4\chi^2 - 2}\right] + \pi. \tag{12.11}$$

Summing sides by sides the relations (12.10), (12.11) the unexpected result: $g_- + 2g + g_+ \equiv \pi$ follows. With the same approach, it can be also shown that: $2\arctan(\sqrt{2}) + \arctan(2\sqrt{2}) = \pi$. It follows that the result (12.9) can be simplified in the following one:

$$\int_A dA(\boldsymbol{x})\, f(\boldsymbol{x}) = \sqrt{2}\pi r \left(\sqrt{2\chi^2 - 1} - \chi\right). \tag{12.12}$$

12.2   Change of Coordinates Inside a Double Integral

The above sample calculation shows the complexity of the evaluation of a double integral on a domain that is not elementary. But if it is possible to transform the domain of integration in a simpler one changing the coordinates inside the integral, its calculation becomes much more easy. For example, coming back to the previous calculation, if the cartesian coordinates are changed in the polar ones (ρ, θ), the hollow disk is transformed in the rectangle $(r, R) \times [0, 2\pi)$ and the integral becomes elementary. But how the integral modifies changing the coordinates?

Consider a one-to-one map with continuous first derivatives which transforms a domain B lying on a cartesian plane (ξ, η) in the domain A of the plane (x, y):

$$\begin{cases} x = x(\xi, \eta) \\ y = y(\xi, \eta), \end{cases} \quad \text{or in vector form:} \quad \boldsymbol{x} = \boldsymbol{x}(\boldsymbol{\xi}). \tag{12.13}$$

In order to understand the effect of the above coordinate change on the integral, a pointwise analysis is appropriate: choose a generic point $\boldsymbol{\xi}_0 \in B$ mapped to the point $\boldsymbol{x}_0 \in A$ by the functions (12.13). In $\boldsymbol{\xi}_0$ the curves $C_1 : \xi \equiv \xi_0$ and $C_2 : \eta \equiv \eta_0$ are considered: they are straight lines parallel to the coordinate axes. The curves $C_{1,2}$ are transformed by the map (12.13) in other two curves in the plane (x, y), both passing through the point $\boldsymbol{x}_0$. They will be named $\mathscr{C}_{1,2}$ hereafter. They are parametrised in η

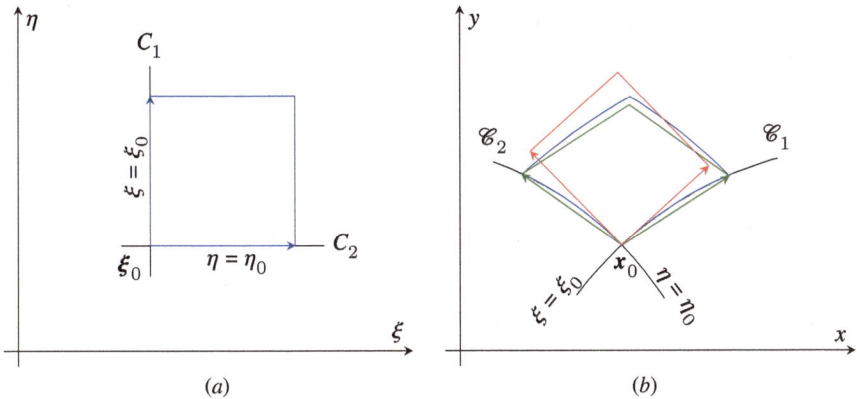

Fig. 12.2 The plane (ξ, η) with the point $\boldsymbol{\xi}_0$ and the rectangle formed by the increments of its coordinates (ξ_0, η_0) (blue line) are drawn in (**a**). In (**b**) the plane (x, y) with the transformed point $\boldsymbol{x}_0$ and the deformed rectangle (blue line) obtained applying the map (12.13) to the rectangle in (**a**) are drawn. Moreover, the parallelograms that have their sides parallel to the increment vectors $\boldsymbol{v}_1$, $\boldsymbol{v}_2$ (green) and to the tangent vectors $\partial_\xi \boldsymbol{x}$, $\partial_\eta \boldsymbol{x}$ (red) are also drawn

and in ξ, respectively, and their equations follow from the map (12.13):

$$\mathscr{C}_1 : \boldsymbol{x} = \boldsymbol{x}(\xi, \eta_0) , \qquad \mathscr{C}_2 : \boldsymbol{x} = \boldsymbol{x}(\xi_0, \eta) . \tag{12.14}$$

The coordinates of the point $\boldsymbol{\xi}_0$ are now increased by two small and positive quantities $\Delta \xi$ and $\Delta \eta$, as shown in Fig. 12.2a. The area of the resulting rectangle is $\Delta \xi \, \Delta \eta =: d A(\boldsymbol{\xi})$. Mapping this rectangle to the plane (x, y), through the coordinate change (12.13), the deformed rectangle shown in Fig. 12.2b is obtained. Its area is now evaluated.

When a point moves on the curve C_1 from ξ_0 to $\xi_0 + \Delta \xi$ the position of the transformed point $\boldsymbol{x}(\boldsymbol{\xi}) \in \mathscr{C}_1$ (12.13) undergoes the vector increment:

$$\boldsymbol{v}_1 = \boldsymbol{x}(\xi_0 + \Delta \xi, \eta_0) - \boldsymbol{x}(\xi_0, \eta_0) = \partial_\xi \boldsymbol{x}(\boldsymbol{\xi}_0)\Delta \xi + O(\Delta \xi^2) ,$$

as well as in correspondence to the motion of a point on the curve C_2 from η_0 to $\eta_0 + \Delta \eta_0$, the transformed point shifts on $\mathscr{C}_2$ of the vector:

$$\boldsymbol{v}_2 = \boldsymbol{x}(\xi_0, \eta_0 + \Delta \eta) - \boldsymbol{x}(\xi_0, \eta_0) = \partial_\eta \boldsymbol{x}(\boldsymbol{\xi}_0)\Delta \eta + O(\Delta \eta^2) .$$

The area of the deformed rectangle is then given by the modulus of the vector product $\boldsymbol{v}_1 \times \boldsymbol{v}_2$, unless terms of order higher than $\Delta \xi \Delta \eta$:

$$d A(\boldsymbol{x}_0) = |\boldsymbol{v}_1 \times \boldsymbol{v}_2| + o(\Delta \xi \Delta \eta)$$
$$= |\partial_\xi \boldsymbol{x}(\boldsymbol{\xi}_0) \times \partial_\eta \boldsymbol{x}(\boldsymbol{\xi}_0)| \, \Delta \xi \Delta \eta + o(\Delta \xi \Delta \eta) . \tag{12.15}$$

In the above formula, the modulus of the vector $\partial_\xi x(\xi_0) \times \partial_\eta x(\xi_0)$, is given by the modulus of the determinant J:

$$\begin{vmatrix} \partial_\xi x(\xi) & \partial_\eta x(\xi) \\ \partial_\xi y(\xi) & \partial_\eta y(\xi) \end{vmatrix} =: \frac{\partial(x, y)}{\partial(\xi, \eta)} =: J(\xi) . \tag{12.16}$$

evaluated in ξ_0. The determinant (12.16) is called "*jacobian*" of the change of coordinates (12.13). This map being one-to-one, it is non-vanishing everywhere, so that the two vectors $\partial_\xi x$ and $\partial_\eta x$ are always a *local* basis on the plane (x, y). The definition (12.16) is inserted into the area element (12.15), that becomes:

$$dA(x_0) = |J(\xi_0)| \, dA(\xi_0) + o(\Delta\xi\Delta\eta) . \tag{12.17}$$

Considering that, when the norm of the decomposition vanishes, terms of order higher than $\Delta\xi\Delta\eta$ give a vanishing contribution to the integral sums, the result (12.17) leads to the following law for changing the area element inside a double integral:

$$dA(x_0) = |J(\xi_0)| \, dA(\xi_0) \tag{12.18}$$

The integral on the domain A of the plane (x, y) can be transformed by means of the map (12.13) in the integral on the domain B of the plane (ξ, η) in the following way:

$$\int_A dA(x) \, f(x) = \int_B dA(\xi) \, |J(\xi)| f[x(\xi)] , \tag{12.19}$$

having used the law of changing of the area element (12.18).

As discussed before, for the integral on the hollow disk of the function (12.1) it is particularly convenient the polar change of coordinates:

$$\begin{cases} x = \rho\cos\theta \\ y = \rho\sin\theta , \end{cases} \tag{12.20}$$

where $r \leq \rho \leq R$ and $0 \leq \theta < 2\pi$. The jacobian determinant (12.16) is specified as:

$$J(\rho, \theta) = \begin{vmatrix} \cos\theta & -\rho\sin\theta \\ \sin\theta & \rho\cos\theta \end{vmatrix} \equiv \rho , \tag{12.21}$$

so that the area element (12.18) results to be $dA(x) = \rho \, d\rho \, d\theta$. The transformed set B reduces in that case to the rectangle $[r, R] \times [0, 2\pi)$, which will be considered as normal with respect to the axis ρ. By means of the change of coordinates (12.20) and of the related jacobian determinant (12.21), the integral becomes:

$$\int_A dA(x) \, f(x) = \int_r^R d\rho \, \rho \int_0^{2\pi} \frac{d\theta}{\rho(\cos\theta + \sin\theta) + 2R}$$

$$= R \int_{1/\chi}^{1} d\mu\, \mu \int_{-\pi}^{+\pi} \frac{d\theta}{\mu(\cos\theta + \sin\theta) + 2}, \qquad (12.22)$$

having introduced the new variable $\mu := \rho/R$. The integral in θ is evaluated by means of the change of variable $t = \tan(\theta/2)$ (this is the reason for integrating on $(-\pi, +\pi)$):

$$\int_{-\pi}^{+\pi} \frac{d\theta}{\mu(\cos\theta + \sin\theta) + 2} = 2 \int_{-\infty}^{+\infty} \frac{dt}{(2 - \mu)t^2 + 2\mu t + (2 + \mu)}.$$

On the latter integral the new change of variable:

$$t = \frac{\sqrt{2(2 - \mu^2)}}{2 - \mu}\, \eta - \frac{\mu}{2 - \mu},$$

is performed, so that it is evaluated as:

$$\int_{-\pi}^{+\pi} \frac{d\theta}{\mu(\cos\theta + \sin\theta) + 2} = \frac{2\pi}{\sqrt{2(2 - \mu)}}.$$

The external integral in μ:

$$\int_A dA(\boldsymbol{x})\, f(\boldsymbol{x}) = \sqrt{2}\pi R \int_{1/\chi}^{1} d\mu\, \frac{\mu}{\sqrt{2 - \mu}}$$

is evaluated by the change of variable: $\mu = \sqrt{2} \sin\omega$ with $\arcsin[1/(\sqrt{2}\chi)] < \omega < \arcsin(1/\sqrt{2})$. It still follows the result (12.12). As a final remark, notice the important simplification induced by the change of coordinates (12.20). In the next lesson, other sample integrals will be evaluated changing their coordinates.

12.3 Suggested Readings

For a complement in the study of the present lesson, may be useful to explore Lax and Terrell (2017, Chap. 6, pp. 245–260) and Adams and Essex (2010, Chap. 14, pp. 808–817).

References

R. A. Adams and C. Essex. *Calculus: Several Variables*. Pearson Education Canada, Toronto, ON, 7th edition, 2010.

P. D. Lax and M. S. Terrell. *Multivariable Calculus with Applications*. Springer, Cham, CH, 2017.

Chapter 13
Two-Dimensional Integration and Area of a Surface

In the present lesson, several examples of two-dimensional integrals are evaluated adopting suitable changes of coordinates. Moreover, the calculus of the area of a (finite) surface is addressed, as a first application of the double integral.

13.1 Examples of Calculus of Two-Dimensional Integrals by Means of the Change of Variables

The evaluation of double integrals by means of the change of coordinates is discussed in sample cases.

Example 13.1 Consider the function:

$$f(x, y) = x^2 y^2 \tag{13.1}$$

and integrate it inside the ellipse A, having semi-axes of lengths a (along x) and b (along y). Assume that $a > b$, without losing generality. Introduced the focal semi-distance $c = (a^2 - b^2)^{1/2}$, the new variables ρ and θ are considered, being related to the cartesian coordinates x and y through the following equations:

$$\begin{cases} x = c \cosh \rho \cos \theta \\ y = c \sinh \rho \sin \theta, \end{cases} \tag{13.2}$$

with $\rho \geq 0$ and $\theta \in [0, 2\pi)$. For $\rho = 0$ the point $\boldsymbol{x}(0, \theta)$ runs twice on the focal segment, starting from the focus at $+c$ ($\theta = 0$) and moving along the decreasing x, until it reaches the focus at $-c$ for $\theta = \pi$. For still increasing θ, the point $\boldsymbol{x}(0, \theta)$ moves along the increasing x and comes back to the focus at $+c$ for $\theta = 2\pi$. If

© The Author(s), under exclusive license to Springer Nature Switzerland AG 2024
G. Riccardi et al., *Multidimensional Differential and Integral Calculus*,
https://doi.org/10.1007/978-3-031-70326-3_13

positive values of ρ are considered, the point $x(\rho, \theta)$ runs on confocal ellipses having semi-axes of lengths $c \cosh \rho$ and $c \sinh \rho$. In correspondence to the value $\overline{\rho} = 1/2 \cdot \log[(a+b)/(a-b)]$, the point $x(\overline{\rho}, \theta)$ moves on the ellipse A. As a consequence, the elliptical domain A is transformed by the inverse of the change of coordinates (13.2) in the rectangle $[0, \overline{\rho}) \times [0, 2\pi)$ of the (ρ, θ)-plane. In Fig. 13.1a the image through the map (13.2) of a cartesian regular grid in $[0, \overline{\rho}) \times [0, 2\pi)$ is drawn in the (x, y)-plane with red lines. The surface $z = f(x, y)$ with $(x, y) \in A$ is drawn in Fig. 13.1b.

The jacobian determinant of the change of coordinates (13.2) is:

$$J = \begin{vmatrix} c \sinh \rho \cos \theta & -c \cosh \rho \sin \theta \\ c \cosh \rho \sin \theta & c \sinh \rho \cos \theta \end{vmatrix} = c^2(\cosh^2 \rho - \cos^2 \theta) = c^2(\sinh^2 \rho + \sin^2 \theta)$$

that is used in evaluating the integral[1]:

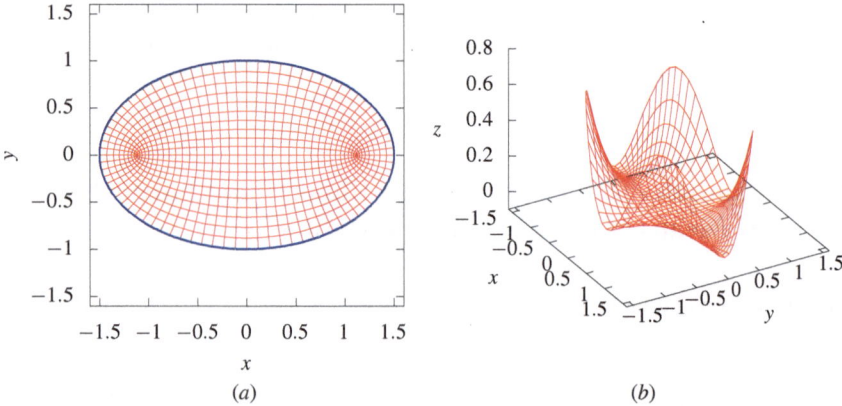

$$(a) \qquad\qquad\qquad\qquad (b)$$

Fig. 13.1 In **a** the image of a cartesian, regular grid 10×80 in $[0, \overline{\rho}) \times [0, 2\pi)$ is drawn with red lines. The boundary ∂A ($a = 1.5$, $b = 1$) is drawn with a blue line. In **b** the surface $z = x^2y^2$ for (x, y) belonging to the grid b inside the ellipse A is drawn with red lines

---

[1] The following results ($n \geq 2$):

$$I_n^c(x) := \int dx \cos^n x = +\frac{1}{n} \sin x \cos^{n-1} x + \frac{n-1}{n} I_{n-2}^c$$

$$I_n^s(x) := \int dx \sin^n x = -\frac{1}{n} \sin^{n-1} x \cos x + \frac{n-1}{n} I_{n-2}^s$$

$$J_n^c(x) := \int dx \cosh^n x = +\frac{1}{n} \sinh x \cosh^{n-1} x + \frac{n-1}{n} J_{n-2}^c$$

$$J_n^s(x) := \int dx \sinh^n x = +\frac{1}{n} \sinh^{n-1} x \cosh x - \frac{n-1}{n} J_{n-2}^s$$

$$\int_A dA(\boldsymbol{x})\, x^2 y^2 = c^6 \left[\int_0^{\overline{\rho}} d\rho\, \cosh^2 \rho\, \sinh^4 \rho \int_0^{2\pi} d\theta\, \cos^2 \theta\, \sin^2 \theta \right.$$

$$\left. + \int_0^{\overline{\rho}} d\rho\, \cosh^2 \rho\, \sinh^2 \rho \int_0^{2\pi} d\theta\, \cos^2 \theta\, \sin^4 \theta \right]$$

$$= c^6 \left\{ \frac{1}{48} \left[\sinh \overline{\rho} \cosh \overline{\rho} \left(8 \sinh^4 \overline{\rho} + 2 \sinh^2 \overline{\rho} - 3 \right) + 3\overline{\rho} \right] \frac{3}{4} \pi \right.$$

$$\left. + \frac{1}{8} \left[\sinh \overline{\rho} \cosh \overline{\rho} \left(1 + 2 \sinh^2 \overline{\rho} \right) - \overline{\rho} \right] \frac{1}{8} \pi \right\}$$

$$= \frac{c^6}{32} \left[\sinh \overline{\rho} \cosh \overline{\rho} \left(4 \sinh^4 \overline{\rho} + 2 \sinh^2 \overline{\rho} - 1 \right) + \overline{\rho} \right]. \qquad (13.3)$$

Example 13.2 Evaluate the integral:

$$\int_A dA(\boldsymbol{x})\, (x^2 - y^2), \qquad (13.4)$$

the domain A being shown in Fig. 13.2. It lies between the two arcs of spiral:

$$\begin{cases} x = a\, (\theta + \theta_0) \cos \theta \\ y = a\, (\theta + \theta_0) \sin \theta \end{cases} \quad \text{and:} \quad \begin{cases} x = b\, (\theta + \theta_0) \cos \theta \\ y = b\, (\theta + \theta_0) \sin \theta \end{cases} \qquad (13.5)$$

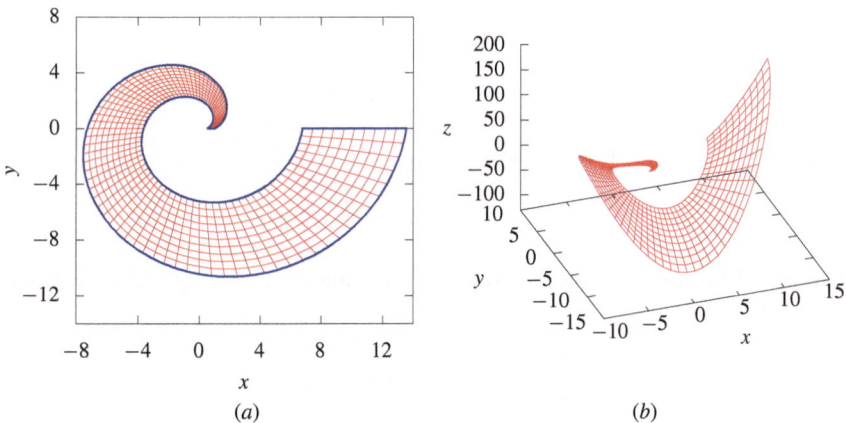

(a)   (b)

Fig. 13.2 For the following values of the parameters in the change of coordinates (13.6): $a = 1$, $b = 2$, $\theta_0 = 1/2$, in **a** the image (red lines) of a cartesian regular grid 10×80 in the (χ, θ)-plane is drawn. The curve ∂A is also drawn with a blue line. The surface $z = x^2 - y^2$ for $(x, y) \in A$ is drawn in **b** with red lines

will be used. In the above formulae, the integrals for $n = 0$ are $I_0^c = I_0^s = J_0^c = J_0^s = x$, whereas the ones for $n = 1$ are $I_1^c = \sin x$, $I_1^s = -\cos x$, $J_1^c = \sinh x$ and $J_1^s = \cosh x$, unless additive constants.

with $0 \le \theta \le 2\pi$, $a < b$ and $b\theta_0 < a(2\pi + \theta_0)$. The shape of the domain A is very complicated: it implies that the evaluation of the integral (13.4) in cartesian coordinates results to be very difficult. However, it becomes simple in the coordinate system (χ, θ) with $0 \le \chi \le 1$ and:

$$\begin{cases} x = \rho(\chi)\,(\theta + \theta_0)\,\cos\theta \\ y = \rho(\chi)\,(\theta + \theta_0)\,\sin\theta\,. \end{cases} \tag{13.6}$$

The simplest choice of the function $\rho(\chi)$ will be adopted here, that is the linear one: $\rho(\chi) = a + (b - a)\chi$. In Fig. 13.2a the image in the (x, y)-plane of a cartesian regular grid in the (χ, θ)-plane is drawn and in Fig. 13.2b the surface $z = x^2 - y^2$ for $(x, y) \in A$ is shown. The jacobian determinant of the change of coordinates (13.6) is evaluated as:

$$\begin{aligned}
J &= \begin{vmatrix} (b-a)(\theta + \theta_0)\cos\theta & \rho(\chi)[\cos\theta - (\theta + \theta_0)\sin\theta] \\ (b-a)(\theta + \theta_0)\sin\theta & \rho(\chi)[\sin\theta + (\theta + \theta_0)\cos\theta] \end{vmatrix} \\
&= (b-a)(\theta + \theta_0)\rho(\chi)\begin{vmatrix} \cos\theta & \cos\theta - (\theta + \theta_0)\sin\theta \\ \sin\theta & \sin\theta + (\theta + \theta_0)\cos\theta \end{vmatrix} \\
&= (b-a)(\theta + \theta_0)^2\rho(\chi)\,.
\end{aligned} \tag{13.7}$$

Using the change of coordinates (13.6) and the jacobian determinant (13.7), the integral (13.4) becomes:

$$\begin{aligned}
\int_A dA(x)\,(x^2 - y^2) &= \int_0^1 d\chi \int_0^{2\pi} d\theta\,(b-a)\rho(\chi)(\theta + \theta_0)^2 \times \\
&\qquad \times [\rho^2(\chi)(\theta + \theta_0)^2\cos^2\theta - \rho^2(\chi)(\theta + \theta_0)^2\sin^2\theta] \\
&= (b-a)\int_0^1 d\chi\,\rho^3(\chi)\int_0^{2\pi} d\theta\,(\theta + \theta_0)^4\cos 2\theta \\
&= \frac{\pi}{4}\,(b-a)\,(a^3 + a^2b + ab^2 + b^3)\,(8\pi^2 + 12\pi\theta_0 + 6\theta_0^2 - 3)\,.
\end{aligned} \tag{13.8}$$

Example 13.3 Consider the integral of the function $f(x, y) = xy^2$ calculated inside the cardioid A:

$$\begin{cases} x = [(1 + \cos\theta)\cos\theta]/2 \\ y = [(1 + \cos\theta)\sin\theta]/2\,, \end{cases}$$

with $0 \le \theta < 2\pi$. The new coordinate system (χ, θ) is built in the following way:

$$\begin{cases} x = [\chi(1 + \cos\theta)\cos\theta]/2 \\ y = [\chi(1 + \cos\theta)\sin\theta]/2\,. \end{cases} \tag{13.9}$$

It maps the rectangle $(0, 1) \times [0, 2\pi)$ on the domain of integration A.
The jacobian determinant of the change of coordinates (13.9) is written as:

$$J = \begin{vmatrix} [(1 + \cos\theta)\cos\theta]/2 & -\chi\sin\theta(1 + 2\cos\theta)/2 \\ [(1 + \cos\theta)\sin\theta]/2 & \chi(\cos\theta + \cos^2\theta - \sin^2\theta)/2 \end{vmatrix}$$

$$= \frac{\chi}{4}(1 + \cos\theta) \begin{vmatrix} \cos\theta & -\sin\theta(1 + 2\cos\theta) \\ \sin\theta & \cos\theta + \cos^2\theta - \sin^2\theta \end{vmatrix}$$

$$= \frac{\chi}{4}(1 + \cos\theta)^2 . \tag{13.10}$$

The integral is calculated as:

$$\int_A dA(x)\,xy^2 =$$

$$= \int_0^1 d\chi \int_0^{2\pi} d\theta \underbrace{\frac{\chi}{4}(1 + \cos\theta)^2}_{J} \underbrace{\frac{\chi}{2}(1 + \cos\theta)\cos\theta}_{x} \underbrace{\frac{\chi^2}{4}(1 + \cos\theta)^2\sin^2\theta}_{y^2}$$

$$= \frac{1}{32} \int_0^1 d\chi\,\chi^4 \int_0^{2\pi} d\theta\,(\cos\theta - \cos^2\theta)(1 + \cos\theta)^6$$

$$= \frac{33}{2048}\pi \simeq 0.05062 . \tag{13.11}$$

13.2 Evaluation of the Surface Area

The calculation of two-dimensional integrals is now applied for evaluating the area of a surface Σ. To this aim, it is given in the following parametric form:

$$\begin{cases} x = x(u, v) \\ y = y(u, v) \\ z = z(u, v) \end{cases} \quad \text{or in vector form:} \quad x = x(u), \tag{13.12}$$

the point in the parameter space $(u, v) = u$ belonging to an open set A of $\mathbb{R}^2$. Note that the vector form of the parametric representation (13.12) gives a first example of a vector function from $\mathbb{R}^2$ to $\mathbb{R}^3$. It is assumed sufficiently smooth: its components x, y and z are partially differentiable with continuous first derivatives. Moreover, the tangent vectors $\partial_u x$ and $\partial_v x$ (to the curves $v \equiv$ const. and $u \equiv$ const. on Σ, respectively) are not parallel, *i.e.* their vector product everywhere satisfies the request:

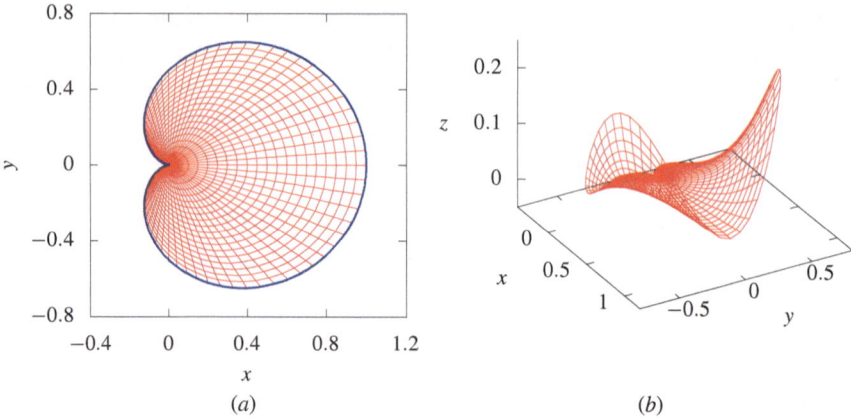

Fig. 13.3 In **a** the image in the (x, y)-plane of a cartesian regular grid 20×80 in the rectangle $(0, 1) \times [0, 2\pi)$ of the (χ, θ)-plane is drawn with red lines. The curve ∂A is also drawn with a blue line. In **b** the surface $z = xy^2$ with $(x, y) \in A$ is drawn with red lines

$$\partial_u x \times \partial_v x \neq 0 . \tag{13.13}$$

The condition (13.13) implies that the vectors $\partial_u x$ and $\partial_v x$ form a *local basis* in the plane which is tangent to Σ at its point x_0. As a consequence, their vector product defines a direction normal to the surface at the same point (Fig. 13.3).

Consider now the points $u_0 \in A$ and $x(u_0) =: x_0 \in \Sigma$ and two *positive* increments $\Delta u, \Delta v$ such that the points $(u_0 + \Delta u, v_0), (u_0, v_0 + \Delta v)$ and $(u_0 + \Delta u, v_0 + \Delta v)$ still belong to A. Noted that to the area element $\Delta u \Delta v$ in the parameter space corresponds the area $|\partial_u x(u_0)\Delta u \times \partial_v x(u_0)\Delta v| + o(\Delta u \Delta v)$ on the surface Σ (remember that $o(\Delta u \Delta v)$ stays for terms which go to 0 faster than $\Delta u \Delta v$, *i.e.* $o(\Delta u \Delta v)/(\Delta u \Delta v) \to 0$ as Δu and Δv vanish), it follows that the area element $dA(u_0) = \Delta u \Delta v = du dv$ in the parameter space generates the area element:

$$dA(x_0) = |\partial_u x(u_0) \times \partial_v x(u_0)| \, dA(u_0) \tag{13.14}$$

on the surface Σ. The modulus of the vector product is often evaluated defining the functions:

$$E(u) := |\partial_u x(u)|^2, \quad G(u) := |\partial_v x(u)|^2, \quad F(u) := \partial_u x(u) \cdot \partial_v x(u) \tag{13.15}$$

by means of the formula:

$$|\partial_u x(u_0) \times \partial_v x(u_0)| = \sqrt{E(u_0)G(u_0) - F^2(u_0)} . \tag{13.16}$$

Examples of area calculations are given below.

Example 13.4 The area of a two-bases spherical cap is evaluated, in the hypothesis that the bases lie on parallel planes. The sphere is assumed with its center at the origin, its radius is named R and the above planes are described by the equations $z = a$ and $z = b$ (with $-R < a < b < +R$), without losing generality. A point x on the sphere is identified by the parameters: *latitude* $\varphi \in [-\pi/2, +\pi/2]$ (angle between x and the equatorial plane $z = 0$) and *longitude* $\theta \in [0, 2\pi)$ (angle formed by the plane through the z-axis and the point x with the plane $y = 0$). With these parameters the functions (13.15) are calculated as: $E \equiv R^2 \cos^2 \varphi$, $G \equiv R^2$ and $F \equiv 0$. Moreover, if x belongs to the spherical cap, its latitude lies between $\arcsin(a/R)$ and $\arcsin(b/R)$. The area of the spherical cap is now evaluated as:

$$|\Sigma| = \int_{\arcsin(a/R)}^{\arcsin(b/R)} d\varphi \int_0^{2\pi} d\theta \, R^2 \cos \varphi = 2\pi R(b - a).$$

It depends on the sphere radius and on the distance between the planes, only. The same does not occur in the next example.

Example 13.5 The area of a two-bases (elliptic) paraboloid cap is now evaluated. The paraboloid is given in the explicit form: $z = x^2 + y^2$ and the two planes are assumed parallel to the (x, y)-plane and given by the equations: $z = a$ and $z = b$, with $0 < a < b$. The orthogonal projection of this surface on the (x, y)-plane is the circular ring with center at the origin and radii $r = \sqrt{a}$ and $R = \sqrt{b} > r$. An interior point u of the circular ring is identified by means of the polar coordinates $\rho \in (r, R)$ and $\theta \in [0, 2\pi)$, so that the following parametric representation of the surface:

$$x(\rho, \theta) = \begin{pmatrix} \rho \cos \theta \\ \rho \sin \theta \\ \rho^2 \end{pmatrix}$$

is obtained. The functions (13.15) are calulated as: $E \equiv 1 + 4\rho^2$, $G \equiv \rho^2$ and $F \equiv 0$. It follows the area of the paraboloid cap:

$$|\Sigma| = 2\pi \int_r^R d\rho \, \rho \sqrt{1 + 4\rho^2} = \frac{\pi}{6} \left[(1 + 4b)^{3/2} - (1 + 4a)^{3/2} \right]. \qquad (13.17)$$

Example 13.6 The evaluation of the area of a surface often leads to double integrals that cannot be analytically calculated. In these cases, a numerical integration is also required. A sample case is obtained modifying the previous example. The two planes π and Π are now assumed not parallel. They still cross the points $(0, 0, a)$ and $(0, 0, b)$ (with $0 < a < b$), but the *unit* vectors normal to these planes (n and N, respectively) are not parallel to the z-axis. The planes are described by the following equations:

$$\pi : \quad x n_x + y n_y + (z - a) n_z = 0$$
$$\Pi : \quad x N_x + y N_y + (z - b) N_z = 0. \qquad (13.18)$$

Hereafter, is will be assumed that n_z and N_z are positive, without losing generality.

The projections on the plane $z = 0$ of the curves along which the planes (13.18) intersect the paraboloid $z = x^2 + y^2$ are two circles having centers and radii given by the following equations:

$$
x_c = -\frac{n_x}{2n_z}, \quad y_c = -\frac{n_y}{2n_z}, \quad r = \left[a + \frac{1}{4}\left(\frac{1}{n_z^2} - 1\right)\right]^{1/2},
$$

$$
X_c = -\frac{N_x}{2N_z}, \quad Y_c = -\frac{N_y}{2N_z}, \quad R = \left[b + \frac{1}{4}\left(\frac{1}{N_z^2} - 1\right)\right]^{1/2}.
$$

(13.19)

Finally, it is assumed that the two circles (13.19) do not meet, *i.e.* the distance d between the two centers is always smaller than the difference of the radii:

$$
d := [(x_c - X_c)^2 + (y_c - Y_c)^2]^{1/2} < R - r. \tag{13.20}
$$

Inserting the values (13.19), this condition is written as:

$$
4n_z N_z \left(\frac{a+b}{2} - rR\right) + (n_x N_x + n_y N_y) > 0, \tag{13.21}
$$

which is assumed always verified below (for example, the constraint (13.20) is satisfied for $n_x = n_y = 0$, or for $N_x = N_y = 0$).

A suitable parametrization of the surface Σ is now built, by means of a coordinate change on the (x, y)-plane which leads to a simple description of the set B given by the intersection between the inside of the circle of radius R and center at X_c and the outside of the circle of radius r and center at x_c. The new parameter $\chi \in [0, 1]$ is introduced, for specifying the position of the center (ξ, belonging to the straight line from x_c to X_c) and the radius (ρ) of a new circle $\mathscr{C}(\chi)$:

$$
\xi = X_c + \chi(x_c - X_c), \quad \rho = R - (R - r)\chi. \tag{13.22}
$$

With the parametrization (13.22), the circle $\mathscr{C}(0)$ lies on the bigger circle, whereas $\mathscr{C}(1)$ lies on the smaller one. For $0 < \chi < 1$ the circle $\mathscr{C}(\chi)$ lies inside B. Indeed, consider a reference axis passing through the centers X_c and x_c, with its origin superimposed to X_c: the points in which the $\mathscr{C}(\chi)$ intersects the axis are in $\chi d \mp \rho$, these abscissae satisfying the inequalities $-R < \chi d - \rho < d - r$ and $d + r < \chi d + \rho < R$.

The surface Σ lies on the paraboloid $z = x^2 + y^2$ between the planes π and Π: a sample is shown in Fig. 13.4. Named as θ_0 the argument of the vector $x_c - X_c$ joining the centers, the following parametrization of Σ is adopted:

Fig. 13.4 Surface Σ on the paraboloid $z = x^2 + y^2$ between the planes π ($a = 1$) and Π ($b = 3$) having normal unit vectors of the form $(\cos\varphi\cos\alpha, \cos\varphi\sin\alpha, \sin\varphi)$ with $\varphi = 80°, \alpha = 20°$ (π) and $\varphi = 30°, \alpha = 120°$ (Π)

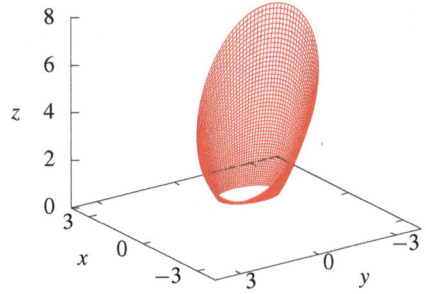

$$\begin{cases} x(\chi,\theta) = X_c + \chi(x_c - X_c) + [R - (R-r)\chi]\cos(\theta + \theta_0) \\ y(\chi,\theta) = Y_c + \chi(y_c - Y_c) + [R - (R-r)\chi]\sin(\theta + \theta_0) \\ z(\chi,\theta) = x^2(\chi,\theta) + y^2(\chi,\theta) \end{cases} \quad (13.23)$$

with $\chi \in [0,1]$ and $\theta \in [0, 2\pi)$. The partial derivatives with respect to the parameters are:

$$\begin{cases} \partial_\chi x = (x_c - X_c) - (R-r)\cos(\theta + \theta_0) \\ \partial_\chi y = (y_c - Y_c) - (R-r)\sin(\theta + \theta_0) \\ \partial_\chi z = 2x\partial_\chi x + 2y\partial_\chi y \end{cases} \begin{cases} \partial_\theta x = -[R - (R-r)\chi]\sin(\theta + \theta_0) \\ \partial_\theta y = +[R - (R-r)\chi]\cos(\theta + \theta_0) \\ \partial_\theta z = 2x\partial_\theta x + 2y\partial_\theta y \end{cases}$$

and they are used in order to calculate the functions E, G and F (13.15). The area of Σ is numerically evaluated integrating with the trapezoidal rule the square root $(EG - F^2)^{1/2}$ on the rectangle $[0,1] \times [0, 2\pi]$. For the sample of surface Σ shown in Fig. 13.4, the area is about 31.93482.

13.3 Suggested Readings

A wide overview on the application of two-dimensional integration to the calculation of surface area is found in: Lax and Terrell (2017, Chap. 6, pp. 205–222; Chap. 7, pp. 310–332), Adams and Essex (2010, Chap. 14, pp. 830–839), Lang (1987, Chap. XII), Apostol (1969, Chap. 12).

References

R. A. Adams and C. Essex. *Calculus: Several Variables*. Pearson Education Canada, Toronto, ON, 7th edition, 2010.

T. M. Apostol. *Calculus, Volume II*. Wiley, Hoboken, NJ, 2nd edition, 1969.

S. Lang. *Calculus of Several Variables*. Springer, New York, 3rd edition, 1987.
P. D. Lax and M. S. Terrell. *Multivariable Calculus with Applications*. Springer, Cham, CH, 2017.

Chapter 14
Vector Functions of Vector Variables

In the previous lessons some basic issues and properties of a scalar function of a vector variable[1] $x \in \mathbb{R}^n$ have been discussed. Tools as limit, continuity, derivative and differential have been extended to such functions. However, these functions are not sufficient for facing many physical applications involving the velocity in a fluid, or the electric field, or the magnetic one acting on a (moving) electric charge. All these applications need *vector* functions (velocity, electric or magnetic fields) of vector variables (usually, the position in the space and, sometimes, the time). From an abstract point of view, a vector function (f) of a vector variable (x) is a map:

$$f : \mathbb{R}^n \longrightarrow \mathbb{R}^m$$
$$x \mapsto f(x), \tag{14.1}$$

having domain an (open) subset of $\mathbb{R}^n$ and range $\mathbb{R}^m$. The dimension of the domain (n) of f (14.1) is, in general, different from the dimension of the range (m). At this stage, it is clear that the scalar functions of vector variables earlier investigated are particular samples of the maps (14.1), having $m = 1$. Hereafter, m will be assumed > 1.

14.1 Vector Functions of Vector Variables

Once a function as f (14.1) has been assigned, the first important issue to be discussed concerns its *existence field*, *i.e.* the subset $\mathscr{E}$ of $\mathbb{R}^n$ such that for any $x \in \mathscr{E}$ the m components of $f(x)$ simultaneously exist. This topic is approached in a similar way to what was done previously in discussing scalar functions of vector variables.

---

[1] A vector $x \in \mathbb{R}^n$ will be represented by means of the list of its components along the reference axes, arranged in a column vector. For typographical needs, the column vector is frequently replaced by a row one, indicating that the transpose vector has to be considered, *e.g.*: $x = (x_1, x_2, \ldots, x_n)^{\mathrm{T}}$.

© The Author(s), under exclusive license to Springer Nature Switzerland AG 2024
G. Riccardi et al., *Multidimensional Differential and Integral Calculus*,
https://doi.org/10.1007/978-3-031-70326-3_14

Indeed, assume that the existence fields $\mathcal{E}_j$ of the m components $f_j(x)$ ($j = 1, 2,$ $\ldots, m$) of the function $f(x)$ have been found. The existence field $\mathcal{E}$ of f is then obtained by means of the intersection of these sets:

$$\mathcal{E} = \bigcap_{j=1}^{m} \mathcal{E}_j . \tag{14.2}$$

The existence field (14.2) being defined, it is now possible to extend the tool of the limit at the point $x_0 \in \mathcal{E}$ to a function as f (14.1). First of all, consider the finite limit, *i.e.* the formula:

$$\lim_{x \to x_0} f(x) = l , \tag{14.3}$$

l being a finite m-dimensional vector. As before, the meaning of the formula (14.3) is that in correspondence to any neighbourhood[2] (arbitrarily small) $A(l) \subset \mathbb{R}^m$ of the point l, it is possible to find a neighbourhood $A(x_0) \subset \mathbb{R}^n$ of x_0 such that f (14.1) maps any point in $A(x_0)$ in a point belonging to $A(l)$, or more shortly $f[A(x_0)] \subseteq A(l)$. In order to simplify the check of the limit (14.3), the neighbourhoods $A(x_0)$ and $A(l)$ are often chosen as spherical, due to the fact that it is very simple to handle their sizes. In this way, the above geometrical meaning corresponds to the following symbolic writing (14.3):

$$\forall \varepsilon > 0 \, \exists \delta_\varepsilon > 0 \mid \forall x \in B_{\delta_\varepsilon}(x_0) - \{x_0\} : \mid f(x) - l \mid < \varepsilon . \tag{14.4}$$

Introduced the limit at the point x_0 (14.3), also the *continuity* at the same point can be easily extended to a function f (14.1). If l is just the value of f at the point x_0, *i.e.* $l = f(x_0)$, this function is said to be continuous at the point x_0. A few of sample cases are discussed below.

Example 14.1 Consider the function ($n = m = 3$):

$$f(x) = \begin{pmatrix} yz \\ x - z \\ x^2 \end{pmatrix} , \tag{14.5}$$

where x, y and z are the three components of x. It is now proved that:

$$\lim_{x \to (0,1,1)^{\mathrm{T}}} f(x) = \begin{pmatrix} 1 \\ -1 \\ 0 \end{pmatrix} . \tag{14.6}$$

Once a positive (small) arbitrary number ε is fixed, in correspondence to it another positive number δ_ε has to be found such that all the points $x = (x, y, z)^{\mathrm{T}} \in B_{\delta_\varepsilon}(x_0)$:

---

[2] Remember that the neighbourhood of a point is an open set containing that point.

$$|x - x_0| < \delta_\varepsilon \quad \Leftrightarrow \quad (x - 0)^2 + (y - 1)^2 + (z - 1)^2 < \delta_\varepsilon^2 \qquad (14.7)$$

(with the only possible exception of x_0) are mapped by f to interior points of $B_\varepsilon(l)$, *i.e.* they verify the inequality:

$$|f(x) - l| < \varepsilon \quad \Leftrightarrow \quad (yz - 1)^2 + (x - z + 1)^2 + (x^2 - 0)^2 < \varepsilon^2. \qquad (14.8)$$

In order to evaluate a possible choice of δ_ε, the inequality (14.8) is transformed and the condition that $x \in B_{\delta_\varepsilon}(x_0)$ (14.7) is used:

$$(yz - 1)^2 + (x - z + 1)^2 + (x^2 - 0)^2 \equiv$$
$$\equiv [(y - 1)(z - 1) + (y - 1) + (z - 1)]^2 + [x - (z - 1)]^2 + (x - 0)^2$$
$$\equiv [|(y - 1)(z - 1) + (y - 1) + (z - 1)|]^2 + [|(x - 0) - (z - 1)|]^2 + (x - 0)^2$$
$$\leq (|y - 1||z - 1| + |y - 1| + |z - 1|)^2 + (|x - 0| + |z - 1|)^2 + (x - 0)^2$$
$$< 14\delta_\varepsilon^2,$$

having assumed $\delta_\varepsilon < 1$. It follows that the choice $\delta_\varepsilon = \varepsilon/4$ proves the limit (14.6).

Example 14.2 It is now shown that the function (14.5) is continuous at any $x_0 \in \mathbb{R}^3$ (despite this is quite obvious, the three components of f being continuous $\mathbb{R}^3$). Once a positive (small) number ε has been fixed, the proof consists in showing that, in correspondence to it, another positive number δ_ε (depending on x_0) can be found, such that all the points belonging to the spherical neighbourhood $B_{\delta_\varepsilon}(x_0)$ are mapped by f (14.5) inside the fixed spherical neighbourhood $B_\varepsilon[f(x_0)]$, *i.e.* $f[B_{\delta_\varepsilon}(x_0)] \subseteq B_\varepsilon[f(x_0)]$. In other terms, given ε it is possible to find a number δ_ε such that, if the point $x = (x, y, z)^\mathsf{T}$ verifies the

$$(x - x_0)^2 + (y - y_0)^2 + (z - z_0)^2 < \delta_\varepsilon^2, \qquad (14.9)$$

its image $f(x)$ belongs to the sphere with center at $f(x_0)$ and radius ε:

$$(yz - y_0 z_0)^2 + [(x - z) - (x_0 - z_0)]^2 + (x^2 - x_0^2)^2 < \varepsilon^2. \qquad (14.10)$$

Taken $\delta_\varepsilon < 1$, the condition (14.10) is rewritten so as to enable the use of the condition $x \in B_{\delta_\varepsilon}(x_0)$ (14.9):

$$(yz - y_0 z_0)^2 + [(x - z) - (x_0 - z_0)]^2 + (x^2 - x_0^2)^2 \equiv$$

$$\equiv \quad |(y - y_0)(z - z_0) + z_0(y - y_0) + y_0(z - z_0)|^2 + |(x - x_0) - (z - z_0)|^2$$
$$+ |x - x_0|^2 |(x - x_0) + 2x_0|^2$$

$$\leq \quad (|y - y_0||z - z_0| + |z_0||y - y_0| + |y_0||z - z_0|)^2 + (|x - x_0| + |z - z_0|)^2$$
$$+ |x - x_0|^2 (|x - x_0| + 2|x_0|)^2$$

$$\leq \delta_\varepsilon^2 [(1 + |z_0| + |y_0|)^2 + 4 + (1 + 2|x_0|)^2].$$

It follows that it is sufficient to chose δ_ε as:

$$\delta_\varepsilon = \frac{\varepsilon}{2[(1 + |z_0| + |y_0|)^2 + 4 + (1 + 2|x_0|)^2]^{1/2}}$$

for the condition (14.10) to be verified.

Example 14.3 Consider now the function ($n = m = 3$):

$$f(x) = \begin{pmatrix} x \cos y \\ z \\ y \cos x \end{pmatrix} \tag{14.11}$$

and show that:

$$\lim_{x \to (\pi, \pi, 0)^{\mathrm{T}}} f(x) = \begin{pmatrix} -\pi \\ 0 \\ -\pi \end{pmatrix}, \tag{14.12}$$

so that the function (14.11) is continuous at the point $(\pi, \pi, 0)^{\mathrm{T}}$. As before, it must be shown that once a (small) positive number ε is fixed, it is possible, in correspondence to it to find a positive number δ_ε such that for any x verifying the condition:

$$(x - \pi)^2 + (y - \pi)^2 + (z - 0)^2 < \delta_\varepsilon^2, \tag{14.13}$$

i.e. x belongs to a sphere with center at $(\pi, \pi, 0)^{\mathrm{T}}$ and radius δ_ε, the following inequality:

$$(x \cos y + \pi)^2 + (z - 0)^2 + (y \cos x + \pi)^2 < \varepsilon^2 \tag{14.14}$$

is satisfied. The first step consists in finding a simpler form of the left hand side of the inequality (14.14) which satisfies the two requests: *I*) it is larger than the present one, even if *II*) it still vanishes as $x \to (\pi, \pi, 0)^{\mathrm{T}}$. Once this form is found, it is constrained to be smaller than ε^2, from which it follows an estimate of δ_ε. Taken $\delta_\varepsilon < 1$, the following simpler form of the left hand side of the inequality (14.14) is found:

$$(x \cos y + \pi)^2 + (z - 0)^2 + (y \cos x + \pi)^2 \equiv$$
$$\equiv \ |(x - \pi)(\cos y + 1) - (x - \pi) + \pi(\cos y + 1)|^2 + z^2$$
$$+ |(y - \pi)(\cos y + 1) - (y - \pi) + \pi(\cos x + 1)|^2 \qquad (14.15)$$
$$\leq \ (|x - \pi||1 + \cos y| + |x - \pi| + \pi|1 + \cos y|)^2 + z^2$$
$$+ (|y - \pi||1 + \cos y| + |y - \pi| + \pi|1 + \cos x|)^2 .$$

The quantity (14.15) is not yet suitable for using the condition (14.13), due to the presence of the terms $|1 + \cos y|$ and $|1 + \cos x|$.

These terms are handled noting that the function $\eta = \cos \xi$ lies below the straight lines $\eta = -2\xi/\pi + 1$ and $\eta = 2\xi/\pi - 3$ for $\pi/2 < \xi < \pi$ and $\pi < \xi < 3\pi/2$, respectively. These two lines are used[3] in order to estimate the distance between the curves $\eta = \cos \xi$ and $\eta = -1$ as follows:

$$|\cos \xi - (-1)| \leq \frac{2}{\pi} |\xi - \pi| . \qquad (14.16)$$

This inequality is verified in the entire interval $(\pi/2, 3\pi/2)$ and is used in order to build a quantity which is simpler than the one in Eq. (14.15), but it still verifies the requests *I* and *II*. Using also the condition (14.13), an inequality stronger but much more simpler with respect to the one (14.14) is obtained:

$$(x \cos y + \pi)^2 + (z - 0)^2 + (y \cos x + \pi)^2 \leq 33 \delta_\varepsilon^2 < \varepsilon^2 .$$

As a consequence, the choice $\delta_\varepsilon = \varepsilon/6$ guarantees that the inequality (14.14) is verified in all the points x belonging to the neighbourhood of $(\pi, \pi, 0)^{\mathrm{T}}$ specified by the condition (14.13).

Example 14.4 Finally, consider the function ($n = 3, m = 2$):

$$f(x) = \begin{pmatrix} x \log(x^2 + y^2) \\ z\sqrt{x^2 + y^2} \end{pmatrix} , \qquad (14.17)$$

that is defined in all $\mathbb{R}^3$ deprived of the z-axis (on which $x = y = 0$). It is now shown that the singularities on the z-axis are eliminable, the limit:

---

[3] Once it is noted that $-1 = \cos \pi$, the inequality (14.16) can be generalised in the following way. Assume that the function $\eta = g(\xi)$ is continuous with continuous and bounded first derivative in the neighbourhood $(\xi_0 - l, \xi_0 + l)$ of the point ξ_0, l being a given positive number. It follows that a finite positive number D exists such that $|g'(\xi)| \leq D$ for all $\xi \in (\xi_0 - l, \xi_0 + l)$. Using the Lagrange theorem, it follows that for any $\Delta\xi \in (-l, +l)$ a point ξ' belonging to the interval $(\xi_0, \xi_0 + \Delta\xi)$ if $\Delta\xi > 0$, or to the one $(\xi_0 + \Delta\xi, \xi_0)$ if $\Delta\xi < 0$, exists such that $g(\xi_0 + \Delta\xi) - g(\xi_0) = g'(\xi')\Delta\xi$. The boundedness of the first derivative finally implies:

$$|g(\xi_0 + \Delta\xi) - g(\xi_0)| \leq D|\Delta\xi| .$$

The function $\eta = \cos \xi$ verifies the above conditions in a neighbourhood of $\xi_0 = \pi$, with $D = 1$. As a consequence of the above discussion, an inequality like (14.16) is quickly obtained, without examining details of the function near the point ξ_0.

$$\lim_{x \to (0,0,z_0)^{\mathrm{T}}} f(x) = \begin{pmatrix} 0 \\ 0 \end{pmatrix} \tag{14.18}$$

existing for any z_0. In this case it is convenient to use cylindrical coordinates (ρ, θ, z) with $\rho \geq 0$ and $\theta \in [0, 2\pi)$ having their axis superimposed to the z-axis. The coordinates on the (x, y)-plane are given by: $x = \rho \cos\theta$ and $y = \rho \sin\theta$. In order to verify the limit (14.18), an arbitrary (small) positive number ε is fixed and, in correspondence to it, a positive number δ_ε (depending on z_0) is found, such that for any point (ρ, θ, z) verifying the condition:

$$|x - (0, 0, z_0)^{\mathrm{T}}|^2 = \rho^2 + (z - z_0)^2 < \delta_\varepsilon^2 \tag{14.19}$$

the following inequality:

$$|f(x) - 0|^2 = 4\rho^2 \log^2 \rho \, \cos^2 \theta + z^2 \rho^2 < \varepsilon^2 \tag{14.20}$$

is satisfied. Assumed $\delta_\varepsilon < 1$ (that implies $\rho < 1$), the key point lies in observing that the inequality:

$$\log \xi < \frac{\xi}{2} \tag{14.21}$$

is verified for any positive ξ (the inequality (14.21) is easily verified plotting both functions). Using the inequality (14.21) with $\xi = \rho^{-1/2}$ and $\rho < 1$, the new inequality: $-\rho \log \rho < \rho^{1/2}$ is obtained. It is used in the inequality (14.20) in order to enforce the simpler condition:

$$4\rho^2 \log^2 \rho \, \cos^2 \theta + z^2 \rho^2 < \{4 + \rho \, [(z - z_0)^2 + 2|z_0||z - z_0| + z_0^2]\} \rho$$
$$< (5 + 2|z_0| + z_0^2) \, \delta_\varepsilon \,,$$

which shows that choosing $\delta_\varepsilon = \varepsilon/(3 + |z_0|)^2$ at all points verifying the inequality (14.19), the condition (14.20) is also satisfied.

The limit is extended now to the cases in which the vector x goes to infinity. If such a limit exists, it is a finite vector $l \in \mathbb{R}^m$ (a), or the function diverges (b):

$$(a): \lim_{x \to \infty} f(x) = l, \qquad (b): \lim_{x \to \infty} f(x) = \infty. \tag{14.22}$$

The symbolic writing (a) means that for any neighbourhood of the point l, it is possible to find a neighbourhood of infinity (i.e., outside a large, closed sphere having its center at the origin) which is mapped inside the above neighbourhood of l by the function f. If the neighbourhood of l is chosen as spherical with radius ε $(B_\varepsilon(l))$, the symbol (a) implies that, in correspondence to any value of ε, a large positive number P_ε exists, such that for any x having modulus larger than P_ε the image point $f(x)$ lies inside $B_\varepsilon(l)$, i.e. $|f(x) - l| < \varepsilon$. In an analogous way, the symbolic writing (b) means that for any neighbourhood of infinity (i.e., outside a large, closed sphere $\overline{B_M(0)}$ having its center at the origin and radius M), another neighbourhood of infinity can be found (i.e., outside $\overline{B_{P_M}(0)}$, having radius P_M depending on M) such that any point lying in this set is mapped outside $\overline{B_M(0)}$ by the function f.

A last case is possible. A function $f : \mathbb{R}^n \rightarrow \mathbb{R}^m$ diverges at a point x_0. This behaviour is described by the symbolic writing:

$$\lim_{x \to x_0} f(x) = \infty, \tag{14.23}$$

which means that, for any neighbourhood of infinity (i.e., outside the closed sphere $\overline{B_M(0)}$ having a large radius M) a neighbourhood of the point x_0 (i.e., the sphere $B_{\delta_M}(x_0)$ having radius δ_M depending on M) exists, such that any point of this neighbourhood is mapped inside the above neighbourhood of infinity by the function f. More formally, once an arbitrary (large) positive number M has been fixed, in correspondence to it a (small) positive number δ_M can be found such that the image of any point x verifying the inequality $|x - x_0| < \delta_M$ satisfies the condition $|f(x)| > M$. A few sample cases are now discussed with $n = 3$ and $x = (x, y, z)^{\mathrm{T}}$.

Example 14.5 It is now shown that the function ($m = 3$):

$$f(x) = \begin{pmatrix} x/(x^2 + y^2 + z^2) \\ (z - x)/(x^2 + y^2 + z^2) \\ z/(x^2 + y^2 + z^2) \end{pmatrix}$$

vanishes at infinity. This behaviour is described by the symbol:

$$\lim_{x \to \infty} f(x) = 0$$

and it is proved choosing an arbitrary (small) positive number ε, in correspondence to which a (large) positive number P_ε is found such that for any x verifying the condition $|x| > P_\varepsilon$ the inequality $|f(x) - 0| < \varepsilon$, i.e.:

$$\frac{x^2 + (z - x)^2 + z^2}{(x^2 + y^2 + z^2)^2} < \varepsilon^2, \tag{14.24}$$

is satisfied. Using the spherical coordinates with center at the origin:

$$\begin{cases} x = r \cos \phi \cos \theta \\ y = r \cos \phi \sin \theta \\ z = r \sin \phi, \end{cases} \tag{14.25}$$

with $r > P_\varepsilon, \phi \in [-\pi/2, +\pi/2]$ and $\theta \in [0, 2\pi)$, the condition (14.24) is rewritten in the following way:

$$\frac{x^2 + (z - x)^2 + z^2}{(x^2 + y^2 + z^2)^2} = \frac{2}{r^2} \left(\cos^2 \phi \cos^2 \theta - \cos \phi \sin \phi \cos \theta + \sin^2 \phi \right) < \frac{6}{P_\varepsilon^2} < \varepsilon^2,$$

which shows that it is sufficient to choose $P_\varepsilon = 3/\varepsilon$ for satisfying the inequality (14.24).

Example 14.6 Consider the function ($m = 2$):

$$f(x) = \begin{pmatrix} (xyz)^{1/3} \\ (x^2 + y^2 + z^2)^{1/2}/2 \end{pmatrix}$$

and show that it diverges for diverging x, as in the case (b) of the formulae (14.22). This behaviour is proved in the following way. A (large) positive number M is fixed, in correspondence of which it is now shown that a positive number P_M can be found such that the image of any x, written in spherical coordinates (14.25) with $r > P_M$, verifies the inequality:

$$|f(x)|^2 = r^2 \left[(\cos^2 \phi \sin \phi \cos \theta \sin \theta)^{2/3} + \frac{1}{4} \right] > \frac{r^2}{4} > \frac{P_M^2}{4}.$$

As a consequence, it is sufficient to choose $P_M = 2M$ for satisfying the inequality $|f(x)| > M$ and prove the limit.

Example 14.7 Consider the function ($m = 3$):

$$f(x) = \begin{pmatrix} 1/(x^2 + y^2 + z^2) \\ 1 + x + z \\ xy \end{pmatrix}$$

and show that it diverges for x going to the origin, *i.e.*

$$\lim_{x \to 0} f(x) = \infty.$$

This behaviour is proved showing that for any (large) positive number M it is possible to find a (small) positive number δ_M such that the image of any $x \in B_{\delta_M}(0)$ satisfies the inequality $|f(x)| > M$. In terms of the spherical coordinates (14.25) with $r < \delta_M$ this condition is rewritten in the following way:

$$|f(x)|^2 = \frac{1}{r^4} + (1 + r \cos \phi \cos \theta + r \sin \phi)^2 + r^4 \cos^4 \phi \sin^2 \theta \cos^2 \theta \geq \frac{1}{r^4} > \frac{1}{\delta_M^4}$$

and it is sufficient to choose $\delta_M = 1/M^{1/4}$ for the condition $|f(x)| > M$ to be verified.

14.2   Divergence and Curl of a Vector Function of a Vector Variable

At this point is quite easy to extend the calculation of a partial derivative to a vector function of a vector variable (14.1). In effect, assumed that all the m components of

f are partially differentiable at a given point $x_0 \in \mathbb{R}^n$, it is possible to evaluate the partial derivative at x_i ($i = 1, 2, \ldots, n$) of any components of f at that point. As a consequence, the partial derivative with respect to x_i of f is defined as the vector of $\mathbb{R}^m$ built with the corresponding derivatives of the components of f:

$$\partial_{x_i} f(x_0) = \begin{pmatrix} \partial_{x_i} f_1(x_0) \\ \partial_{x_i} f_2(x_0) \\ \vdots \\ \partial_{x_i} f_m(x_0) \end{pmatrix}. \tag{14.26}$$

Hereafter two particular *differential operators* (*i.e.*, operators involving partial derivatives) acting on the function f will play important roles. They are defined for $n = m$, and the second one will be used here for $n = m = 3$, only. The first operator acts on the partially differentiable function f and gives a scalar function of x, according to the rule:

$$\sum_{i=1}^{n} \partial_{x_i} f_i(x) =: \nabla \cdot f(x), \tag{14.27}$$

the left hand side of which is often written in the simpler form: $\partial_{x_i} f_i(x)$, having used the sum rule. The differential operator (14.27) is called *divergence of f*, its action on f produces a scalar function of x. In addition to the symbol at the right hand side of the Eq. (14.27), the divergence is often indicated by means of the symbol $\mathrm{div}(f)$. According to the definition (14.27), the divergence of a function f (14.1) with $n = m$ is evaluated adding the derivative along x_1 of the first component f_1, derivative along x_2 of the second component f_2 and so on, up to the derivative along x_n of the nth component f_n. As a sample case, the divergence of the function (14.5) trivially vanishes:

$$\nabla \cdot f(x) = \partial_x(yz) + \partial_y(x - z) + \partial_z(x^2) = 0,$$

whereas the divergence of the function (14.11) is:

$$\nabla \cdot f(x) = \partial_x(x \cos y) + \partial_y(z) + \partial_z(y \cos x) = \cos y.$$

Exercise 14.1 Show that, if $g : \mathbb{R}^3 \to \mathbb{R}$ is twice partially differentiable, the rule:

$$\nabla \cdot \nabla g = \nabla^2 g$$

follows from the definition (14.27) of the divergence.

Exercise 14.2 Evaluate the divergence of the following functions ($x \in \mathbb{R}^3$ in the cases (a), (b), (c), whereas $x \in \mathbb{R}^4$ in the case (d)):

$$(a)\ \begin{pmatrix} x^2 y \\ xyz \\ yz^2 \end{pmatrix}, \quad (b)\ \begin{pmatrix} \cos(xy) \\ \sin(z) \\ \sin(yz) \end{pmatrix}, \quad (c)\ \begin{pmatrix} y \log x \\ xz \\ x \log z \end{pmatrix}, \quad (d)\ \begin{pmatrix} \cosh(xu) \\ \sinh(yu) \\ xy \\ \cosh(xyzu) \end{pmatrix}.$$

The second differential operator acts on partially differentiable functions $f : \mathbb{R}^3 \to \mathbb{R}^3$, only (it can be generalised to functions with $n = m > 3$, but this issue goes beyond the aims of the present course). It produces another vector function of a vector variable, according to the symbolic rule:

$$\begin{vmatrix} e_1 & e_2 & e_3 \\ \partial_{x_1} & \partial_{x_2} & \partial_{x_3} \\ f_1(x) & f_2(x) & f_3(x) \end{vmatrix} = \begin{pmatrix} \partial_{x_2} f_3(x) - \partial_{x_3} f_2(x) \\ \partial_{x_3} f_1(x) - \partial_{x_1} f_3(x) \\ \partial_{x_1} f_2(x) - \partial_{x_2} f_1(x) \end{pmatrix} =: \nabla \times f(x),$$

(14.28)

$e_{1,2,3}$ being the unit vectors of the axis 1 (x), 2 (y) and 3 (z), respectively. The operator (14.28) is called *curl of f*. In addition to the symbol used in the definition (14.28), the curl of f is also indicated with the symbol curl(f). As sample cases, the curl of the function (14.5) is evaluated as:

$$\nabla \times f(x) = \begin{vmatrix} e_x & e_y & e_z \\ \partial_x & \partial_y & \partial_z \\ yz & x - z & x^2 \end{vmatrix} = \begin{pmatrix} 1 \\ y - 2x \\ 1 - z \end{pmatrix},$$

whereas the curl of the function (14.11) is:

$$\nabla \times f(x) = \begin{vmatrix} e_x & e_y & e_z \\ \partial_x & \partial_y & \partial_z \\ x \cos y & z & y \cos x \end{vmatrix} = \begin{pmatrix} \cos x - 1 \\ y \sin x \\ x \sin y \end{pmatrix}.$$

Exercise 14.3 Show that, if $g : \mathbb{R}^3 \to \mathbb{R}$ has continuous second partial derivatives, the curl of its gradient, *i.e.* $\nabla \times \nabla g$, vanishes. Moreover, prove that, if $f : \mathbb{R}^3 \to \mathbb{R}^3$ has continuous second partial derivatives, the divergence of its curl, *i.e.* $\nabla \cdot \nabla \times f$, vanishes.

Exercise 14.4 Evaluate the curl of the following vector functions:

$$(a)\ \begin{pmatrix} yz \\ z^2 \\ xy \end{pmatrix}, \quad (b)\ \begin{pmatrix} xz \cos(y) \\ y \sin(y) \\ xy \cos(z) \end{pmatrix}, \quad (c)\ \begin{pmatrix} z \log(y) \\ -x \log(y) + y \log(x) \\ \log(z) \end{pmatrix},$$

$$(d)\ \begin{pmatrix} x/(x^2 + z^2) \\ z/(x^2 + y^2) \\ xyz \end{pmatrix}, \quad (e)\ \begin{pmatrix} -\cosh(yz) \\ \cosh(xz) \\ -\cosh(xy) \end{pmatrix}, \quad (f)\ \begin{pmatrix} -e^{x-z} \\ xze^y \\ e^{z-x} \end{pmatrix}.$$

Chapter 15
Line Integral and Flux of Vector Functions

In the present lesson, two important kinds of integrals of vector functions of vector variables will be discussed. They are used in many physical applications and will be connected by means of a famous theorem in a successive lesson.

One of the most important physical tools is the *work* ($\mathcal{L}$) done by a force field f on a pointwise particle. From a mathematical point of view, the force field f : $\mathbb{R}^3 \times [0, +\infty) \to \mathbb{R}^3$ is a vector function of the position ($x \in \mathbb{R}^3$) and, in general, of the time ($t \in [0, +\infty)$), which specifies the force acting in the place x and at time t due to the presence of an external entity (*e.g.*, a massive body attracting a pointwise mass, or an electric charge acting on a charged particle), being not influenced by the motion of the particle.

In correspondence to an elementary (infinitely small) displacement of the particle from x to $x + dx$, the force field makes an elementary work $d\mathcal{L}$ given by the product of the component of the force along dx times the length of the displacement, *i.e.* $d\mathcal{L} = f \cdot dx$. The particle moves from x to $x + dx$ in an elementary time dt, with a velocity $\dot{x} = dx/dt$. As a consequence, the elementary displacement dx can be also rewritten in terms of the time as $\dot{x}\, dt$.

Assume now that the particle motion takes place for a finite time, *e.g.* the interval $[0, T]$ with T finite. During this time, the particle moves along a curve $\mathscr{C}$, which is automatically parametrised in time (as well known, $\mathscr{C}$ is the trajectory of the particle), and the force field does the work:

$$\mathcal{L} = \int_0^T dt\, \dot{x}(t) \cdot f[x(t), t]\,. \tag{15.1}$$

The work (15.1) is evaluated calculating the force *on the trajectory*, that is the scalar product of force and elementary displacement dx (written in terms of the parameter t as $\dot{x}\, dt$) and finally integrating. It will be seen below that this is the standard procedure to evaluate the line integral of the force field f on the curve $\mathcal{L}$.

G. Riccardi et al., *Multidimensional Differential and Integral Calculus*,
https://doi.org/10.1007/978-3-031-70326-3_15

Coming back to the physical meaning of the integral at the right hand side of Eq. (15.1), it is useful to consider that a particle of mass m (constant in time) moves in presence of the force field f according to the equation:

$$\frac{d}{dt} m\dot{x} = f ,\qquad(15.2)$$

which means that the variation in time of the particle momentum $m\dot{x}$ is due to the acting force f. Taking the scalar product of $dx = \dot{x}\, dt$ for both sides of Eq. (15.2), the relation:

$$d\left(\frac{1}{2} m|\dot{x}|^2\right) = \dot{x}\, dt \cdot m \frac{d\dot{x}}{dt} = dx \cdot f = d\mathcal{L}\qquad(15.3)$$

is obtained. Equation (15.3) has an important physical meaning: it says that the work done by the force field along the trajectory leads to an equal increase of the kinetic energy $m|\dot{x}|^2/2$ of the particle. This remark highlights the importance of the line integral (15.1), as it will be now discussed in a more general context.

15.1 Circulation of a Vector Function on a Curve

Given a function $f : \mathbb{R}^n \to \mathbb{R}^n$ which is continuous and a curve $\mathscr{C}$, *i.e.* a vector function of a scalar variable (the parameter σ)

$$\begin{aligned} x : (a, b) &\to \mathbb{R}^n \\ \sigma &\mapsto x(\sigma) \end{aligned}\qquad(15.4)$$

continuous in (a, b) with continuous first derivative, the following integral:

$$\int_a^b d\sigma\, x'(\sigma) \cdot f[x(\sigma)] =: \int_{\mathscr{C}} dx \cdot f(x) ,\qquad(15.5)$$

is now considered. It is called *line integral of f on the curve $\mathscr{C}$*. In order to familiarise with the tool (15.5), some sample cases are now discussed.

Example 15.1 Consider the function from $\mathbb{R}^2$ to $\mathbb{R}^2$:

$$f(x) = \begin{pmatrix} x - y \\ xy \end{pmatrix}\qquad(15.6)$$

and the curve $\mathscr{C}$ in the plane (x, y):

$$x(\sigma) = \begin{pmatrix} \sigma \sin\sigma \\ \sigma \cos 2\sigma \end{pmatrix} ,\qquad(15.7)$$

for $\sigma \in (0, 2\pi)$. The curve (15.7) is drawn in Fig. 15.1 with a green dashed line. In a certain number n ($n = 30$ in figure) of points (specified by small yellow circles) $\boldsymbol{x}(\sigma_k)$, the vectors $d\boldsymbol{x} = \boldsymbol{x}'(\sigma_k)d\sigma$ with $d\sigma = 2\pi/n$ (red lines) and $\boldsymbol{f}[\boldsymbol{x}(\sigma_k)]$ (black, with a scale factor $1/5$) are superimposed to the curve.

The integral (15.5) is specified using the definitions of the vector function $\boldsymbol{f}$ (15.6) and of the curve $\mathscr{C}$ (15.7) in the following form:

$$\int_0^{2\pi} d\sigma \begin{pmatrix} \sin\sigma + \sigma\cos\sigma \\ \cos 2\sigma - 2\sigma\sin 2\sigma \end{pmatrix} \cdot \begin{pmatrix} \sigma\sin\sigma - \sigma\cos 2\sigma \\ \sigma\sin\sigma\,\sigma\cos 2\sigma \end{pmatrix} =$$

$$= \frac{1}{4} \int_0^{2\pi} d\sigma\,\sigma \Big[\; 2\,(1 + \sin\sigma - \cos 2\sigma - \sin 3\sigma)$$

$$+ \sigma\,(2\sin\sigma - 2\cos\sigma + 2\sin 2\sigma - \sin 3\sigma - 2\cos 3\sigma + \sin 5\sigma)$$

$$+ \sigma^2\,(\cos 5\sigma - \cos 3\sigma)\Big]$$

$$= \frac{1}{4} \Big(4\pi^2 + 2I_{1,1}^s - 2I_{2,1}^c + 2I_{2,1}^s - 2I_{1,2}^c + 2I_{2,2}^s - 2I_{1,3}^s - 2I_{2,3}^c - I_{2,3}^s$$

$$- I_{3,3}^c + I_{2,5}^s + I_{3,5}^c\Big)$$

$$= -26\pi \left(\frac{1}{9} + \frac{2\pi}{25}\right), \tag{15.8}$$

by means of the following integrals (k and m positive integers):

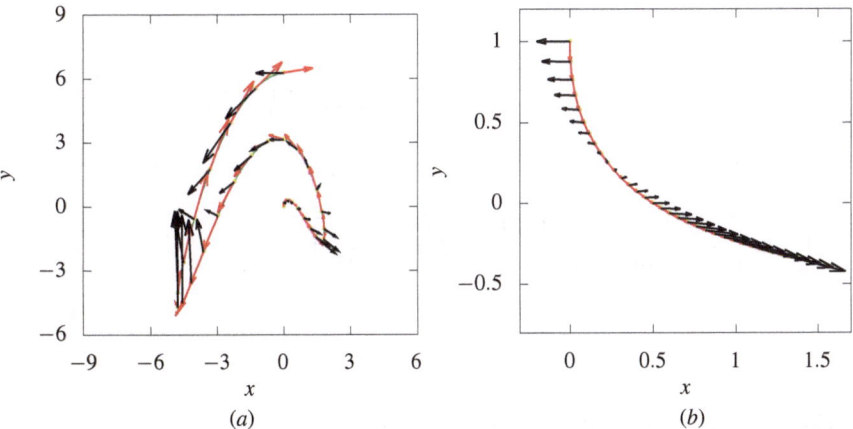

(a)   (b)

Fig. 15.1 In **a** the curve (15.7) is drawn with a green dashed line. On a certain number of points (small yellow circles) of the curve the vectors $d\boldsymbol{x}$ (red) and $\boldsymbol{f}(\boldsymbol{x})$ (black) are also drawn. With the same rules about the colours, in **b** the curve (15.10) and the corresponding vectors $d\boldsymbol{x}$ and $\boldsymbol{f}(\boldsymbol{x})$ are drawn

$$I^c_{k,m} := \int_0^{2\pi} d\sigma\, \sigma^k \cos m\sigma = -\frac{k}{m} I^s_{k-1,m}$$

$$\text{(15.9)}$$

$$I^s_{k,m} := \int_0^{2\pi} d\sigma\, \sigma^k \sin m\sigma = -\frac{(2\pi)^k}{m} + \frac{k}{m} I^c_{k-1,m}\,,$$

starting from $I^c_{0,m} = I^s_{0,m} = 0$.

Example 15.2 Consider the same vector function (15.6) on a different curve:

$$x(\sigma) = \begin{pmatrix} \sigma^2/(1+\sigma) \\ (1-\sigma)/(1+\sigma) \end{pmatrix},$$

$$\text{(15.10)}$$

with $\sigma \in (0, a)$ and $a > 0$. The line integral (15.5) becomes:

$$\int_0^a d\sigma \begin{pmatrix} \sigma(\sigma+2)/(\sigma+1)^2 \\ -2/(\sigma+1)^2 \end{pmatrix} \cdot \begin{pmatrix} (\sigma^2+\sigma-1)/(\sigma+1) \\ \sigma^2(1-\sigma)/(\sigma+1)^2 \end{pmatrix} =$$

$$= \int_0^a d\sigma \left[\frac{\sigma(\sigma+2)(\sigma^2+\sigma-1)}{(\sigma+1)^3} - 2\frac{\sigma^2(1-\sigma)}{(\sigma+1)^4} \right]. \quad \text{(15.11)}$$

In order to calculate the two integrals in the formula (15.11) the fractions have to be decomposed in simpler ones using the following identities:

$$\frac{\sigma(\sigma+2)(\sigma^2+\sigma-1)}{(\sigma+1)^3} \equiv \sigma - \frac{2}{\sigma+1} + \frac{1}{(\sigma+1)^2} + \frac{1}{(\sigma+1)^3}$$

$$\frac{\sigma^2(1-\sigma)}{(\sigma+1)^4} \equiv -\frac{1}{\sigma+1} + \frac{4}{(\sigma+1)^2} - \frac{5}{(\sigma+1)^3} + \frac{2}{(\sigma+1)^4}\,,$$

which leads to the result:

$$\int_0^a d\sigma \left[\frac{\sigma(\sigma+2)(\sigma^2+\sigma-1)}{(\sigma+1)^3} - 2\frac{\sigma^2(1-\sigma)}{(\sigma+1)^4} \right] =$$

$$= -\frac{17}{6} + \frac{a^2}{2} + \frac{7}{a+1} - \frac{11}{2}\frac{1}{(a+1)^2} + \frac{4}{3}\frac{1}{(a+1)^3}\,.$$

Exercise 15.1 Evaluate the line integral of the vector functions f on the specified curves $\mathscr{C}$ (see Fig. 15.2a, b and Fig. 15.3a, b):

$$\text{(a)} \quad f(x) = \begin{pmatrix} y-z \\ z-x \\ x-y \end{pmatrix} \qquad x(\sigma) = \begin{pmatrix} R\cos\sigma \\ R\sin\sigma \\ z_0 + (h/2)\sin 3\sigma \end{pmatrix} \quad \text{with } \sigma \in [0, 2\pi)$$

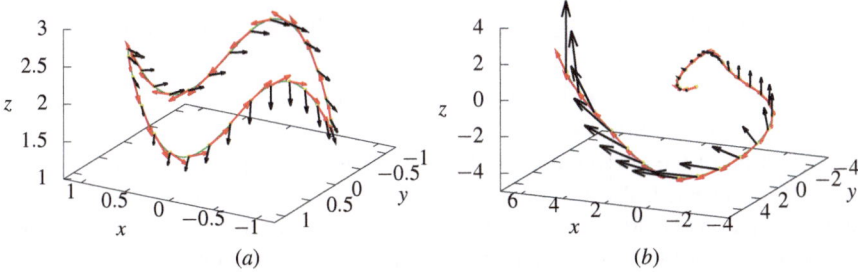

Fig. 15.2 In **a** the curve $\mathscr{C}$ and the vector function $\boldsymbol{f}$ of the Exercise 15.1a ($R = 1, h = 1, z_0 = 2$) are drawn. The same is made in **b** for the Exercise 15.1b. The vectors $\boldsymbol{f}(\boldsymbol{x})$ are scaled by a factor $1/10$

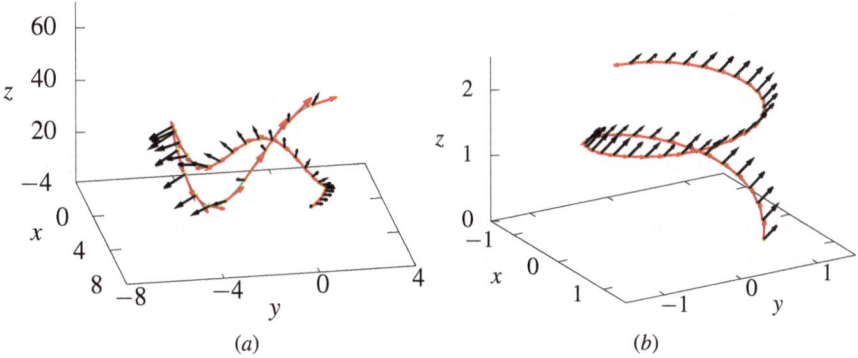

Fig. 15.3 In **a** the curve $\mathscr{C}$ and the vector function $\boldsymbol{f}$ of the Exercise 15.1c are drawn. The same is made in **b** for the Exercise 15.1d. The vectors $\boldsymbol{f}(\boldsymbol{x})$ are scaled by a factor $1/5$

(b) $\boldsymbol{f}(\boldsymbol{x}) = \begin{pmatrix} z^2 \\ xyz \\ x^2 \end{pmatrix}$ $\qquad \boldsymbol{x}(\sigma) = \begin{pmatrix} \sigma \cos \sigma \\ \sin 3\sigma \\ \sigma \sin \sigma \end{pmatrix}$ with $\sigma \in (0, 2\pi)$

(c) $\boldsymbol{f}(\boldsymbol{x}) = \begin{pmatrix} x \\ y \\ x^2 - y^2 \end{pmatrix}$ $\qquad \boldsymbol{x}(\sigma) = \begin{pmatrix} \sigma \cos \sigma \\ \sigma \sin \sigma \\ \sigma^2(1 + 0.5 \cos 4\sigma) \end{pmatrix}$ with $\sigma \in (0, 2\pi)$

(d) $\boldsymbol{f}(\boldsymbol{x}) = \begin{pmatrix} x^2 \\ y^2 \\ 1 \end{pmatrix}$ $\qquad \boldsymbol{x}(\sigma) = \begin{pmatrix} \cos \sigma \\ \sin \sigma \\ \log \sigma \end{pmatrix}$ with $\sigma \in (1, 3\pi)$

Solution 15.1

(a) $\quad -2\pi R^2$

(b) $\quad \dfrac{16}{3} \pi^3 + 8\pi^2 - \dfrac{1015}{128} \pi \simeq 219.41180$

(c) $\quad 2\pi^2(1 + 9\pi^2) \simeq 1773.10285$

$$(4) \quad \log(3\pi) - \frac{1}{3} - \frac{1}{12}(3\sin 1 + 3\cos 1 - \sin 3 + \cos 3) \simeq 1.65882.$$

The line integral (15.5) plays a special role when the curve $\mathscr{C}$ is closed. In this case, it is called *circulation of f on the curve $\mathscr{C}$*. It is an easy consequence of the definition of the circulation that it changes sign, if the positive direction of the curve $\mathscr{C}$ is reversed. As it will be shown below, vector functions exist such that their circulations on arbitrary closed curves vanish. As a sample of these functions, consider a vector function f which is the gradient of a scalar one, *i.e.* $g : \mathbb{R}^n \to \mathbb{R}$. Obviously, for a given f infinitely many functions g exist, differing from each other by additive constants. The line integral (15.5) on the closed curve $\mathscr{C}$ becomes in this case:

$$\int_{\mathscr{C}} dx \cdot f(x) = \int_{\mathscr{C}} dx \cdot \nabla g(x) = \int_{\mathscr{C}} dg(x) = g(x_0^+) - g(x_0^-) = 0, \quad (15.12)$$

x_0 being an arbitrary point of the curve that is considered as starting (x_0^+) and ending (x_0^-) point of the integration path. It will be shown below that also other vector functions have vanishing circulations, even if they are not gradients of scalar functions.

Vice versa, if the circulation of the continuous vector function $f : \mathbb{R}^n \to \mathbb{R}^n$ vanishes on every closed curve, the line integral:

$$\int_{x_0}^{x} dy \cdot f(y) \tag{15.13}$$

does not depend on the integration path used for moving from x_0 to x. For this reason, the integral (15.13) defines a new scalar function of the vector variable x, here called $F(x)$.

Exercise 15.2 Show that the function $F(x)$ (15.13) is continuous (to this aim, it would be sufficient to take f as bounded).

Assuming the function f to be continuous, it becomes possible to evaluate the partial derivative $\partial_i F$ ($i = 1, 2, \ldots, n$). Once the incremental ratio in the ith coordinate direction for the function F (15.13) is written, the Lagrange theorem leads to the following result (in order to indicate that there is not sum on i, this index is put inside brackets):

$$\frac{F(x + e_{(i)}\Delta x_{(i)}) - F(x)}{\Delta x_{(i)}} = \frac{1}{\Delta x_{(i)}} \int_{x}^{x + e_{(i)}\Delta x_{(i)}} dy_{(i)} \, f_{(i)}(y) = f_{(i)}(x + e_{(i)}\xi_{(i)}),$$

ξ_i being a suitable point belonging to the interval $(0, \Delta x_i)$. The limit as $\Delta x_i \to 0$ leads to the relation $\partial_i F = f_i$, or in vector form:

$$\nabla F = f. \tag{15.14}$$

Therefore, F is called potential associated to the field f. As it is easily seen from Eq. (15.14), the potential is not unique, but two potentials associated to the same f differ just by a function of the time (or, in particular, by a constant).

15.2 Flux of a Vector Function Through a Surface

The second kind of integral of a vector function (f) of a vector variable (x) is the *flux of f through a surface*. A particularly meaningful example of flux comes from the Fluid Mechanics. Consider a confined, steady flow (for example, in a pipe) of a fluid having constant and uniform density and evaluate the volume of fluid that crosses a given surface Σ in a unit time. The velocity of the fluid is a three dimensional vector function of a vector variable: $u : \mathbb{R}^3 \to \mathbb{R}^3$, so that at the place x the fluid particle moves at velocity $u(x)$. The surface Σ is oriented in a congruent way with respect to the flow, defining at each point $x \in \Sigma$ the normal unit vector $n(x)$. Due to the fact that each area element $dS(x)$ of the surface Σ is crossed by a volume of fluid $dS(x)\,n(x) \cdot u(x)$ in the time unit, the volume of fluid that crosses the entire surface during a unitary time interval is written as:

$$\int_{\Sigma} dS(x)\,n(x) \cdot u(x) \,. \tag{15.15}$$

The integral (15.15) is often called the (volumetric) *flow rate* through the surface Σ.

In more conceptual terms, assume that a surface Σ is given in a parametrical form and name (u, v) the parameters. The position x on the surface is specified as a partially differentiable vector function of a vector variable: $x : A \subset \mathbb{R}^2 \to \mathbb{R}^3$, having continuous partial derivatives $\partial_u x$ and $\partial_v x$. The components of a vector $u = (u, v)^{\mathsf{T}} \in A$ are called *parameters*. Furthermore, to any straight line parallel to a coordinate axis in A, *i.e.* $v \equiv v_0$ or $u \equiv u_0$, corresponds a *coordinate curve*, $x = x(u, v_0)$ or $x = x(u_0, v)$, lying on Σ. It is also assumed that any point of Σ is crossed by two coordinate curves, so that the tangent vector $\partial_u x$ is never parallel to the tangent vector $\partial_v x$. As a consequence, the normal direction $\partial_u x \times \partial_v x$ (or its opposite $\partial_v x \times \partial_u x$) results to be defined everywhere on Σ. Moreover, due to the fact that $(dA(u, v) = du\,dv)$:

$$dS(u, v) = |\partial_u x \times \partial_v x|\,dA(u, v)\,, \quad n(u, v) = \frac{\partial_u x \times \partial_v x}{|\partial_u x \times \partial_v x|}\,,$$

the product $dS\,n$ is written in the following form:

$$dS(u, v)\,n(u, v) = \partial_u x \times \partial_v x\,dA(u, v)\,. \tag{15.16}$$

This result is particularly meaningful, showing that $dS\,n$ does not depend on the actual parametrization of Σ.

Exercise 15.3 Prove that $dS\,\boldsymbol{n}$ is independent of the choice of the parameters u and v. In other words, if these parameters are changed in $\tilde{u}$ and $\tilde{v}$ (without changing the orientation of $\boldsymbol{n}$) the same value of $dS\,\boldsymbol{n}$ (15.16) is obtained. What happens if $\partial(\tilde{u}, \tilde{v})/\partial(u, v) < 0$?

Suggestion: use the relation: $\boldsymbol{x}(u, v) \equiv \tilde{\boldsymbol{x}}[\tilde{u}(u, v), \tilde{v}(u, v)]$ for writing $\partial_u \boldsymbol{x} \times \partial_v \boldsymbol{x}$ by means of $\partial_{\tilde{u}} \tilde{\boldsymbol{x}} \times \partial_{\tilde{v}} \tilde{\boldsymbol{x}} \ldots$

Once the surface element and the normal unit vector (or, better, their product) have been defined, a continuous vector function $\boldsymbol{f}$ of the vector variable $\boldsymbol{x}$ is defined in a suitable set $\mathscr{E} \subset \mathbb{R}^3$, containing the above surface Σ. The flux of $\boldsymbol{f}$ through the surface Σ is then given by the following integral:

$$\int_{\Sigma} dS(\boldsymbol{x})\,\boldsymbol{n}(\boldsymbol{x}) \cdot \boldsymbol{f}(\boldsymbol{x}) = \int_A dA(u, v)\,\partial_u \boldsymbol{x}(u, v) \times \partial_v \boldsymbol{x}(u, v) \cdot \boldsymbol{f}[\boldsymbol{x}(u, v)] \quad (15.17)$$

which is a usual two-dimensional integral in the space of the parameters A. A sample case is now discussed.

Example 15.3 Consider the flux of the function $\boldsymbol{f}(\boldsymbol{x}) = (x, y, xy)^{\mathsf{T}}$ through the surface of the (elliptic) paraboloid $z = x^2 + y^2$ between the planes $z = 0$ and $z = 1$, oriented outward.

The surface is parametrised in polar form:

$$x = \rho \cos\theta, \quad y = \rho \sin\theta, \quad z = \rho^2$$

with $\rho \in (0, 1)$ and $\theta \in [0, 2\pi)$. The derivatives with respect to the parameters of the position on the surface are:

$$\partial_\rho \boldsymbol{x} = \begin{pmatrix} \cos\theta \\ \sin\theta \\ 2\rho \end{pmatrix}, \quad \partial_\theta \boldsymbol{x} = \begin{pmatrix} -\rho \sin\theta \\ \rho \cos\theta \\ 0 \end{pmatrix},$$

and their vector product (oriented outward) is evaluated as:

$$\partial_\theta \boldsymbol{x} \times \partial_\rho \boldsymbol{x} = \rho \begin{pmatrix} 2\rho \cos\theta \\ 2\rho \sin\theta \\ -1 \end{pmatrix}.$$

The flux through the surface follows from the relation (15.17):

$$\int_{\Sigma} dS(\boldsymbol{x})\,\boldsymbol{n}(\boldsymbol{x}) \cdot \boldsymbol{f}(\boldsymbol{x}) = \int_0^1 d\rho\,\rho^2 \int_0^{2\pi} d\theta \begin{pmatrix} 2\rho \cos\theta \\ 2\rho \sin\theta \\ -1 \end{pmatrix} \cdot \begin{pmatrix} \cos\theta \\ \sin\theta \\ \rho \sin\theta \cos\theta \end{pmatrix}$$

$$= 4\pi \int_0^1 d\rho\,\rho^3$$

$$= \pi .$$

15.3 Suggested Readings

An extensive treatment of line integrals and their applications to vector fields is given in Lax and Terrell (2017, Chap. 7, pp. 279–310), Adams and Essex (2010, Chap. 15, pp. 842–885), Lang (1987, Chap. VIII), Apostol (1969, Chap. 10).

References

R. A. Adams and C. Essex. *Calculus: Several Variables*. Pearson Education Canada, Toronto, ON, 7th edition, 2010.

T. M. Apostol. *Calculus, Volume II*. Wiley, Hoboken, NJ, 2nd edition, 1969.

S. Lang. *Calculus of Several Variables*. Springer, New York, 3rd edition, 1987.

P. D. Lax and M. S. Terrell. *Multivariable Calculus with Applications*. Springer, Cham, CH, 2017.

Chapter 16
Triple Integrals and Coordinate Changes

In the present lesson several issues about multi-dimensional integrals are discussed. In particular, the tool of the flow across a surface of a vector function, that has been introduced in the previous lesson, is here applied to many sample cases (see Sect. 16.1). It has a great importance in several physical applications, from Electromagnetism to Fluid Mechanics. The integrals in $\mathbb{R}^3$ are then introduced in Sect. 16.2 and the reduction of a triple integral to a sequence of three one-dimensional integrals is discussed, too. Finally, in Sect. 16.3 the issue of the change of coordinates in a triple integral is faced: it often happens, indeed, that a difficult triple integral strongly simplifies, once it is rewritten in a suitable coordinate system. For this reason, the coordinate change is a powerful tool in handling multiple integrals.

16.1 Flux of a Vector Function Across a Surface

In the previous lesson the flux of a continuous vector function $f : \mathbb{R}^3 \to \mathbb{R}^3$ across a surface Σ lying inside the existence field of f has been defined. The surface is given in a parametric form, *i.e.* three functions of the two variables (u, v), called parameters, are given in such a way that, when the point (u, v) belongs to a domain A in the parameter plane, the vector $x = (x, y, z)^\mathsf{T}$:

$$\begin{cases} x = x(u, v) \\ y = y(u, v) \\ z = z(u, v) \end{cases} \tag{16.1}$$

moves on the surface Σ. In the following, the functions 16.1 are assumed continuous with continuous first derivatives in A. Moreover, in order to define at each point of Σ the normal direction, it is also required that they satisfy the condition:

© The Author(s), under exclusive license to Springer Nature Switzerland AG 2024
G. Riccardi et al., *Multidimensional Differential and Integral Calculus*,
https://doi.org/10.1007/978-3-031-70326-3_16

$$\partial_u x \times \partial_v x \neq 0 \tag{16.2}$$

at all points of A.[1] The area element on the surface has been written in the following form:

$$dS = |\partial_u x \times \partial_v x| \, du \, dv, \tag{16.3}$$

which is equivalent to the well known form $dS = \sqrt{EG - F^2} \, du \, dv$, once the scalar functions $E = |\partial_u x|^2$, $G = |\partial_v x|^2$ and $F = \partial_u x \cdot \partial_v x$ have been introduced. Σ is considered an *oriented surface* defining at every point of it normal unit vector:

$$v = \frac{\partial_u x \times \partial_v x}{|\partial_u x \times \partial_v x|}, \tag{16.4}$$

that is a continuous vector function of the parameters (u, v). The positive face is the one v points to, whereas the other face is conventionally named as negative. It has been stressed that the product $dS\, v$, evaluated by means of the relations (16.3), (16.4), is written as:

$$dS\, v = \partial_u x \times \partial_v x \, du \, dv, \tag{16.5}$$

the above form showing that it does not depend on the choice of the parameters (u, v). In other words, if the new parameters $u' = u'(u, v)$ and $v' = v'(u, v)$ are defined (with $\partial(u', v')/\partial(u, v) > 0$), the product $dS\, v$ is written in a form which is identical to the one (16.5), but with u' and v' replacing u and v, respectively.

The flux of the continuous function $f : \mathbb{R}^3 \to \mathbb{R}^3$ across Σ is then defined and evaluated as the two-dimensional integral:

$$\int_\Sigma dS(x)\, v(x) \cdot f(x) = \int_A du\,dv\, \partial_u x \times \partial_v x \cdot f[x(u, v)]. \tag{16.6}$$

It is important to note that inside the double integral on Σ (at the left hand side) the function f is evaluated at the point $x \in \Sigma$, *i.e. on the surface* Σ. The tool (16.6) is now applied to sample cases.

Example 16.1 Consider the flux of the function $f(x) = (x, y, xy)^\mathsf{T}$ across the segment of paraboloid $z = 1 - x^2 - y^2$ having z positive. The surface is oriented taking the normal unit vector as coming out the paraboloid. The simplest choice of the parameters lies in the use the polar coordinates (ρ, θ) inside the unit circle, which is also the basis of this segment of the paraboloid. The surface results to be parametrised as $x = (\rho \cos \theta, \rho \sin \theta, 1 - \rho^2)^\mathsf{T}$ and the product $dS\, v$ becomes:

$$dS(x)\, v(x) = \rho \, d\rho \, d\theta \begin{pmatrix} 2\rho \cos \theta \\ 2\rho \sin \theta \\ 1 \end{pmatrix}.$$

---

[1] The condition (16.2) implies that the two tangent vectors $\partial_u x$ (tangent to the curves $v \equiv$ const.) and $\partial_v x$ (tangent to the curves $u \equiv$ const.) are not parallel, so that they form a basis on the tangent plane.

The flux is now evaluated as follows:

$$\int_{\Sigma} dS(x)\, v(x) \cdot f(x) = \int_0^1 d\rho\, \rho \int_0^{2\pi} d\theta \left(2\rho^2 + \rho^2 \sin\theta \cos\theta\right) = \pi .$$

Example 16.2 On a sphere of radius R having its center at a point $z_c \in (0, R)$ of the z-axis consider the segment with $x > 0$, $y > 0$ and $z > 0$. This part of the spherical surface is parametrised with the latitude $\varphi \in (\varphi_m, \pi/2)$, with $\varphi_m = -\arcsin(z_c/R)$, and the longitude $\theta \in (0, \pi/2)$:

$$x(\varphi, \theta) = \begin{pmatrix} R \cos\varphi \cos\theta \\ R \cos\varphi \sin\theta \\ z_c + R \sin\varphi \end{pmatrix} .$$

The surface is oriented defining the normal unit vector v outward the sphere, so that the product $dS\, v$ is:

$$dS(x)\, v(x) = R^2 \cos\varphi\, d\varphi\, d\theta \begin{pmatrix} \cos\varphi \cos\theta \\ \cos\varphi \sin\theta \\ \sin\varphi \end{pmatrix} .$$

The flux of the function $f(x) = (x^2, y^2, z)^{\mathsf{T}}$ across this surface is now evaluated by means of the above result:

$$\int_{\Sigma} dS(x)\, v(x) \cdot f(x) = R^4 \int_{\varphi_m}^{\pi/2} d\varphi\, \cos^4\varphi \left(\int_0^{\pi/2} d\theta\, \cos^3\theta + \int_0^{\pi/2} d\theta\, \sin^3\theta \right)$$
$$+ \frac{\pi}{2} R^2 \left(z_c \int_{\varphi_m}^{\pi/2} d\varphi\, \sin\varphi \cos\varphi + R \int_{\varphi_m}^{\pi/2} d\varphi\, \sin^2\varphi \cos\varphi \right) ,$$

where the following integrals:

$$\int_0^{\pi/2} d\theta\, \cos^3\theta = \int_0^{\pi/2} d\theta\, \sin^3\theta = \frac{2}{3}$$

$$\int_{\varphi_m}^{\pi/2} d\varphi\, \cos^4\varphi = \frac{3}{8}\left(\frac{\pi}{2} - \varphi_m\right) + \frac{1}{8}\frac{z_c}{R}\sqrt{1 - \frac{z_c^2}{R^2}}\left(5 - 2\frac{z_c^2}{R^2}\right)$$

$$\int_{\varphi_m}^{\pi/2} d\varphi\, \sin\varphi \cos\varphi = \frac{1}{2}\left(1 - \frac{z_c^2}{R^2}\right)$$

$$\int_{\varphi_m}^{\pi/2} d\varphi\, \sin^2\varphi \cos\varphi = \frac{1}{3}\left(1 + \frac{z_c^3}{R^3}\right)$$

are used.

Example 16.3 Consider the surface Σ given in explicit form $z = x^2 y$ with $x^2 + y^2 < 1$ and evaluate the flux across Σ of the vector function $f(x) = (y^2, x^2, xy)^{\mathsf{T}}$. Once the unit circle in the (x, y)-plane is parametrised in polar coordinates $x = \rho \cos \theta$, $y = \rho \sin \theta$ with $0 \le \rho \le 1$ and $0 \le \theta < 2\pi$, the surface Σ and the function f *evaluated on this surface* are written in the parametric forms:

$$x(\rho, \theta) = \begin{pmatrix} \rho \cos \theta \\ \rho \sin \theta \\ \rho^3 \sin \theta \cos^2 \theta \end{pmatrix}, \qquad f(\rho, \theta) = \begin{pmatrix} \rho^2 \sin^2 \theta \\ \rho^2 \cos^2 \theta \\ \rho^2 \sin \theta \cos \theta \end{pmatrix}.$$

The flux across the surface is calculated by means of the integral:

$$\int_{\Sigma} dS(x)\, v(x) \cdot f(x) = \int_0^1 d\rho \int_0^{2\pi} d\theta \left(-2\rho^5 \sin^3 \theta \cos \theta - \rho^5 \cos^4 \theta + \rho^3 \sin \theta \cos \theta \right)$$

$$= -\frac{\pi}{8}.$$

16.2 Triple Integrals

In previous lessons the one-dimensional integral has been extended to functions from $\mathbb{R}^2$ to $\mathbb{R}^1$ and bounded sets of $\mathbb{R}^2$, by means of the tool of the double integral. Important applications of this new tool have been discussed, with the aim to evaluate the area of a surface, or the flux of a vector function across a surface. In the present lesson, the integral is extended to functions from $\mathbb{R}^3$ to $\mathbb{R}$ and bounded three dimensional domains, due to the boundedness of D.

Consider a bounded domain $D \subset \mathbb{R}^3$. Due to the fact that D is bounded, in a suitable cartesian reference system (x, y, z) exists a parallelepiped $\Pi = (a_x, b_x) \times (a_y, b_y) \times (a_z, b_z)$ that includes D. Intersecting Π with planes parallel to the coordinate ones, it is divided in N small parallelepiped π_i and the following non-empty sets are defined:

$$a_i = \pi_i \cap D \quad \text{for} \quad i = 1, 2, \ldots, N.$$

The volume of the set a_i will be indicated by $|a_i|$, whereas its diameter[2] will be called d_i. The maximum among the diameters $\{d_i, \ i = 1, 2, \ldots, N\}$ will be called δ. The family $\{a_i, \ i = 1, 2, \ldots, N\}$ of subsets of D is named as *coordinate partition of D*, whereas δ is its *norm*. Retracing the above definitions in the opposite sense, it is possible to take δ as an independent variable and to consider the number of subsets N of D as a (rather complicated) function of δ. This choice results to be advantageous, the limit as $\delta \to 0$ ($|a_i| \to 0$) implying the one as $N \to +\infty$ (the converse is not true).

---

[2] The diameter d of a set A is defined as:

$$d = \sup_{x', x'' \in A} |x' - x''|.$$

Once an arbitrary point $\boldsymbol{\xi}_i$ is chosen in each subset a_i, the *Riemann sum S*:

$$S(\delta) := \sum_{i=1}^{N(\delta)} |a_i| f(\boldsymbol{\xi}_i) \tag{16.7}$$

is introduced. As stressed in the definition (16.7), this sum depends on the above partition of D, and in particular on its norm δ. The use of the notation $S(\delta)$ is a rather coarse simplification, due to the fact that S depends on all the details of the adopted partition of D and, in addition, on δ. The definition (16.7) satisfies an important property. For decreasing δ, the number N of the subsets of the partition increases as δ^{-3}. At the same time, the volumes of these subsets reduce in the same way as δ^3. If the function f is continuous in D, as supposed before, the limit of the sum (16.7) as $\delta \to 0$ exists and is finite. This limit is indicated with the symbol:

$$\int_D dV(\boldsymbol{x}) \, f(\boldsymbol{x}), \tag{16.8}$$

which is named as *triple integral of f on the domain* $D \subset \mathbb{R}^3$. $dV(\boldsymbol{x})$ is the *volume element* at the point $\boldsymbol{x}$ in $\mathbb{R}^3$ and it could be directly thought as the product of the three differentials of the cartesian coordinates: if $\boldsymbol{x} = (x, y, z)^\mathsf{T}$, it is possible to write $dV(\boldsymbol{x}) = dx \, dy \, dz$, too.

The Reader has to be advised that, as for the double integrals, different notations are also employed for indicating the limit as $\delta \to 0$ of the Riemann sum (16.7). In particular, it is often used the symbol:

$$\iiint_D dx \, dy \, dz \, f(x, y, z),$$

which stresses the fact that the integral is triple, *i.e.* it is made on a subset of $\mathbb{R}^3$.

As it has been previously discussed for the double integrals, also the triple integrals are evaluated partitioning the domain D in the union of subdomains that are normal with respect to some coordinate direction. In order to clarify this important point, assume that two continuous functions $\alpha = \alpha(x, y)$ and $\beta = \beta(x, y)$ exist such that any point $\boldsymbol{x} = (x, y, z)^\mathsf{T} \in D$ has the first two coordinates belonging to a certain bounded set $A \subset \mathbb{R}^2$, whereas the third one satisfies the relations: $\alpha(x, y) \leq z \leq \beta(x, y)$. In other words, the orthogonal projection of D on the plane (x, y) determines the set A and D is composed by all the points of $\mathbb{R}^3$ lying between the surfaces $z = \alpha(x, y)$ and $z = \beta(x, y)$. In this case, D is said to be *normal with respect to the axis z*. In order to evaluate the integral (16.8) for the present D, it is appropriate to use the following *reduction formula*:

$$\int_D dV(\boldsymbol{x}) \, f(\boldsymbol{x}) = \int_A dA(x, y) \int_{\alpha(x,y)}^{\beta(x,y)} dz \, f(x, y, z). \tag{16.9}$$

In applying the formula (16.9), the first step consists in evaluating the one dimensional integral in z. The resulting function depends on the two coordinates x and y, only. As a consequence, in the second step it is integrated on the domain A of the plane (x, y).

The domains which are normal with respect to the other coordinate directions are defined in a completely analogous way. Examples of calculation of triple integrals on normal domains in which the reduction formula (16.9), or the analogous one along a different coordinate direction, is applied are summarised below.

Example 16.4 Calculate the integral of the function $f(x) = z^2(x^2 - y^2)$ on the domain $D := \{(y, z) \in B_1(0), 2[(y^2 + z^2) - 1] \leq x \leq 1 - (y^2 + z^2)\}$ ($B_1(0)$ is the circle with center on the origin and radius 1), which results to be normal with respect to the x-direction. Using the reduction formula (16.9) along x and the polar coordinates (ρ, θ) (where $0 \leq \rho \leq 1$ and $0 \leq \theta < 2\pi$) in the (y, z)-plane, the integral becomes:

$$
\int_D dV(x)\, f(x) = \int_{B_1(z)} dA(y, z)\, z^2 \int_{2[(y^2+z^2)-1]}^{1-(y^2+z^2)} dx\, (x^2 - y^2)
$$

$$
= \int_0^1 d\rho\, \rho^3 \int_0^{2\pi} d\theta\, \sin^2\theta \int_{2(\rho^2-1)}^{1-\rho^2} dx\, (x^2 - \rho^2 \cos^2\theta)
$$

$$
= \int_0^1 d\rho\, \rho^3 \int_0^{2\pi} d\theta\, \sin^2\theta\, \Big\{ \frac{1}{3}[(1 - \rho^2)^3 - 8(\rho^2 - 1)^3]
$$

$$
- \rho^2 \cos^2\theta\, [(1 - \rho^2) - 2(\rho^2 - 1)] \Big\}
$$

$$
= 3\pi \int_0^1 d\rho\, \rho^3 \Big[(-\rho^6 + 3\rho^4 - 3\rho^2 + 1) + \frac{1}{4}\rho^2(\rho^2 - 1) \Big]
$$

$$
= \frac{7}{160}\pi \simeq 0.13744 .
$$

Example 16.5 Calculate the integral of the function $f(x) = z/(1 + x^2 y^2)$ on the sector of the elliptical paraboloid $z = 4 - (x^2 + y^2)$ having positive z. This domain is normal with respect to the axis z and the reduction formula (16.9) can be directly applied. Using polar coordinates inside the circle $B_2(0)$ on the (x, y)-plane and changing coordinates from ρ to $\eta := \rho^2/4$, the integral is calculated in the following way:

$$
\int_D dV(x)\, f(x) = \int_{B_2(0)} \frac{dA(x, y)}{1 + x^2 y^2} \int_0^{4-(x^2+y^2)} dz\, z
$$

$$
= 16 \int_0^1 d\eta\, (1 - \eta)^2 \int_0^{2\pi} \frac{d\theta}{1 + 4\eta^2 \sin^2\theta} . \tag{16.10}
$$

The inner integral in θ is evaluated changing $\sin^2\theta$ in $(1 - \cos 2\theta)/2$ and setting $\beta^2 = 2\eta^2/(1 + 2\eta^2) \in (0, 1)$:

$$\int_0^{2\pi} \frac{d\theta}{1 + 4\eta^2 \sin^2 \theta} = \frac{1}{1 + 2\eta^2} \int_{-\pi}^{+\pi} \frac{d\theta}{1 - \beta^2 \cos \theta},$$

where the classical change of coordinate $t = \tan(\theta/2)$ leads to the result:

$$\int_0^{2\pi} \frac{d\theta}{1 + 4\eta^2 \sin^2 \theta} = \frac{2\pi}{\sqrt{1 + 4\eta^2}}. \tag{16.11}$$

The value (16.11) is put inside the reduction formula (16.10) and the variable η is changed in the new one ξ according to the formula: $2\eta = \sinh \xi$. The integral follows as:

$$\int_D dV(\boldsymbol{x}) \, f(\boldsymbol{x}) = 32\pi \int_0^1 d\eta \, \frac{(1 - \eta)^2}{\sqrt{1 + 4\eta^2}}$$

$$= 4\pi \int_0^{\log(2+\sqrt{5})} d\xi \, (2 - \sinh \xi)^2$$

$$= 2\pi \, [7 \log(2 + \sqrt{5}) - 6\sqrt{5} + 8] \simeq 29.46211 \, .$$

Example 16.6 Finally, calculate the integral of the function $f(\boldsymbol{x}) = z \sin \sqrt{x^2 + y^2}$ on the sector with $0 \le z \le 1$ of the right circular cone having the circle $B_1(\boldsymbol{0})$ of the (x, y)-plane as base and vertex in the point $(0, 0, 1)^\mathsf{T}$. This domain is normal with respect to the axis z, *i.e.* for any $(x, y) \in B_1(\boldsymbol{0})$, the z-coordinate verifies the conditions: $0 \le z \le 1 - \sqrt{x^2 + y^2}$. Changing the coordinates from cartesian to polar (ρ, θ) with $0 \le \rho \le 1$ and $0 \le \theta < 2\pi$ inside $B_1(\boldsymbol{0})$, the integral is evaluated as:

$$\int_D dV(\boldsymbol{x}) \, f(\boldsymbol{x}) = \int_0^1 d\rho \, \rho \sin \rho \int_0^{2\pi} d\theta \int_0^{1-\rho} dz \, z$$

$$= \pi \int_0^1 d\rho \, (\rho - 2\rho^2 + \rho^3) \sin \rho$$

$$= 2\pi \, (\cos 1 + 2 - 3 \sin 1) \simeq 0.09984,$$

Using the integrals:

$$\int_0^1 d\rho \, \rho \sin \rho = \sin 1 - \cos 1 \, ,$$

$$\int_0^1 d\rho \, \rho \cos \rho = \sin 1 + \cos 1 - 1 \, ,$$

$$\int_0^1 d\rho \, \rho^2 \sin \rho = \cos 1 - 2(1 - \sin 1) \, ,$$

$$\int_0^1 d\rho \, \rho^2 \cos \rho = 2 \cos 1 - \sin 1 \, ,$$

$$\int_0^1 d\rho \, \rho^3 \sin \rho = 5 \cos 1 - 3 \sin 1 \, .$$

16.3   Change of Variables in the Triple Integrals

Also for the triple integrals, it often occurs that the domain of integration (D) has a simple geometrical representation in a system of coordinates different from the cartesian ones. In the present section, the consequences of the change from the cartesian coordinates $x = (x, y, z)^\mathsf{T}$ to the new ones $u = (u, v, w)^\mathsf{T}$ is investigated. This change is achieved by means of a suitable *vector function of a vector variable*, that is called *transformation*:

$$x = x(u). \tag{16.12}$$

The function (16.12) is assumed to be bijective, *i.e.* onto (surjective) and one-to-one (injective), continuous and having continuous first derivatives. The image of the cartesian domain of integration D will be called $u(D) =: T$ and is, in general, a simpler domain. Finally, it is required that the *Jacobian determinant* of the transformation (16.12):

$$J = \frac{\partial(x, y, z)}{\partial(u, v, w)} := \begin{vmatrix} \partial_u x & \partial_v x & \partial_w x \\ \partial_u y & \partial_v y & \partial_w y \\ \partial_u z & \partial_v z & \partial_w z \end{vmatrix}, \tag{16.13}$$

viewed as a scalar function of the vector variable u, does not change its sign for any $u \in T$. In other words, $J(u)$ (16.13) is positive, or negative, in T. It vanishes in subsets of T having zero volume, as points, curves or surfaces.

Fixed a point u_0 belonging to the inside of T, consider the positive increments Δu, Δv and Δw of the coordinates u_0, v_0 and w_0 so small that all the points $(u_0 + \Delta u, v_0, w_0)^\mathsf{T}$, $(u_0, v_0 + \Delta v, w_0)^\mathsf{T}$, $(u_0, v_0, w_0 + \Delta w)^\mathsf{T}$, $(u_0 + \Delta u, v_0 + \Delta v, w_0)^\mathsf{T}$ and so on still belong to the domain T. When a point u moves from $(u_0, v_0, w_0)^\mathsf{T}$ to $(u_0 + \Delta u, v_0, w_0)^\mathsf{T}$ keeping v and w fixed to v_0 and w_0, the corresponding increment of the cartesian position x occurs along the curve $x = x(u, v_0, w_0)$ so that:

$$x(u_0 + \Delta u, v_0, w_0) - x(u_0, v_0, w_0) = \partial_u x_0 \Delta u + O(\Delta u)^2, \tag{16.14}$$

where the symbol $\partial_u x_0$ means $\partial_u x(u_0, v_0, w_0)$ and the modulus of the vector $O(\Delta u)^2$ vanishes at least as $(\Delta u)^2$ for $\Delta u \to 0$. Analogous formulae can be easily found for the increments along the other two coordinate directions v and w. Remembering that the volume of the parallelepiped formed by the vectors a, b and c is given by the absolute value of their mixed product $a \cdot b \times c$, to the element of volume in the new coordinates $\Delta u \Delta v \Delta w = dV(u_0)$ corresponds the cartesian element:

$$dV(x_0) = |\partial_u x_0 \cdot \partial_v x_0 \times \partial_w x_0| \, dV(u_0) + \text{higher order terms.} \tag{16.15}$$

Note that the triple product $\partial_u x_0 \cdot \partial_v x_0 \times \partial_w x_0$ is just the Jacobian determinant (16.13) evaluated in u_0. Moreover, the higher order terms in the formula (16.15) come from the corresponding terms which appear in writing the increments along the coordinate directions u (16.14), v and w. These latter are due to the non-straight shapes of the curves $x = x(u, v_0, w_0)$, $x = x(u_0, v, w_0)$ and $x = x(u_0, v_0, w)$.

A key observation is now pointed out about the role played in building the integral (16.8) by the higher order terms in the volume element (16.15). Indeed, due to the fact that they are of higher order with respect to δ^3 as $\delta \to 0$, their contribution to the Riemann sum (16.7) vanishes in that limit. As a consequence, the volume element in the cartesian coordinates x is related to the analogous element in the coordinates u by the following rule:

$$dV(x) = |J(u)| \, dV(u) . \tag{16.16}$$

Before showing some sample cases, it is worth noting that the physical dimensions of the volume element $dV(u)$ can differ from the cube of a length, if the physical dimension of some new coordinate (u, v or w) is not a length. Important examples are the following ones.

Consider the *spherical coordinates* (ρ, φ, θ) related to the cartesian ones by the following rules:

$$\begin{cases} x = \rho \cos \varphi \cos \theta \\ y = \rho \cos \varphi \sin \theta \\ z = \rho \sin \varphi \end{cases} \tag{16.17}$$

where $\rho \in [0, +\infty)$ is the radial distance, $\varphi \in [-\pi/2, +\pi/2]$ is the latitude ($\varphi = 0$ on the equatorial (x, y)-plane, $+\pi/2$ at the north pole and $-\pi/2$ at the south pole) and $\theta \in [0, 2\pi)$ is the longitude. This latter angle is usually measured starting from the (x, z)-plane, where $\theta = 0$ is assumed, and it is taken as positive by an observer placed along the positive z-axis in the counterclockwise direction. Due to the fact that two of the three new coordinates are angles (φ and θ), the volume element $dV(u) = d\rho d\varphi d\theta$ has the physical dimension of a length. The Jacobian determinant (16.13) of the transformation (16.17) is calculated as:

$$J = \frac{\partial(x, y, z)}{\partial(\rho, \theta, \varphi)}$$

$$= \begin{vmatrix} \cos \varphi \cos \theta & -\rho \cos \varphi \sin \theta & -\rho \sin \varphi \cos \theta \\ \cos \varphi \sin \theta & \rho \cos \varphi \cos \theta & -\rho \sin \varphi \sin \theta \\ \sin \varphi & 0 & \rho \cos \varphi \end{vmatrix}$$

$$= \rho^2 \cos \varphi \begin{vmatrix} \cos \varphi \cos \theta & -\sin \theta & -\sin \varphi \cos \theta \\ \cos \varphi \sin \theta & \cos \theta & -\sin \varphi \sin \theta \\ \sin \varphi & 0 & \cos \varphi \end{vmatrix}$$

$$= \rho^2 \cos \varphi .$$

Note that J is positive at any point outside the z-axis ($\varphi = \pm\pi/2$), where it vanishes. The law for changing the volume element (16.16) becomes:

$$dV(x) = \rho^2 \cos \varphi \, d\rho \, d\varphi \, d\theta . \tag{16.18}$$

As stated above, the physical dimension of the volume element in the new coordinates $(d\rho\, d\varphi\, d\theta)$ is a length and J gives the remaining $(\text{length})^2$, needed to reach the physical dimensions of $dV(x)$.

Consider now the transformation from cartesian to *toroidal coordinates*. The axis z is assumed as the symmetry axis of the torus. The circle $\mathscr{C}$ on the plane parallel to the plane (x, y) with center at the point z_0 of the z-axis and radius R is fixed. In order to identify the toroidal coordinates of a point $x \in \mathbb{R}^3$, the plane $\pi(x)$ through the z-axis and the point x (here it is assumed that x does not belong to the z-axis) it is firstly considered. The longitude angle φ between the coordinate (x, z)-plane and $\pi(x)$ is the first toroidal coordinate. On $\pi(x)$, the cartesian reference system having axes (ξ, η), with ξ defined by the intersection between $\pi(x)$ and the plane $z = 0$ and η coinciding with z, is now considered. Being the distance from the symmetry axis, the coordinate ξ is assumed non-negative.

In this coordinate system, the intersection C between $\pi(x)$ and $\mathscr{C}$ lies in the point (R, z_0). Finally, the position of x on $\pi(x)$ is specified by means of polar coordinates (ρ, θ) with center on the point C: $\xi = R + \rho\cos\theta, \eta = z_0 + \rho\sin\theta$, having assumed $\theta = 0$ on the half-line $\xi = R + \sigma, \eta = z_0$ with $\sigma > 0$. As a consequence, the point x has toroidal coordinates (φ, ρ, θ):

$$\begin{cases} x = (R + \rho\cos\theta)\cos\varphi \\ y = (R + \rho\cos\theta)\sin\varphi \\ z = z_0 + \rho\sin\theta\,, \end{cases} \tag{16.19}$$

for all the angles θ such that $R + \rho\cos\theta \geq 0$.

The Jacobian determinant is:

$$J = \frac{\partial(x, y, z)}{\partial(\rho, \theta, \varphi)} \tag{16.20}$$

$$= \begin{vmatrix} \cos\theta\cos\varphi & -\rho\sin\theta\cos\varphi & -(R + \rho\cos\theta)\sin\varphi \\ \cos\theta\sin\varphi & -\rho\sin\theta\sin\varphi & (R + \rho\cos\theta)\cos\varphi \\ \sin\theta & \rho\cos\theta & 0 \end{vmatrix} \tag{16.21}$$

$$= \rho(R + \rho\cos\theta)\begin{vmatrix} \cos\theta\cos\varphi & -\sin\theta\cos\varphi & -\sin\varphi \\ \cos\theta\sin\varphi & -\sin\theta\sin\varphi & \cos\varphi \\ \sin\theta & \cos\theta & 0 \end{vmatrix} \tag{16.22}$$

$$= -\rho(R + \rho\cos\theta)\,, \tag{16.23}$$

as a consequence the relation (16.16) between the volume element in the system (ρ, θ, φ) and the one in the cartesian space is written as:

$$dV(x) = \rho(R + \rho\cos\theta)\, d\rho\, d\theta\, d\varphi\,. \tag{16.24}$$

Also in this case, the volume element in the new coordinates $(d\rho\, d\theta\, d\varphi)$ has the physical dimensions of a length, whereas the one of the Jacobian is the square of a length.

16.4 Suggested Readings

A comprehensive presentation of triple integrals and their solution using change of coordinates is found in Lax and Terrell (2017, Chap. 6, pp. 261–278), Adams and Essex (2010, Chap. 14, pp. 818–829), Lang (1987, Chap. XI, Chap. XVII, pp. 478–481), Apostol (1974, Chap. 14), Apostol (1969, Chap. 11, pp. 405–413).

References

R. A. Adams and C. Essex. *Calculus: Several Variables*. Pearson Education Canada, Toronto, ON, 7th edition, 2010.

T. M. Apostol. *Calculus, Volume II*. Wiley, Hoboken, NJ, 2nd edition, 1969.

T. M. Apostol. *Mathematical Analysis: A Modern Approach to Advanced Calculus*. Pearson Education US, Hoboken, NJ, 2nd edition, 1974.

S. Lang. *Calculus of Several Variables*. Springer, New York, 3rd edition, 1987.

P. D. Lax and M. S. Terrell. *Multivariable Calculus with Applications*. Springer, Cham, CH, 2017.

Chapter 17
Green's Formulae for the Integral Calculus

In the present lesson some important relations used in evaluating multi-dimensional integrals are deduced and applied. These relations are at the bases of two theorems—attributed to C.F. Gauss[1] and G.G. Stokes[2]—that play a fundamental role in the multi-dimensional integration and in its applications to the Electromagnetism and to the Continuum Mechanics.

17.1 Green's Formulae

Green's formulae can be viewed as a multi-dimensional extension of the one-dimensional integration by parts. They establish the equality between the integral on a (bounded) domain D with smooth (and bounded) boundary ∂D of the derivative along the direction $\boldsymbol{d}$ of a continuous function $f : \mathbb{R}^n \to \mathbb{R}$ ($n = 2$ or 3) having continuous partial derivatives and the integral on ∂D of the same function multiplied by the component along $\boldsymbol{d}$ of the outward unit normal vector. Besides having a great usefulness in the calculations, these formulae are at the basis of the analytical handling of Electromagnetism and Continuum Mechanics. The discussion starts from the two-dimensional case ($n = 2$).

Consider a continuous function $f : \mathbb{R}^2 \to \mathbb{R}$ having continuous partial derivatives and the integral over the (bounded) domain D of its derivative along x:

$$\int_D dA(\boldsymbol{x})\, \partial_x f(x, y). \tag{17.1}$$

The simplest case is the one in which D is normal with respect to the y-axis, as in Fig. 17.1a: two functions (α and β) of y exist, such that any point $(x, y) \in D$ has

---

[1] https://mathshistory.st-andrews.ac.uk/Biographies/Gauss/.

[2] https://mathshistory.st-andrews.ac.uk/Biographies/Stokes/.

© The Author(s), under exclusive license to Springer Nature Switzerland AG 2024   205
G. Riccardi et al., *Multidimensional Differential and Integral Calculus*,
https://doi.org/10.1007/978-3-031-70326-3_17

$a \leq y \leq b$ and, in correspondence to the fixed value of y, $\alpha(y) \leq x \leq \beta(y)$. In this case, the integral (17.1) can be reduced to a one-dimensional integral:

$$\int_a^b dy \int_{\alpha(y)}^{\beta(y)} dx\, \partial_x f(x, y) = \int_a^b dy\, \{f[\beta(y), y] - f[\alpha(y), y]\}$$

$$= \int_a^b dy\, f[\beta(y), y] + \int_b^a dy\, f[\alpha(y), y]. \quad (17.2)$$

The boundary of D is now oriented in the counterclockwise direction. It follows that ∂D is the union of four smooth curves: the curve $(\beta(y), y)$ for y running from a to b; the straight segment (it can collapse to a point) (x, b) for x from $\beta(b)$ to $\alpha(b)$; the curve $(\alpha(y), y)$ for y from b to a and the straight segment (again, it can collapse to a point) (x, a) for x from $\alpha(a)$ to $\beta(a)$. However, due to the fact that dy vanishes on the two segments, the integral (17.2) can be rewritten as:

$$\int_a^b dy \int_{\alpha(y)}^{\beta(y)} dx\, \partial_x f(x, y) = \int_{\partial D} dy\, f(x, y). \quad (17.3)$$

The integral (17.3) is now put in a more significant form.

As well known (see lesson 2), naming with s the curvilinear abscissa on the curve $(\beta(y), y)$, the derivative in s of the coordinate y is written as:

$$\frac{dy}{ds} = \frac{1}{\sqrt{1 + \beta'^2}}. \quad (17.4)$$

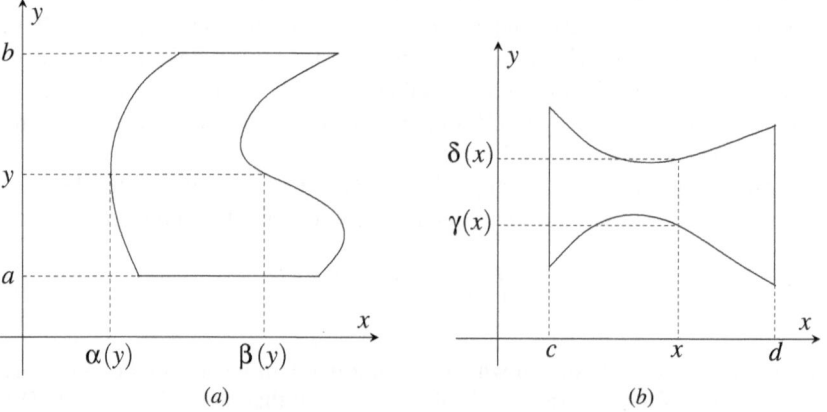

Fig. 17.1 Bounded domains with respect to the y-axis (**a**) and to the x-axis (**b**)

Due to the fact that the tangent unit vector τ is:

$$\tau = \frac{1}{\sqrt{1+\beta'^2}} \begin{pmatrix} \beta' \\ 1 \end{pmatrix},$$

the corresponding outward normal unit vector is obtained considering the opposite of the orthogonal of τ (or, the orthogonal vector obtained by means of a rotation of $\pi/2$ in the clockwise direction):

$$\nu = \frac{1}{\sqrt{1+\beta'^2}} \begin{pmatrix} 1 \\ -\beta' \end{pmatrix}. \tag{17.5}$$

Comparing Eqs. (17.4) and (17.5) it follows that $\nu_x = dy/ds$ and then:

$$dy = \nu_x \, ds. \tag{17.6}$$

Analogous considerations are also valid for the curve $(\alpha(y), y)$. The use of the differential (17.6) inside the integral at the right hand side of Eq. (17.3) leads to a first Green's formula:

$$\int_D dA(\boldsymbol{x}) \, \partial_x f(x, y) = \int_{\partial D} ds \, \nu_x(s) \, f[x(s), y(s)], \tag{17.7}$$

where it must be recalled that the domain boundary ∂D is counterclockwise oriented and the normal unit vector is taken outwards D.

Assume now that the domain D is normal with respect to the x-axis, as shown in Fig. 17.1b. In this case two functions $\gamma(x)$ and $\delta(x)$ exist such that any point $(x, y) \in D$ satisfies the relations: $c \le x \le d$ and, in correspondence to the fixed value of x, $\gamma(x) \le y \le \delta(x)$. In this case, the integral (17.1) is rewritten as:

$$\int_D dA(\boldsymbol{x}) \, \partial_x f(x, y) = \int_c^d dx \int_{\gamma(x)}^{\delta(x)} dy \, \partial_x f(x, y). \tag{17.8}$$

The integrals at the right hand side of the above relation are now handled using the result (11.2) about the derivative of a definite integral having variable limits, deduced in Sect. 11.1. In the present case, it is specified as:

$$\int_{\gamma(x)}^{\delta(x)} dy \, \partial_x f(x, y) = \frac{d}{dx} \int_{\gamma(x)}^{\delta(x)} dy \, f(x, y) - \delta'(x) \, f[x, \delta(x)] + \gamma'(x) \, f[x, \gamma(x)]. \tag{17.9}$$

Inserting the result (17.9) in the form (17.8) of the integral, it can be rewritten in the significant form:

$$\int_D dA(\boldsymbol{x})\, \partial_x f(x, y) = \int_{\gamma(d)}^{\delta(d)} dy\, f(d, y) - \int_{\gamma(c)}^{\delta(c)} dy\, f(c, y)$$

$$- \int_c^d dx\, \delta'(x)\, f[x, \delta(x)] + \int_c^d dx\, \gamma'(x)\, f[x, \gamma(x)]$$

$$= \int_{\gamma(d)}^{\delta(d)} dy\, f(d, y) + \int_{\delta(c)}^{\gamma(c)} dy\, f(c, y)$$

$$+ \int_d^c dx\, \delta'(x)\, f[x, \delta(x)] + \int_c^d dx\, \gamma'(x)\, f[x, \gamma(x)],$$

$$(17.10)$$

which still leads to the form (17.3), once it is noted that $dy = \gamma'(x)\, dx$ on the curve $y = \gamma(x)$ and similarly on the curve $y = \delta(x)$. Moreover, on the two segments parallel to the y-axis the x-component of the outward unit normal vector is $+1$ in correspondence to $x = d$ and -1 for $x = c$. As a consequence, on the above segments $dy = -ds = v_x ds$ for $x = c$ and $dy = ds = v_x ds$ in correspondence to $x = d$. In this way the form (17.7) of the integral (17.1) is recovered, also in the case in which D is normal with respect to the x-axis.

As shown in Fig. 17.2 in a sample case, any (bounded) domain $D \subset \mathbb{R}^2$ can be viewed as the union of sub-domains that are normal at least with respect to one of the two axes. Once the boundaries of these domains are oriented in the same direction of the boundary of D, the integral at the right hand side of Eq. (17.7) can be written as the sum of the corresponding integrals evaluated on the boundaries of the sub-domains. The sides of these boundaries not lying on ∂D give a vanishing contribution to that sum. Indeed, consider two sub-domains having a side in common: in the sum this side gives two opposite contributions, being considered two times with opposite orientations. As a consequence, in the sum of the integrals on the boundaries of the sub-domains the only contributions that do not cancel are the ones made on arcs

Fig. 17.2 The (bounded) domain $D \subset \mathbb{R}^2$ can be viewed as the union of the domains A_1 (normal with respect to the y-axis), A_2 (y), A_3 (x) and A_4 (x). The boundary of D is oriented counterclockwise, as well as the boundaries of the sub-domains A_k ($k = 1, 2, 3$ and 4). As a consequence, the integral on ∂D is the sum of the integrals on ∂A_k, the contributions due to the common sides appearing twice, but with opposite signs

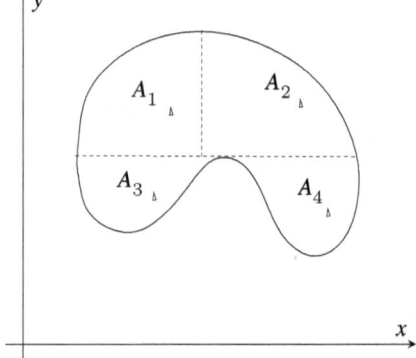

lying on ∂D so that the formula (17.7) is valid for any (bounded) domain D, with smooth and finite boundary ∂D.

Examine what happens to the integral (17.1) if the partial derivative along x is changed with the one along y:

$$\int_D dA(x)\, \partial_y f(x, y). \qquad (17.11)$$

Using the same approach of the previous case, the following relation

$$\int_D dA(x)\, \partial_y f(x, y) = -\int_{\partial D} dx\, f(x, y) = \int_{\partial D} ds\, v_y(s)\, f[x(s), y(s)], \qquad (17.12)$$

is found. Note that it is analogous to the formula (17.7), with the y-component of the normal unit vector in place of the one along x. At this stage, one can conclude, at least in the two-dimensional case, that to the derivative along the direction x (or y) in the left hand side corresponds the component along x (or y) of the normal unit vector at the right hand side. The relations (17.7), (17.12) are named as *Green's formulae* in the plane. These formulae were presented in 1828 in a paper published by George Green.[3] At the same time, they were discovered by Mikhail Ostrogradski.[4]

Assume now that the integral of a function f on a (bounded) three-dimensional domain D having a bounded boundary ∂D:

$$\int_D dV(x)\, \partial_x f(x) \qquad (17.13)$$

is given. In the integral (17.13) the function f is continuous with continuous first partial derivatives. The domain D is now assumed as normal with respect to the axis z: a bounded domain A in the (x, y)-plane exists as well as two continuous functions $\alpha(x, y)$ and $\beta(x, y)$ with continuous first partial derivatives such that for any $x \in D$, $(x, y) \in A$ and $\alpha(x, y) \leq z \leq \beta(x, y)$. As a consequence, the integral (17.13) can be decomposed as follows:

$$\int_D dV(x)\, \partial_x f(x) = \int_A dA(x, y) \int_{\alpha(x,y)}^{\beta(x,y)} dz\, \partial_x f(x) \qquad (17.14)$$

and the relation (11.2) about the derivative of a definite integral having variable limits is still applied:

---

[3] https://mathshistory.st-andrews.ac.uk/Biographies/Green/.

[4] https://mathshistory.st-andrews.ac.uk/Biographies/Ostrogradski/.

$$\int_{\alpha(x,y)}^{\beta(x,y)} dz\, \partial_x f(\boldsymbol{x}) = \partial_x \int_{\alpha(x,y)}^{\beta(x,y)} dz\, f(\boldsymbol{x})$$

$$- \partial_x \beta(x,y)\, f[x,y,\beta(x,y)] + \partial_x \alpha(x,y)\, f[x,y,\alpha(x,y)]\,.$$

Inserting the above relation inside the integral (17.14), the reduction formula:

$$\int_D dV(\boldsymbol{x})\, \partial_x f(\boldsymbol{x}) = \int_A dA(x,y)\left[\partial_x \int_{\alpha(x,y)}^{\beta(x,y)} dz\, f(\boldsymbol{x})\right]$$

$$- \int_A dA(x,y)\, \partial_x \beta(x,y)\, f[x,y,\beta(x,y)]$$

$$+ \int_A dA(x,y)\, \partial_x \alpha(x,y)\, f[x,y,\alpha(x,y)] \quad (17.15)$$

is obtained. The first integral at the right hand side can be further reduced by means of the two-dimensional Green's formula (17.7), whereas the calculation of the second and third integrals is more complicated.

In order to obtain a final form of the second integral at the right hand side of Eq. (17.15) in which the normal unit vector appears, The coordinates x and y are considered as parameters for the surface $z = \beta(x,y)$. As a consequence, a point on this surface is represented as $(x,y,\beta(x,y))$ and, using the results of the previous lesson (16), the product $\boldsymbol{v}\, dA$ is written as:

$$\boldsymbol{v}(x,y)\, dA(x,y) = \begin{vmatrix} \boldsymbol{e}_x & \boldsymbol{e}_y & \boldsymbol{e}_z \\ 1 & 0 & \partial_x \beta \\ 0 & 1 & \partial_y \beta \end{vmatrix} dA(x,y)\,,$$

the x-component of which is:

$$v_x(x,y)\, dA(x,y = -\partial_x \beta\, dA(x,y)\,. \quad (17.16)$$

For the third integral at the right hand side of Eq. (17.15) the surface $z = \alpha(x,y)$ is considered and the same product is written as:

$$v_x(x,y)\, dA(x,y) = -(-\partial_x \alpha)\, dA(x,y)\,, \quad (17.17)$$

the sign change being necessary in order to have an outward vector. Accounting for the relations (17.16), (17.17), the right hand side of Eq. (17.15) is rewritten as follows:

$$\int_D dV(\boldsymbol{x})\, \partial_x f(\boldsymbol{x}) = \int_{\partial A} ds\, v_x(s) \int_{\alpha(x(s),y(s))}^{\beta(x(s),y(s))} dz\, f[x(s),y(s),z]$$

$$+ \int_A dA(x, y) \, v_x(x, y) \, f[x, y, \beta(x, y)] +$$

$$+ \int_A dA(x, y) \, v_x(x, y) \, f[x, y, \alpha(x, y)]$$

$$= \int_{\partial D} dA(\boldsymbol{x}) \, v_x(\boldsymbol{x}) \, f(\boldsymbol{x}). \tag{17.18}$$

The result (17.18) is the three-dimensional extension of the Green's formula (17.18). A completely analogous result is obtained considering a domain D that is normal with respect to the axis x (or y) and, as a consequence, for an arbitrary (bounded) domain D having a bounded, smooth boundary.

The previous Green's formulae can be summarised by means of the following vector formula:

$$\int_D dV(\boldsymbol{x}) \, \nabla f(\boldsymbol{x}) = \int_{\partial D} dA(\boldsymbol{x}) \, \boldsymbol{v}(\boldsymbol{x}) \, f(\boldsymbol{x}). \tag{17.19}$$

Finally, through a scalar multiplication by a (fixed and arbitrary) unit vector $\boldsymbol{d}$ of both sides of the relation (17.19), the reduction formula for the volume integral of the derivative along the direction $\boldsymbol{d}$ of f is obtained. This integral is therefore written as the integral on the boundary of the domain (∂D) of the function f times the component along $\boldsymbol{d}$ of the unit outward vector $\boldsymbol{v}$.

17.2 Examples of Green's Formulae Application

Some applications of the two-dimensional formulae deduced in Sect. 17.1 are now discussed.

Consider the domain D shown in Fig. 17.3a and the integral:

$$\int_D dA(\boldsymbol{x}) \, \partial_x \frac{y}{x^2 + 2y^2 + 1}. \tag{17.20}$$

Due to the fact that the shape of D and the integrand function are quite simple, the integral (17.20) can be directly calculated once the derivative is evaluated:

$$\int_D dA(\boldsymbol{x}) \, \partial_x \frac{y}{x^2 + 2y^2 + 1} = -2 \int_D dA(\boldsymbol{x}) \, \frac{xy}{(x^2 + 2y^2 + 1)^2}$$

$$= -2 \int_{-1}^{+1} dx \, x \int_{(x-3)/2}^{1-x^2} dy \, \frac{y}{(x^2 + 2y^2 + 1)^2}$$

$$= \frac{1}{2} \int_{-1}^{+1} dx \, \frac{x}{2x^4 - 3x^2 + 3} - \int_{-1}^{+1} dx \, \frac{x}{3x^2 - 6x + 11}$$

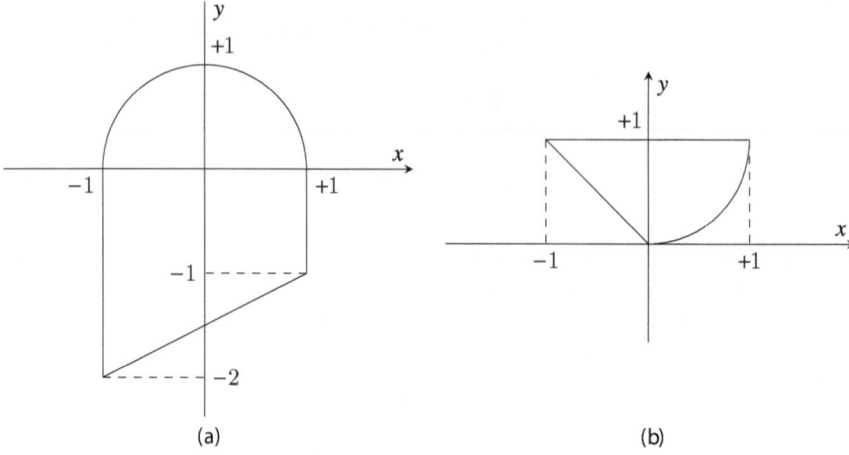

(a)   (b)

Fig. 17.3 In **a** and **b** the domains D of the integrals (17.20) and (17.23) are drawn. The curved parts of the boundaries are arcs of the parabolas $y = 1 - x^2$ (**a**) and $y = x^2$ (**b**). The domain **a** is normal with respect to the x-axis, whereas the domain **b** is taken as normal with respect to the y-axis

$$= \frac{1}{6} \log \frac{5}{2} - \frac{1}{4}\sqrt{\frac{2}{3}} \arctan \sqrt{\frac{3}{2}}. \tag{17.21}$$

The same integral (17.20) is now evaluated using Green's formula. Noted that $v_x ds = dy$, it follows:

$$\int_D dA(x) \, \partial_x \frac{y}{x^2 + 2y^2 + 1}$$

$$= \int_{-1}^{+1} \frac{dx}{2} \frac{(x-3)/2}{x^2 + 2[(x-3)/2]^2 + 1} + \int_{-1}^{0} dy \frac{y}{(+1)^2 + 2y^2 + 1}$$

$$+ \int_{+1}^{-1} (-2x\,dx) \frac{1 - x^2}{x^2 + 2(1 - x^2)^2 + 1} + \int_{0}^{-2} dy \frac{y}{(-1)^2 + 2y^2 + 1}$$

$$= \frac{1}{2} \int_{-1}^{+1} dx \frac{x-3}{3x^2 - 6x + 11} - \frac{\log 2}{4} + \frac{\log 5}{4}. \tag{17.22}$$

Once the integral in Eq. (17.22) is calculated by means of the change of variable $x = 2\sqrt{2/3}\,\xi + 1$, the result (17.21) still follows.

The second sample case consists in the calculation of the integral:

$$\int_D dA(x) \, \partial_x \frac{1 + xy}{1 + x + y} \tag{17.23}$$

on the domain drawn in Fig. 17.3b. This domain is normal with respect to both coordinate axes, but in order to directly evaluate the integral (17.23) it is more convenient to take it as normal with respect to the y-axis. It follows:

$$\int_D dA(\boldsymbol{x}) \, \partial_x \frac{1+xy}{1+x+y} = \int_0^1 dy \, (y^2 + y - 1) \int_{-y}^{\sqrt{y}} \frac{dx}{(1+x+y)^2}$$

$$= -\frac{1}{6} - 2 \int_0^1 d\eta \, \frac{\eta^5 + \eta^3 - \eta}{\eta^2 + \eta + 1} , \qquad (17.24)$$

in which the change of variable $\eta = \sqrt{y}$ has been used in the last integral. Decomposing the fraction in the following way:

$$\frac{\eta^5 + \eta^3 - \eta}{\eta^2 + \eta + 1} \equiv -\frac{2\eta + 1}{\eta^2 + \eta + 1} + \frac{1}{\eta^2 + \eta + 1} + \eta^3 - \eta^2 + \eta$$

the integral in Eq. (17.24) becomes:

$$\int_D dA(\boldsymbol{x}) \, \partial_x \frac{1+xy}{1+x+y} = -1 + 2\log 3 - 2 \int_0^1 \frac{d\eta}{\eta^2 + \eta + 1} . \qquad (17.25)$$

By means of the change of variable $\eta = (\sqrt{3}/2)\xi - 1/2$, the integral (17.25) is easily evaluated:

$$\int_D dA(\boldsymbol{x}) \, \partial_x \frac{1+xy}{1+x+y} = -1 + 2\log 3 - \frac{4}{\sqrt{3}} \arctan \frac{\sqrt{3}}{3} \simeq -0.01197 . \quad (17.26)$$

The same integral is now evaluated using the Green's formula and the relation $v_x ds = dy$:

$$\int_D dA(\boldsymbol{x}) \, \partial_x \frac{1+xy}{1+x+y} = \int_0^1 (2x\,dx) \frac{1+xx^2}{1+x+x^2} + \int_{-1}^0 (-dx) \frac{1+x(-x)}{1+x+(-x)}$$

$$= -\frac{2}{3} + 2 \int_0^1 dx \, \frac{x^4 + x}{x^2 + x + 1} \qquad (17.27)$$

and the fraction is decomposed in the following way:

$$\frac{x^4 + x}{x^2 + x + 1} \equiv \frac{2x + 1}{x^2 + x + 1} - \frac{1}{x^2 + x + 1} + x^2 - x .$$

Using the above decomposition inside the integral (17.27) and going forward as in the previous integration, the result (17.26) is still obtained in a simpler way.

17.3   Suggested Readings

Derivation of Green's formulae and their application in vector calculus are found in Lax and Terrell (2017, Chap. 8, pp. 333–345), Adams and Essex (2010, Chap. 16, pp. 888–924), Lang (1987, Chap. X, pp. 269–280, Chap. XVII, pp. 474–477), Apostol (1969, Chap. 11, pp. 378–391).

References

R. A. Adams and C. Essex. *Calculus: Several Variables*. Pearson Education Canada, Toronto, ON, 7th edition, 2010.

T. M. Apostol. *Calculus, Volume II*. Wiley, Hoboken, NJ, 2nd edition, 1969.

S. Lang. *Calculus of Several Variables*. Springer, New York, 3rd edition, 1987.

P. D. Lax and M. S. Terrell. *Multivariable Calculus with Applications*. Springer, Cham, CH, 2017.

Chapter 18
Application of Green's Formulae

In the present lesson more samples of application of Green's formulae will be given, in order to familiarise with this essential tool of the multidimensional integration.

18.1 Samples of Using Green's Formulae in the Calculation of Integrals

Example 18.1 Consider the domain $D \subset \mathbb{R}^3$ bounded by the surface of the (elliptic) paraboloid $z = x^2 + y^2$ (drawn with red lines) and by the plane $z = 1$ (blue), as shown in Fig. 18.1a. On this domain the following integral:

$$\int_D dV(\boldsymbol{x}) \, \partial_x \frac{xy^2}{1+z} . \tag{18.1}$$

has to be evaluated. The integral (18.1) is directly calculated, evaluating the derivative and integrating:

$$
\begin{aligned}
\int_D dV(\boldsymbol{x}) \, \partial_x \frac{xy^2}{1+z} &= \int_{B_1(0)} dA(x,y) \int_{x^2+y^2}^1 dz \, \frac{y^2}{1+z} \\
&= \int_{B_1(0)} dA(x,y) \, y^2 \left[\log 2 - \log \left(1 + x^2 + y^2\right) \right] \\
&= \frac{\pi}{2} \int_0^1 d\rho \, \frac{\rho^5}{1+\rho^2} , \tag{18.2}
\end{aligned}
$$

using polar coordinates in the unit circle $B_1(0)$ (sphere having center at the origin and radius 1) of the (x, y)-plane. The resulting integral in Eq. (18.2) is easily evaluated:

$$\int_0^1 d\rho \, \frac{\rho^5}{1+\rho^2} = \frac{\log 2}{2} - \frac{1}{4} \tag{18.3}$$

© The Author(s), under exclusive license to Springer Nature Switzerland AG 2024   215
G. Riccardi et al., *Multidimensional Differential and Integral Calculus*,
https://doi.org/10.1007/978-3-031-70326-3_18

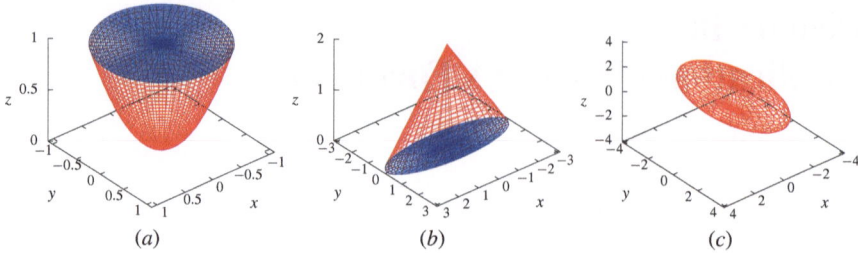

Fig. 18.1 Domains D used in the Examples 18.1, 18.2, 18.3 **a**; 18.4, 18.5 **b** with $a = 3, b = 1$ and $h = 2$; 18.6 **c** with $a = 2, b = 4, c = 1$

and it is used inside Eq. (18.2):

$$\int_D dV(\mathbf{x})\,\partial_x \frac{xy^2}{1+z} = \frac{\pi}{4}\left(\log 2 - \frac{1}{2}\right).\tag{18.4}$$

The same integral (18.1) is now evaluated by means of Green's formula:

$$\int_D dV(\mathbf{x})\,\partial_x \frac{xy^2}{1+z} = \int_{\partial D} dS(\mathbf{x})\,v_x(\mathbf{x})\,\frac{xy^2}{1+z},\tag{18.5}$$

v_x being the x-component of the outward unit normal vector. Due to the fact that the normal to the plane $z = 1$ is parallel to the z-axis, its component along the coordinate direction x vanishes. As a consequence, to the integral at the right hand side of Eq. (18.5) only the paraboloid gives a non-vanishing contribution. Using the polar coordinates inside the unit circle in order to parametrise this surface:

$$\mathbf{x}(\rho, \theta) = \begin{pmatrix} \rho \cos \theta \\ \rho \sin \theta \\ \rho^2 \end{pmatrix},$$

leads to the form of the product $dS\,\mathbf{v}$:

$$dS(\mathbf{x})\,\mathbf{v}(\mathbf{x}) = \partial_\theta \mathbf{x} \times \partial_\rho \mathbf{x}\, d\theta\, d\rho = \begin{pmatrix} 2\rho^2 \cos \theta \\ 2\rho^2 \sin \theta \\ -\rho \end{pmatrix}.\tag{18.6}$$

Using the result (18.6) inside the integral at the right hand side of Eq. (18.5):

$$\int_{\partial D} dS(\mathbf{x})\,v_x(\mathbf{x})\,\frac{xy^2}{1+z} = \int_0^1 d\rho \int_0^{2\pi} d\theta\, 2\rho^2 \cos \theta\, \frac{\rho \cos \theta\, \rho^2 \sin^2 \theta}{1+\rho^2}$$

$$= 2\int_0^1 d\rho\, \frac{\rho^5}{1+\rho^2} \int_0^{2\pi} \cos^2 \theta \sin^2 \theta = \frac{\pi}{4}\left(\log 2 - \frac{1}{2}\right),$$

in which the integral (18.3) has been still used. Obviously, this result is identical to the previous one (18.4).

Example 18.2 The integral (18.1) is slightly modified:

$$\int_D dV(x)\, \partial_z \frac{x^2 y^2}{1+z}, \tag{18.7}$$

where D is the same domain used in the previous exercise and shown in Fig. 18.1a. Due to the fact that the domain D is normal with respect to the z-axis, a direct calculation gives:

$$\int_D dV(x)\, \partial_z \frac{x^2 y^2}{1+z} = \int_{B_1(0)} dA(x,y)\, x^2 y^2 \int_{x^2+y^2}^{1} dz\, \partial_z \frac{1}{1+z}$$

$$= \int_0^1 d\rho\, \rho^5 \left(\frac{1}{2} - \frac{1}{1+\rho^2}\right) \int_0^{2\pi} d\theta\, \cos^2\theta \sin^2\theta$$

$$= \frac{\pi}{4}\left(\frac{1}{3} - \frac{\log 2}{2}\right), \tag{18.8}$$

in which the result (18.3) has been used. The same integral (18.7) is now evaluated by means of the suitable Green's formula. It is worth noting that, due to the presence of the z-component of the normal unit vector, in the present case also the part of ∂D lying in the plane $z=1$ gives a contribution. Writing the contributions of the paraboloid (in which the form (18.6) of the product $dS\, v$ is used) and of the planar circular disk:

$$\int_D dV(x)\, \partial_z \frac{x^2 y^2}{1+z} = \int_{\partial D} dS(x)\, v_z(x)\, \frac{x^2 y^2}{1+z}$$

$$= \int_0^1 d\rho \int_0^{2\pi} d\theta\, (-\rho)\, \frac{\rho^2 \cos^2\theta\, \rho^2 \sin^2\theta}{1+\rho^2}$$

$$+ \int_0^1 d\rho \int_0^{2\pi} d\theta\, \frac{\rho^2 \cos^2\theta\, \rho^2 \sin^2\theta}{1+1},$$

which still gives the result (18.8).

Example 18.3 In the same domain D used in the first two sample cases (see Fig. 18.1a), consider now the integral:

$$\int_D dV(x)\, \partial_z \log\left[1 + z\left(x^2 + y^2\right)\right]. \tag{18.9}$$

It is directly evaluated accounting for that the domain D is normal with respect to the z-axis:

$$\int_D dV(x)\, \partial_z \log\left[1 + z\left(x^2 + y^2\right)\right] = \int_{B_1(0)} dA(x,y) \int_{x^2+y^2}^{1} dz\, \frac{x^2 + y^2}{1 + z(x^2 + y^2)}$$

$$= 2\pi \int_0^1 d\rho\, \rho \left[\log(1 + \rho^2) - \log(1 + \rho^4)\right]$$

$$= \pi \left(\log 2 + 1 - \frac{\pi}{2}\right). \tag{18.10}$$

The same integral is now evaluated using the proper Green's formula:

$$\int_D dV(\boldsymbol{x})\, \partial_z \log\left[1 + z\left(x^2 + y^2\right)\right] = \int_0^1 d\rho \int_0^{2\pi} d\theta\, (-\rho) \log(1 + \rho^4)$$

$$+ \int_0^1 d\rho\, \rho \int_0^{2\pi} d\theta\, \log(1 + \rho^2),$$

which still leads to the result (18.10).

Example 18.4 Consider the truncated cone shown in Fig. 18.1b. Its height is h and its generic section with a plane $z = $ const. is an ellipse having axes parallel to the coordinate directions x and y. Naming as a and b the lengths of the semi-axes along x and y of the section at $z = 0$, the lengths of the corresponding axes of a section with a plane parallel to the (x, y)-plane at height $0 \le z \le h$ are $a(1 - z/h)$ and $b/(1 - z/h)$. As a consequence, the coordinates x, y and z of any point belonging to the lateral surface of the cone are related by the equation:

$$\frac{x^2}{a^2(1 - z/h)^2} + \frac{y^2}{b^2(1 - z/h)^2} = 1$$

which gives, using the formulae $x = r\, a \cos\theta$ and $y = r\, b \sin\theta$, the third coordinate $z = (1 - r)h$, for $0 < r < 1$. It follows that the lateral surface of the truncated cone can be parametrised as:

$$\boldsymbol{x}(r, \theta) = \begin{pmatrix} r\, a \cos\theta \\ r\, b \sin\theta \\ (1 - r)\, h \end{pmatrix}, \tag{18.11}$$

the parameters (r, θ) belonging to the rectangle $(0, 1) \times (0, 2\pi)$. The differential quantity $dS(\boldsymbol{x})\, \boldsymbol{v}(\boldsymbol{x})$ is written as:

$$dS\, \boldsymbol{v} = \partial_r \boldsymbol{x} \times \partial_\theta \boldsymbol{x}\, dr\, d\theta$$

$$= \begin{pmatrix} a \cos\theta \\ b \sin\theta \\ -h \end{pmatrix} \times \begin{pmatrix} -r\, a \sin\theta \\ r\, b \cos\theta \\ 0 \end{pmatrix} dr\, d\theta$$

$$= r \begin{pmatrix} hb \cos\theta \\ ha \sin\theta \\ ab \end{pmatrix} dr\, d\theta, \tag{18.12}$$

choosing $\boldsymbol{v}$ outward the lateral surface.

Inside the truncated cone it is now evaluated the volume integral of the derivative:

$$\partial_y \left[z \log \left(1 + \frac{y}{b} \right) \right]. \tag{18.13}$$

This calculation is firstly performed in a direct way, *i.e.* evaluating the volume integral, then it is shown that the use of the suitable Green formula leads to simpler integrals.

The volume integral is calculated parametrizing the truncated cone by means of the new coordinates $r \in (0, 1)$, $\theta \in [0, 2\pi)$ and $z \in (0, (1 - r)h)$ as $(ra \cos \theta, rb \sin \theta, z)$, The jacobian determinant of this change of coordinates is written as:

$$\frac{\partial(x, y, z)}{\partial(r, \theta, z)} = \begin{vmatrix} a \cos \theta & b \sin \theta & 0 \\ -ra \sin \theta & rb \cos \theta & 0 \\ 0 & 0 & 1 \end{vmatrix} = abr. \tag{18.14}$$

It follows the volume integral of the derivative (18.13):

$$\frac{1}{b} \int_D dV(x) \frac{z}{1 + y/b} = a \int_0^1 dr\, r \int_0^{(1-r)h} dz\, z \int_0^{2\pi} \frac{d\theta}{1 + r \sin \theta}$$

$$= \frac{ah^2}{2} \int_0^1 dr\, r(1 - r)^2 \int_0^{2\pi} \frac{d\theta}{1 + r \sin \theta} \tag{18.15}$$

The integral in θ in the formula (18.15) is evaluated moving the integration interval to $(-\pi, +\pi)$ (the integral does not change, due to the periodicity of the integrand function) and performing the changes of variable from θ to $t = \tan(\theta/2)$ and from t to $\xi = (t + r)/(1 - r^2)^{1/2}$:

$$\int_0^{2\pi} \frac{d\theta}{1 + r \sin \theta} = \int_{-\infty}^{+\infty} \underbrace{\frac{2dt}{t^2 + 1}}_{d\theta} \frac{1}{1 + r \frac{2t}{t^2 + 1}} = \frac{2}{\sqrt{1 - r^2}} \int_{-\infty}^{+\infty} \frac{d\xi}{\xi^2 + 1} = \frac{2\pi}{\sqrt{1 - r^2}}. \tag{18.16}$$

The result (18.16) is inserted into the integral (18.15) and the integration variable is changed in α, with $\cos \alpha = r$:

$$\frac{1}{b} \int_D dV(x) \frac{z}{1 + y/b} = \pi ah^2 \int_0^1 dr \frac{r(1 - r)^2}{\sqrt{1 - r^2}} = \frac{\pi(10 - 3\pi)}{6} ah^2 \simeq 0.30119\, ah^2. \tag{18.17}$$

The same volume integral on D of the function (18.13) is now evaluated by means of the suitable Green's formula. Due to the fact that the normal to the basis of the truncated cone has vanishing component along y, it follows:

$$\int_{\partial D} dS(x)\, v_y(x)\, z \log\left(1 + \frac{y}{b}\right) = \int_0^1 \int_0^{2\pi} \underbrace{dr\, d\theta\, rha\sin\theta}_{dS\, v_y}\; \underbrace{(1-r)h}_{z}\; \log(\underbrace{1 + r\sin\theta}_{y/b})$$

$$= ah^2 \int_0^1 dr\, r^2(1-r) \int_0^{2\pi} d\theta\, \frac{\cos^2\theta}{1 + r\sin\theta}$$

$$= ah^2 \int_0^1 dr\, (1-r) \left[2\pi - (1 - r^2) \int_0^{2\pi} \frac{d\theta}{1 + r\sin\theta} \right]$$

$$= \pi ah^2 \left[1 - 2 \int_0^1 dr\, (1-r)\sqrt{1 - r^2} \right],$$

in which the result (18.16) has been used. By means of the same change of variable from r to $\cos\alpha = r$, the result (18.17) follows once again.

Example 18.5 Consider the function:

$$\partial_z \frac{1}{2 + x/a + y/b + z/h} \tag{18.18}$$

and its integral on the domain D (defined in the previous example) written in the coordinate system (r, θ, ζ) with $\zeta := z/h$:

$$\int_D dV(x)\, \partial_z \frac{1}{2 + x/a + y/b + z/h} =$$

$$= -\frac{1}{h} \int_D \frac{dV(x)}{(2 + x/a + y/b + z/h)^2}$$

$$= ab \int_0^1 dr\, r \int_0^{2\pi} d\theta \int_0^{1-r} d\zeta \left\{ -\frac{1}{[2 + r(\cos\theta + \sin\theta) + \zeta]^2} \right\}$$

$$= ab \int_0^1 dr\, r \int_0^{2\pi} d\theta \left[\frac{1}{3 - r + r(\cos\theta + \sin\theta)} - \frac{1}{2 + r(\cos\theta + \sin\theta)} \right]. \tag{18.19}$$

The first internal integral in the formula (18.19) is evaluated by means of the changes of variable from θ to $t = \tan(\theta/2)$ and from t to ξ:

$$t = \frac{\sqrt{9 - 6r - r^2}}{3 - 2r}\, \xi - \frac{r}{3 - 2r},$$

so that the integral is evaluated as:

$$\int_0^{2\pi} \frac{d\theta}{3-r+r(\cos\theta+\sin\theta)} = \int_{-\pi}^{+\pi} \frac{d\theta}{3-r+r(\cos\theta+\sin\theta)}$$

$$= \int_{-\infty}^{+\infty} \frac{dt}{(3-2r)t^2+2rt+3}$$

$$= \frac{1}{\sqrt{9-6r-r^2}} \int_{-\infty}^{+\infty} \frac{d\xi}{\xi^2+1}$$

$$= \frac{\pi}{\sqrt{9-6r-r^2}}. \tag{18.20}$$

The second internal integral is calculated in an analogous way, but with:

$$t = \frac{\sqrt{2(2-r^2)}}{2-r}\xi - \frac{r}{2-r},$$

so that one obtains:

$$\int_0^{2\pi} \frac{d\theta}{2+r(\cos\theta+\sin\theta)} = \int_{-\pi}^{+\pi} \frac{d\theta}{2+r(\cos\theta+\sin\theta)}$$

$$= \int_{-\infty}^{+\infty} \frac{dt}{(2-r)t^2+2rt+(2+r)}$$

$$= \frac{1}{\sqrt{2(2-r^2)}} \int_{-\infty}^{+\infty} \frac{d\xi}{\xi^2+1}$$

$$= \frac{\pi}{\sqrt{2(2-r^2)}}. \tag{18.21}$$

Using the results (18.20) and (18.21) inside the integral (18.19), this latter becomes:

$$\int_D dV \, \partial_z \frac{1}{2+x/a+y/b+z/h} = 2\pi ab \left[\int_0^1 dr \, \frac{r}{\sqrt{9-6r-r^2}} - \int_0^1 dr \, \frac{r}{\sqrt{2(2-r^2)}} \right].$$

The first integral on the right-hand side is evaluated by means of the change of variables $r = 3(\sqrt{2}\sin\eta - 1)$, whereas $r = \sqrt{2}\sin\eta$ is used in the second one. The final result follows as:

$$\int_D dV(x) \, \partial_z \frac{1}{2+x/a+y/b+z/h} = 2\pi ab \left[2 - \frac{\sqrt{2}}{2} + \frac{3\pi}{4} - 3 \arcsin\left(\frac{2\sqrt{2}}{3}\right) \right]$$

$$\simeq -0.27514 \, ab. \tag{18.22}$$

The same volume integral of the derivative (18.18) can be evaluated by means of the suitable Green's formula. Due to the presence of the derivative along z, the basis of the truncated cone (on which $n_z \equiv -1$) gives now a non-vanishing contribution. It is evaluated by means of the change of variables $x = r \, a \cos\theta$, $y = r \, b \sin\theta$, which gives $dS = rab\, dr d\theta$. Using the relation (18.12) on the lateral surface, the integral leads directly to the relation (18.19):

$$
\int_D dV(x)\, \partial_z \frac{1}{2 + x/a + y/b + z/h} = \underbrace{ab \int_0^1 dr\, r \int_0^{2\pi} \frac{d\theta}{3 - r + r(\cos\theta + \sin\theta)}}_{\text{contribution of the lateral surface}}
$$

$$
\underbrace{-ab \int_0^1 dr\, r \int_0^{2\pi} \frac{d\theta}{2 + r(\cos\theta + \sin\theta)}}_{\text{contribution of the basis}} .
$$

Example 18.6 Consider now the ellipsoid drawn in Fig. 18.1c having semi-axes a, b and c along the coordinate directions x, y and z, respectively, and integrate inside this domain (called D, for short) the derivative:

$$
\partial_x (xy^2z^2) . \tag{18.23}
$$

This calculation is firstly performed as a volume integral. The cartesian coordinates (x, y, z) are changed in (r, φ, θ) by means of the following relation:

$$
x(\rho, \varphi, \theta) = r \begin{pmatrix} a \cos\varphi \cos\theta \\ b \cos\varphi \sin\theta \\ c \sin\varphi \end{pmatrix} . \tag{18.24}
$$

The coordinates (18.24) belong to the parallelepiped $(r, \varphi, \theta) \in (0, 1) \times (-\pi/2, \pi/2) \times (0, 2\pi)$ and the related jacobian determinant is:

$$
J = abc \begin{vmatrix} \cos\varphi \cos\theta & \cos\varphi \sin\theta & \sin\varphi \\ -r \cos\varphi \sin\theta & r \cos\varphi \cos\theta & 0 \\ -r \sin\varphi \cos\theta & -r \sin\varphi \sin\theta & r \cos\varphi \end{vmatrix} = abc\, r^2 \cos\varphi .
$$

The volume integral on D is now evaluated as:

$$
\int_D dV\, \partial_x(xy^2z^2)
$$

$$
= \int_D dV\, y^2z^2
$$

$$
= \int_0^1 dr \int_{-\pi/2}^{+\pi/2} d\varphi \int_0^{2\pi} d\theta\, \underbrace{abc\, r^2 \cos\varphi}_{J}\, \underbrace{(br \cos\varphi \sin\theta)^2}_{y^2}\, \underbrace{(cr \sin\varphi)^2}_{z^2}
$$

$$
= \frac{4\pi}{105} ab^3c^3 . \tag{18.25}
$$

The same integral is now evaluated by means of the suitable Green's formula. The position on the surface ∂D is given by the relation (18.24) written in correspondence to $r = 1$ and the vector $dS\, \nu$ with ν outward is written:

$$dS\,v = \partial_\theta x \times \partial_\varphi x\, d\varphi\, d\theta = \begin{pmatrix} bc\,\cos^2\varphi\,\cos\theta \\ ac\,\cos^2\varphi\,\sin\theta \\ ab\,\sin\varphi\,\cos\varphi \end{pmatrix} d\varphi\, d\theta\,. \tag{18.26}$$

Using the x-component of the vector (18.26), Green's formula along x is written as:

$$\int_D dV\,\partial_x(xy^2z^2) = \int_{\partial D} dS\,v_x\,xy^2z^2$$

$$= \int_{-\pi/2}^{+\pi/2} d\varphi \int_0^{2\pi} d\theta\, bc\,\cos^2\varphi\,\cos\theta\,\underbrace{a\,\cos\varphi\,\cos\theta}_{x} \times$$

$$\times\,\underbrace{(b\,\cos\varphi\,\sin\theta)^2}_{y^2}\,\underbrace{(c\,\sin\varphi)^2}_{z^2}\,,$$

which still leads to the result (18.25) once the integrals:

$$\int_{-\pi/2}^{+\pi/2} d\varphi\,\cos^5\varphi\,\sin^2\varphi = \frac{16}{105}\,, \qquad \int_0^{2\pi} d\theta\,\cos^2\theta\,\sin^2\theta = \frac{\pi}{4}$$

are used.

Chapter 19
Gauss and Stokes Theorems

In this lesson two powerful consequences of Green's formulae are deduced and applied in some sample cases. The first one (Sect. 19.1) is called *Gauss'* (or *divergence*) *theorem* and relates the volume integral on a finite domain of the divergence of a vector field to the flux of the same field across the boundary of the domain. The second consequence (Sect. 19.2) is called *Stokes' theorem* and enables to write the flux of the rotor of a vector field on a surface with the circuitation of the same field on the boundary of the surface itself.

19.1 Gauss' Theorem

Consider a bounded domain D having a smooth and finite boundary ∂D and the integral on D of the divergence of a continuous vector function $\boldsymbol{f} : \mathbb{R}^n \to \mathbb{R}^n$ (in particular, the values $n = 2$ and $n = 3$ will be considered in the examples) having continuous first derivatives. Applying Green's formulae the volume integral of the divergence is transformed into the flux of the vector field:

$$
\begin{aligned}
\int_D dV(\boldsymbol{x}) \, \nabla \cdot \boldsymbol{f}(\boldsymbol{x}) &= \int_D dV(\boldsymbol{x}) \, \partial_k f_k(\boldsymbol{x}) \\
&= \int_{\partial D} dS(\boldsymbol{x}) \, v_k(\boldsymbol{x}) \, f_k(\boldsymbol{x}) \\
&= \int_{\partial D} dS(\boldsymbol{x}) \, \boldsymbol{v}(\boldsymbol{x}) \cdot \boldsymbol{f}(\boldsymbol{x}) \tag{19.1}
\end{aligned}
$$

(the summation convention is used). The relation (19.1) enables to transform the volume integral on D of the divergence of $\boldsymbol{f}$ into the flux of the same $\boldsymbol{f}$ across the boundary of D, and viceversa. It plays an important and ubiquitous role in all the theories involving continuous media as the Electromagnetism or the Fluid Dynamics. It is known as *Gauss'* (or *divergence*) *theorem*. Its practical application is now discussed in few sample cases.

Example 19.1 Consider the doman D inside the paraboloid $z = x^2 + y^2$ between the two planes $z = 0$ and $z = h$ and the function $f : \mathbb{R}^3 \to \mathbb{R}^3$ continuous with continuous partial derivatives $f(x) = (x, x^2 y z^2, z)^T$. The volume integral of the divergence $\nabla \cdot f = 2 + x^2 z^2$ is evaluated as:

$$
\int_D dV(x)\, \nabla \cdot f(x) = \int_D dV(x,y,z)\,\left(2 + x^2 z^2\right)
$$
$$
= \int_0^{\sqrt{h}} d\rho\, \rho \int_0^{2\pi} d\theta \int_{\rho^2}^h dz\,\left(2 + \rho^2 \cos^2\theta\, z^2\right)
$$
$$
= \pi h^2 \left(1 + \frac{h^3}{20}\right). \tag{19.2}
$$

This result is compared with the integral of the flux across the boundary of D, writing firstly the contribution on the section of the paraboloid boundary between the two planes $z = 0$ and $z = h$ and then the contribution of the disk on the plane $z = h$ ($f \cdot v$ identically vanishes on the disk on the plane $z = 0$):

$$
\int_{\partial D} dS(x)\, v(x) \cdot f(x) = \int_0^{\sqrt{h}} d\rho \int_0^{2\pi} d\theta \begin{pmatrix} 2\rho^2 \cos\theta \\ 2\rho^2 \sin\theta \\ -\rho \end{pmatrix} \cdot \begin{pmatrix} \rho\cos\theta \\ \rho^7 \cos^2\theta \sin\theta \\ \rho^2 \end{pmatrix}
$$
$$
+ \int_0^{\sqrt{h}} d\rho\, \rho \int_0^{2\pi} d\theta \begin{pmatrix} 0 \\ 0 \\ 1 \end{pmatrix} \cdot \begin{pmatrix} \rho\cos\theta \\ h^2 \rho^3 \cos^2\theta \sin\theta \\ h \end{pmatrix},
$$

which still leads to the result (19.2) (the integral on the first row gives $\pi h^5/20$ and the one in the second row πh^2).

Example 19.2 The domain D is now the torus generated by the rotation of a circle about the z-axis. The circle generated by the centers of the torus sections lies on the plane $z = 0$ and has radius R. The section of the torus with a symmetry plane passing through the z-axis is a circle of radius $r < R$, so that D is a *ring torus* (a doubly connected subset of $\mathbb{R}^3$). it is firstly considered the integral of the divergence of the function $f(x) = (xy^2, yz^2, z^3/3)^T$, continuous with continuous partial derivatives, inside D. In order to calculate the volume integral, the toroidal coordinates:

$$
x(\rho, \varphi, \theta) = \begin{pmatrix} (R + \rho\cos\varphi)\cos\theta \\ (R + \rho\cos\varphi)\sin\theta \\ \rho\sin\varphi \end{pmatrix} \tag{19.3}
$$

are used. When the new coordinates (ρ, φ, θ) belong to the parallelepiped $(0, r) \times [0, 2\pi) \times [0, 2\pi)$, the point (19.3) moves inside D. The following results:

$$|J(\rho, \varphi, \theta)| = \rho(R + \rho \cos \varphi)$$

$$dS(\boldsymbol{x})\boldsymbol{v}(\boldsymbol{x}) = \partial_\theta \boldsymbol{x}(r, \varphi, \theta) \times \partial_\varphi \boldsymbol{x}(r, \varphi, \theta) d\varphi \, d\theta = r(R + r \cos \varphi) \begin{pmatrix} \cos \varphi \cos \theta \\ \cos \varphi \sin \theta \\ \sin \varphi \end{pmatrix} d\varphi \, d\theta$$

$$(19.4)$$

will be used below. In the second row, the normal unit vector $\boldsymbol{v}$ on ∂D points outward D.

The volume integral of the divergence $\nabla \cdot \boldsymbol{f} = y^2 + 2z^2$ is evaluated by means of the jacobian determinant (19.4):

$$\int_D dV(\boldsymbol{x}) \, \nabla \cdot \boldsymbol{f}(\boldsymbol{x}) = \int_D dV(x, y, z) \left(y^2 + 2z^2\right)$$

$$= \int_0^r d\rho \int_0^{2\pi} d\varphi \int_0^{2\pi} d\theta \underbrace{\rho(R + \rho \cos \varphi)}_{J} \times$$

$$\times \left[\underbrace{(R + \rho \cos \varphi)^2 \sin^2 \theta}_{y^2} + 2 \underbrace{\rho^2 \sin^2 \varphi}_{z^2} \right]$$

$$= \frac{\pi^2}{4} R r^2 \left(4R^2 + 7r^2\right). \qquad (19.5)$$

The flux of $\boldsymbol{f}$ across ∂D is given by the surface integral:

$$\int_{\partial D} dS(\boldsymbol{x}) \, \boldsymbol{v}(\boldsymbol{x}) \cdot \boldsymbol{f}(\boldsymbol{x}) =$$

$$= \int_0^{2\pi} d\varphi \int_0^{2\pi} d\theta \, r(R + r \cos \varphi) \begin{pmatrix} \cos \varphi \cos \theta \\ \cos \varphi \sin \theta \\ \sin \varphi \end{pmatrix} \cdot \begin{pmatrix} (R + r \cos \varphi)^3 \sin^2 \theta \cos \theta \\ r^2(R + r \cos \varphi) \sin^2 \varphi \sin \theta \\ r^3/3 \sin^3 \varphi \end{pmatrix},$$

in which the quantity $dS\boldsymbol{v}$ has been evaluated in Eq. (19.4). The integrals inside the above formula are evaluated by means of the following ones:

$$\int_0^{2\pi} dx \, \cos^2 x = \pi, \quad \int_0^{2\pi} dx \, \sin^2 x \cos^2 x = \frac{\pi}{4}, \quad \int_0^{2\pi} dx \, \sin^4 x = \int_0^{2\pi} dx \, \cos^4 x = \frac{3}{4}\pi$$

and the result (19.5) is still obtained.

Example 19.3 Consider now the quite different problem of approximating the area $|D|$ of a bounded, simply connected domain $D \subset \mathbb{R}^2$ having a complicated shape, once a large number (n) of points $\{\boldsymbol{x}_k, k = 1, 2, \ldots, n\}$ (the definition $\boldsymbol{x}_{n+1} := \boldsymbol{x}_1$ is also adopted below, for short) on its boundary has been given. Using Gauss' theorem (19.1) and the identity $\nabla \cdot \boldsymbol{x} \equiv 2$, the area of D is evaluated as:

$$|D| = \frac{1}{2} \int_D dA(\boldsymbol{x}) \, \nabla \cdot \boldsymbol{x} = \frac{1}{2} \int_{\partial D} ds \, \boldsymbol{x} \cdot \boldsymbol{v} \,. \tag{19.6}$$

The result (19.6) can easily be approximated. The arc of ∂D between the point $\boldsymbol{x}_k$ and its cuccessive one $\boldsymbol{x}_{k+1}$ is approximated with a straight segment $\boldsymbol{x}(s) \simeq \boldsymbol{x}_k + \boldsymbol{\tau}_k s$ for $s \in (0, l_k)$, where $\boldsymbol{\tau}_k = (\boldsymbol{x}_{k+1} - \boldsymbol{x}_k)/l_k$ approximates the tangent unit vector ($\boldsymbol{v}_k = -\boldsymbol{\tau}_k^{\perp}$ is the corresponding approximation of the outward normal unit vector) and $l_k = |\boldsymbol{x}_{k+1} - \boldsymbol{x}_k|$ is the length of the segment. The integral (19.6) is calculated as:

$$\int_{\partial D} \boldsymbol{x} \cdot \boldsymbol{v} \, ds \simeq \sum_{k=1}^{n} \int_0^{l_k} (\boldsymbol{x}_k + \boldsymbol{\tau}_k s) \cdot \boldsymbol{v}_k \, ds = \sum_{k=1}^{n} \boldsymbol{x}_k \cdot \boldsymbol{v}_k \, l_k \,,$$

but $\boldsymbol{v}_k \, l_k = -(\boldsymbol{x}_{k+1} - \boldsymbol{x}_k)^{\perp}$ and then $\boldsymbol{x}_k \cdot \boldsymbol{v}_k \, l_k = -\boldsymbol{x}_k \cdot \boldsymbol{x}_{k+1}^{\perp} = \boldsymbol{x}_k^{\perp} \cdot \boldsymbol{x}_{k+1}$. The approximation of the area $|D|$:

$$|D| \simeq \frac{1}{2} \sum_{k=1}^{n} \boldsymbol{x}_k^{\perp} \cdot \boldsymbol{x}_{k+1} \tag{19.7}$$

follows. The practical use of the formula (19.7) shows that it gives a satisfactory approximation of $|D|$ if many points are used in high curvature regions of ∂D.

Exercise 19.1 Using completely analogous approaches, approximate the position of the centroid of D and its moments of inertia:

$$I_{xx} = \int_D dA(x, y) \, x^2 \,, \quad I_{xy} = \int_D dA(x, y) \, xy \,, \quad I_{yy} = \int_D dA(x, y) \, y^2 \,.$$

19.2   Stokes' Theorem

A second application of Green's formulae leads to a useful relation between the flux of the curl of a vector function $\boldsymbol{f} : \mathbb{R}^3 \to \mathbb{R}^3$ across a (bounded and smooth) surface $S \subset \mathbb{R}^3$ and the circulation of the same function on the boundary of this surface. This closed (smooth) curve is also called the *edge* of S and it will be indicated by ∂S hereafter. Moreover, $\boldsymbol{f}$ is assumed continuous with continuous first derivatives in an open set containing S.

In order to point out this relation, the orientation of S is specified defining the direction of the normal unit vector $\boldsymbol{v}$ at all of its points, so that the flux of a vector field across this surface can be evaluated. The edge ∂S of S is then oriented in a congruent way: it is required that an observer placed at a point $\boldsymbol{x} \in S$, having $\boldsymbol{v}(\boldsymbol{x})$ directed from his foots to his head, sees a point moving in the positive direction along the edge as rotating counterclockwise. This situation is schematised in Fig. 19.1.

Fig. 19.1 Positive direction
on the edge of a surface S. If
an observer in $x \in S$, with
$v(x)$ directed from his foots
to his head, sees a point
moving in the positive
direction on ∂S as rotating
counterclockwise

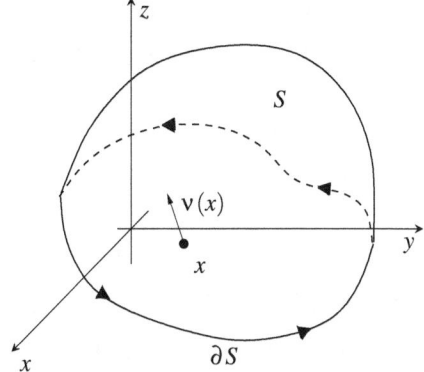

Once the surface S and its edge ∂S have been oriented, Stokes' theorem states
that:

$$\int_S dS(x)\, v(x) \cdot \nabla \times f(x) = \int_{\partial S} dx \cdot f(x), \qquad (19.8)$$

or the flux of the curl across the surface is equal to the circulation on its edge. The
relation (19.8) is now proved. The surface S is parametrised with the two parameters
u and v belonging to the (bounded) domain $\Sigma \subset \mathbb{R}^2$. For short, the parameters
are collected in the vector $u = (u, v)$. The differential quantity $dS\, v$ is written as
$\partial_u x \times \partial_v x\, dA$ and the flux of the curl of f is written in the following form:

$$\int_S dS(x)\, v(x) \cdot \nabla \times f(x) = \int_\Sigma dA(u)\, \partial_u x(u) \times \partial_v x(u) \cdot \nabla \times f[x(u)]. \quad (19.9)$$

The scalar quantity inside the integral at the right hand side can be rewritten in a
simpler form, reorganizing the vector products. For short, the derivatives will be
indicated by subscripts separated by commas, e.g. $\partial_z f_y = f_{y,z}$ or $\partial_u y = y_{,u}$. Once
the two products are written by components:

$$\partial_u x \times \partial_v x = \begin{pmatrix} y_{,u} z_{,v} - z_{,u} y_{,v} \\ z_{,u} x_{,v} - x_{,u} z_{,v} \\ x_{,u} y_{,v} - y_{,u} x_{,v} \end{pmatrix}, \quad \nabla \times f = \begin{pmatrix} f_{z,y} - f_{y,z} \\ f_{x,z} - f_{z,x} \\ f_{y,x} - f_{x,y} \end{pmatrix},$$

their scalar product is evaluated[1] as:

$$\partial_u x \times \partial_v x \cdot \nabla \times f =$$

---

[1] The calculation in Eq. (19.10) is less tedious if it is performed using indices:

$$\partial_u x \times \partial_v x \cdot \nabla \times f = \varepsilon_{ijk} \partial_u x_j\, \partial_v x_k\, \varepsilon_{ilm} \partial_{x_l} f_m$$

and the rule: $\varepsilon_{ijk}\varepsilon_{ilm} = \delta_{jl}\delta_{km} - \delta_{jm}\delta_{kl}$.

$$
\begin{aligned}
&= \left[x_{,v}(f_{x,y}y_{,u} + f_{x,z}z_{,u}) + y_{,v}(f_{y,x}x_{,u} + f_{y,z}z_{,u}) + z_{,v}(f_{z,x}x_{,u} + f_{z,y}y_{,u}) \right] + \\
&\quad - \left[x_{,u}(f_{x,y}y_{,v} + f_{x,z}z_{,v}) + y_{,u}(f_{y,x}x_{,v} + f_{y,z}z_{,v}) + z_{,u}(f_{z,x}x_{,v} + f_{z,y}y_{,v}) \right] \\
&= \left[\; x_{,v}(f_{x,x}x_{,u} + f_{x,y}y_{,u} + f_{x,z}z_{,u}) + y_{,v}(f_{y,x}x_{,u} + f_{y,y}y_{,u} + f_{y,z}z_{,u}) + \right. \\
&\quad \left. + z_{,v}(f_{z,x}x_{,u} + f_{z,y}y_{,u} + f_{z,z}z_{,u}) \right] + \\
&\quad - \left[\; x_{,u}(f_{x,x}x_{,v} + f_{x,y}y_{,v} + f_{x,z}z_{,v}) + y_{,u}(f_{y,x}x_{,v} + f_{y,y}y_{,v} + f_{y,z}z_{,v}) + \right. \\
&\quad \left. + z_{,u}(f_{z,x}x_{,v} + f_{z,y}y_{,v} + f_{z,z}z_{,v}) \right] \\
&= \left(x_{,v}f_{x,u} + y_{,v}f_{y,u} + z_{,v}f_{z,u} \right) - \left(x_{,u}f_{x,v} + y_{,u}f_{y,v} + z_{,u}f_{z,v} \right) \\
&= \partial_u \left(x_{,v}f_x + y_{,v}f_y + z_{,v}f_z \right) - \partial_v \left(x_{,u}f_x + y_{,u}f_y + z_{,u}f_z \right) \\
&= \partial_u \left(\boldsymbol{f} \cdot \partial_v \boldsymbol{x} \right) - \partial_v \left(\boldsymbol{f} \cdot \partial_u \boldsymbol{x} \right) .
\end{aligned}
\tag{19.10}
$$

The result (19.10) suggests the use of Green's formulae inside the integral on the domain Σ in the parameter plane. Once the curve $\partial \Sigma$ is parametrised with the curvilinear abscissa s, it is described by a smooth function $\boldsymbol{u} = \boldsymbol{u}(s)$ that verifies the differential relation:

$$
ds\, \boldsymbol{v} = \begin{pmatrix} dv \\ -du \end{pmatrix} .
$$

An application of Green's formulae to the right hand side of the relation (19.9) with the integrand function written in the form (19.10) leads to the desired result:

$$
\begin{aligned}
\int_\Sigma dA(\boldsymbol{u})\, [\partial_u \boldsymbol{x}(\boldsymbol{u}) \times \partial_v \boldsymbol{x}(\boldsymbol{u})] \cdot \nabla \times \boldsymbol{f}[\boldsymbol{x}(\boldsymbol{u})] \; &= \\
= \int_\Sigma dA(\boldsymbol{u}) \left\{ \partial_u \left[\boldsymbol{f}[\boldsymbol{x}(\boldsymbol{u})] \cdot \partial_v \boldsymbol{x}(\boldsymbol{u}) \right] - \partial_v \left[\boldsymbol{f}[\boldsymbol{x}(\boldsymbol{u})] \cdot \partial_u \boldsymbol{x}(\boldsymbol{u}) \right] \right\} \\
= \int_{\partial\Sigma} \left\{ \boldsymbol{f}[\boldsymbol{x}(\boldsymbol{u})] \cdot \partial_v \boldsymbol{x}(\boldsymbol{u})\, dv + \boldsymbol{f}[\boldsymbol{x}(\boldsymbol{u})] \cdot \partial_u \boldsymbol{x}(\boldsymbol{u})\, du \right\} \\
= \int_{\partial\Sigma} \left[\partial_u \boldsymbol{x}(\boldsymbol{u})\, du + \partial_v \boldsymbol{x}(\boldsymbol{u})\, dv \right] \cdot \boldsymbol{f}[\boldsymbol{x}(\boldsymbol{u})] \\
= \int_{\partial S} d\boldsymbol{x} \cdot \boldsymbol{f}(\boldsymbol{x}) .
\end{aligned}
$$

Stokes' theorem will be now applied in few sample cases.

Example 19.4 Consider the surface S defined as the part of the paraboloid $z = x^2 + y^2$ included between the planes $z = 0$ and $z = h$ (with $h > 0$). Once this surface is oriented taking the normal vector outward the paraboloid, the calculation of the flux of the curl of the following vector function:

$$
\boldsymbol{f}(\boldsymbol{x}) = \begin{pmatrix} yz \\ (z^2 - x^2)/2 \\ xyz \end{pmatrix} \qquad \text{and then:} \qquad \nabla \times \boldsymbol{f}(\boldsymbol{x}) = \begin{pmatrix} (x-1)z \\ y(1-z) \\ -x - z \end{pmatrix}
\tag{19.11}
$$

across S is required.

The surface is parametrised by means of the parameters ρ and θ, being ρ the radius of the circle resulting intersecting the paraboloid with the plane $z = \rho$ for $0 < \rho < \sqrt{h}$, so that $x(\rho, \theta) = (\rho \cos \theta, \rho \sin \theta, \rho^2)^{\mathrm{T}}$. Taking the normal vector outward the paraboloid, it follows:

$$dS\, v = \partial_\theta x \times \partial_\rho x \, d\rho \, d\theta = \rho \begin{pmatrix} 2\rho \cos \theta \\ 2\rho \sin \theta \\ -1 \end{pmatrix} d\rho \, d\theta$$

and the flux of the curl of f across S is evaluated as:

$$\int_S dS(x)\, v(x) \cdot \nabla \times f(x) = \int_0^{\sqrt{h}} d\rho\, \rho^2 \int_0^{2\pi} d\theta \begin{pmatrix} 2\rho \cos \theta \\ 2\rho \sin \theta \\ -1 \end{pmatrix} \cdot \begin{pmatrix} \rho(\rho \cos \theta - 1) \\ (1 - \rho^2) \sin \theta \\ -\cos \theta - \rho \end{pmatrix} = \pi h^2.$$

$$(19.12)$$

The circulation of f on the edge of S, *i.e.* on the circle belonging to the plane $z = h$ parametrised in terms of the angle $\theta \in [0, 2\pi)$ as:

$$x(\theta) = \begin{pmatrix} \sqrt{h} \cos \theta \\ \sqrt{h} \sin \theta \\ h \end{pmatrix} \quad \text{from which:} \quad dx(\theta) = \sqrt{h} \begin{pmatrix} -\sin \theta \\ \cos \theta \\ 0 \end{pmatrix} d\theta$$

(note that the orientation of this circle has been chosen in such a way that an observer placed on S in the normal direction sees a point moving in the positive direction on ∂S as rotating counterclockwise). The circulation of f is:

$$\int_{\partial S} dx \cdot f(x) = -\sqrt{h} \int_0^{2\pi} d\theta \begin{pmatrix} -\sin \theta \\ \cos \theta \\ 0 \end{pmatrix} \cdot \begin{pmatrix} h^{3/2} \sin \theta \\ h(h - \cos^2 \theta)/2 \\ h^2 \sin \theta \cos \theta \end{pmatrix},$$

which still gives the result (19.12).

Example 19.5 Consider the circular torus having the centers of its circular sections on the circle of the plane $z = 0$ with center at the orgin and radius R. This circle is oriented counterclockwise and parametrised with the angle $\theta \in [0, 2\pi)$, with $\theta = 0$ on the x-axis. Each circular section has radius r ($< R$), is oriented counterclockwise and is parametrised with the angle $\varphi \in [0, 2\pi)$, with $\varphi = 0$ on the plane $z = 0$. The surface S is the part of the above torus having $\pi/2 \leq \theta \leq 2\pi$. Its edge ($\partial S$) is formed by the two circles $\mathscr{C}_{y,x}$ having centers on the points $(R, 0, 0)$, $(0, R, 0)$ and lying in the planes $y = 0$ ($\mathscr{C}_y$) and $x = 0$ ($\mathscr{C}_x$), respectively. If the normal v on S is chosen outward the torus, the circles $\mathscr{C}_{y,x}$ have to be oriented clockwise. The position of a point $x \in S$ is given by:

$$x(\theta, \varphi) = \begin{pmatrix} (R + r \cos \varphi) \cos \theta \\ (R + r \cos \varphi) \sin \theta \\ r \sin \varphi \end{pmatrix},$$

for $\pi/2 \le \theta \le 2\pi$ and $0 \le \varphi < 2\pi$. It follows:

$$dS\, \boldsymbol{v} = \partial_\theta \boldsymbol{x} \times \partial_\varphi \boldsymbol{x}\, d\theta d\varphi = r(R + r \cos \varphi) \begin{pmatrix} \cos \varphi \cos \theta \\ \cos \varphi \sin \theta \\ \sin \varphi \end{pmatrix}. \tag{19.13}$$

The flux of the curl of the vector function (19.11) across S is now evaluated using Eq. (19.13):

$$\int_S dS\, \boldsymbol{v} \cdot \nabla \times \boldsymbol{f} =$$

$$= r \int_{\pi/2}^{2\pi} d\theta \int_0^{2\pi} d\varphi (R + r \cos \varphi) \begin{pmatrix} \cos \varphi \cos \theta \\ \cos \varphi \sin \theta \\ \sin \varphi \end{pmatrix} \cdot \begin{pmatrix} [(R + r \cos \varphi) \cos \theta - 1]r \sin \varphi \\ (R + r \cos \varphi) \sin \theta (1 - r \sin \varphi) \\ -(R + r \cos \varphi) \cos \theta - r \sin \varphi \end{pmatrix}$$

$$= \pi r^2 R \int_{\pi/2}^{2\pi} d\theta\, (2 \sin^2 \theta - 1) = 0. \tag{19.14}$$

The circulation of $\boldsymbol{f}$ on ∂S also vanishes, being zero both the integrals on the circles $\mathscr{C}_{y,x}$ (first and second row, in the equation below), as it is easily recognised evaluating the curve elements and the function (19.11) on these curves:

$$d\boldsymbol{x}(\varphi) = -r \begin{pmatrix} -\sin \varphi \\ 0 \\ \cos \varphi \end{pmatrix} d\varphi \quad \boldsymbol{f}(\varphi) = \frac{1}{2} \begin{pmatrix} 0 \\ r^2 \sin^2 \varphi - (R + r \cos \varphi)^2 \\ 0 \end{pmatrix}$$

$$d\boldsymbol{x}(\varphi) = -r \begin{pmatrix} 0 \\ -\sin \varphi \\ \cos \varphi \end{pmatrix} d\varphi \quad \boldsymbol{f}(\varphi) = \frac{1}{2} \sin \varphi \begin{pmatrix} 2(R + r \cos \varphi) \\ r \sin \varphi \\ 0 \end{pmatrix}$$

(remember that the orientation of these circles is opposite to the one of any section of the torus with a plane passing through the z-axis).

19.3 Suggested Readings

An introduction to theoretical aspects of Gauss and Stokes theorems, complemented by samples of their applications, is found in Lax and Terrell (2017, Chap. 8, pp. 346–367), Adams and Essex (2010, Chap. 16, pp. 907–917), Lang (1987, Chap. XII, pp. 345–364), Apostol (1969, Chap. 12, pp. 438–465).

References

R. A. Adams and C. Essex. *Calculus: Several Variables*. Pearson Education Canada, Toronto, ON, 7th edition, 2010.

T. M. Apostol. *Calculus, Volume II*. Wiley, Hoboken, NJ, 2nd edition, 1969.

S. Lang. *Calculus of Several Variables*. Springer, New York, 3rd edition, 1987.

P. D. Lax and M. S. Terrell. *Multivariable Calculus with Applications*. Springer, Cham, CH, 2017.

References

Chapter 20
Partial Differential Equations

This lectures starts the study of one of the most important chapters of Mathematical Analysis: the partial differential equations. The importance of this topic is immediately clear when facing the study of phenomena taking place in multi-dimensional spaces. Most examples of the great interest for engineering application of such equations comes from problems in Continuum Mechanics and Electromagnetism.

20.1 Definition and Properties of Partial Differential Equations

A *partial differential equation* of order n is an equation for a function $f : \mathbb{R}^m \to \mathbb{R}$, where at least a derivative of order n of such function appears. With $D^k f$ is denoted the set of functions from $\mathbb{R}^m$ to $\mathbb{R}$ consisting of[1] the partial derivatives of order k of the function f, that is:

$$D^k f := \left\{ \partial^k_{x_{i_1} x_{i_2} \cdots x_{i_k}} f \text{ with } i_j \in \{1, 2, \ldots, m\} \text{ and } j = 1, 2, \ldots, k \right\}. \qquad (20.2)$$

In what follows it will be always assumed $i_1 \le i_2 \le \cdots \le i_k$. In general, using the notation (20.2) one can think of a partial differential equation of order n written as follows:

$$F\left(D^n f, D^{n-1} f, \ldots, D^1 f, f; \boldsymbol{x} \right) = 0, \qquad (20.3)$$

---

[1] When one derivative is excluded from the list (20.2), say in particular $\partial^k_{x_{i'_1} x_{i'_2} \cdots x_{i'_k}} f$, the sets (20.2) deprived of this derivative are defined as:

$$D^k_{x_{i'_1} x_{i'_2} \cdots x_{i'_k}} f := D^k f - \{ \partial^k_{x_{i'_1} x_{i'_2} \cdots x_{i'_k}} f \}. \qquad (20.1)$$

G. Riccardi et al., *Multidimensional Differential and Integral Calculus*,
https://doi.org/10.1007/978-3-031-70326-3_20

where suitable assumptions have been made on the regularity of the function F. The definition (20.3) has a general character and allows one to define classes of partial differential equations of paramount importance in the applications.

An Eq. (20.3) is said to be *in normal form* when one of the highest order (n) derivatives is isolated on one side of the equation, moving all other terms to the other side. In such case, there exists a function G such that Eq. (20.3) can be rewritten, with the notation (20.1), as follows:

$$\partial^n_{x_{i'_1} x_{i'_2} \cdots x_{i'_n}} f = G \left(t D^n_{x_{i'_1} x_{i'_2} \cdots x_{i'_n}} f, D^{n-1} f, \ldots, D^1 f, f; \boldsymbol{x} \right), \qquad (20.4)$$

for a given n-tuple of numbers $i'_1, i'_2, \ldots, i'_n$ within the set $\{1, 2, \ldots, m\}$. An Eq. (20.3) is said to be *linear* if F is a linear function of f and of all its derivatives. It is called, instead, *quasi-linear* if F is linear only with respect to the highest order derivatives.

20.2   Reduction to a System of First Order Equations

Similarly to the case of ordinary differential equations, a partial differential equation can be reduced to a system of first order partial differential equations. The following example is in order. Suppose one has an equation for $n = 2$ ed $m = 2$:

$$\partial_{x_1} f \partial^2_{x_1 x_1} f + \partial^2_{x_1 x_2} f = h, \qquad (20.5)$$

where h is a given (continuous) function of x_1 and x_2. Equation (20.5) is quasi-linear and, moreover, it can be easily rewritten in normal form, with respect to the derivative $\partial^2_{x_1 x_2} f$. This aspect will prove of great importance in what follows. If one introduces the two functions:

$$\begin{cases} \partial_{x_1} f := g_1 \\ \partial_{x_2} f := g_2 \end{cases} \qquad (20.6)$$

Equation (20.5) becomes:

$$\partial_{x_1} g_2 = -g_1 \partial_{x_1} g_1 + h, \qquad (20.7)$$

where only the two functions g_1 and g_2 appear. Notice that in Eq. (20.7) the function f does no longer appear! To form a system of two equations in the two unknowns g_1 and g_2 it needs to add a first order equation that be satisfied by the functions g_1 and g_2 (of course, this equation should not contain f). The relation between g_1 and g_2 that can be used to this end is the *compatibility* equation:

$$\partial_{x_2} g_1 = \partial_{x_1} g_2, \qquad (20.8)$$

following the requirement of continuity of the second order derivatives of f, according to Schwarz's theorem. One can then replace the second order Eq. (20.5), with the equivalent system consisting of the new form (20.7) of Eq. (20.5) and of the compatibility Eq. (20.8) (which maintains the quasi-linear structure and the normal form):

$$\begin{cases} \partial_{x_2} g_1 = \partial_{x_1} g_2 \\ \partial_{x_1} g_2 = -g_1 \partial_{x_1} g_1 + h \,. \end{cases} \tag{20.9}$$

A solution to the system (20.9), that is a pair of functions $g_{1,2}$ which satisfies it, allows one to find the actual solution f of Eq. (20.5) through integration of the system (20.6).

A more complicated example, for $n = 3$ ed $m = 2$, is given in the following. Suppose one has the equation in normal form:

$$\partial^3_{x_1 x_1 x_2} f = f + \partial_{x_1} f + \left(\partial^3_{x_1 x_1 x_1} f\right)^2 + \left(\partial^2_{x_1 x_2} f\right)^3 \partial^2_{x_2 x_2} f + h \,, \tag{20.10}$$

where, as usual, h is a (continuous) function of the two real variables x_1 and x_2. As done previously, one introduces the new functions:

$$\partial_{x_1} f =: g_1 \,, \quad \partial_{x_2} f =: g_2 \tag{20.11}$$

for the first order derivatives of f, and also three more functions for the second order derivatives:

$$\partial_{x_1} g_1 = g_{11} \,, \quad \partial_{x_2} g_1 = g_{12} \,, \quad \partial_{x_2} g_2 = g_{22} \,. \tag{20.12}$$

Using the new functions (20.11), and (20.12), Eq. (20.10) can be rewritten as a first order equation in normal form:

$$\partial_{x_1} g_{12} = f + g_1 + \left(\partial_{x_1} g_{11}\right)^2 + g^3_{12} g_{22} + h \,, \tag{20.13}$$

in the five unknown functions: f, g_1, g_{11}, g_{12} and g_{22}. In order to obtain a system of 5 first order equations equivalent to Eq. (20.10), 4 first order equations have to be added to (20.13), so as to relate the 5 function (f, g_1, g_{11}, g_{12} and g_{22}) to each other. Observe, however, that the relation between f and g_{22} is given by the second order equation: $\partial^2_{x_2 x_2} f = g_{22}$ and then, to maintain the system of the first order, it is required to introduce a new unknown function g_2 with a corresponding increase of one in the number of (unknowns and) equations of the system. This can now be written as follows:

$$\begin{cases} \partial_{x_1} g_{12} = f + g_1 + \left(\partial_{x_1} g_{11}\right)^2 + g_{12}^3 g_{22} + h & \text{equation (20.13)} \\ \partial_{x_1} f = g_1 & \text{definition of } g_1 \text{ (20.11)} \\ \partial_{x_2} f = g_2 & \text{definition of } g_2 \text{ (20.11)} \\ \partial_{x_1} g_1 = g_{11} & \text{definition of } g_{11} \text{ (20.12)} \\ \partial_{x_2} g_1 = g_{12} & \text{definition of } g_{12} \text{ (20.12)} \\ \partial_{x_2} g_2 = g_{22} & \text{definition of } g_{22} \text{ (20.12)} \end{cases} \quad (20.14)$$

What would happen in Eq. (20.10) if the first term in the right-hand side (f) were missing? The transformed Eq. (20.13) would be written with the only 4 unknown functions g_1, g_{11}, g_{12} and g_{22}, and one could not use the definitions of g_1 and g_2, that is the second and third equation of the system (20.14). The missing equation would be provided by the compatibility condition (20.8). One would obtain, instead of the system (20.14), a system of 5 equations in 5 unknowns g_1, g_2, g_{11}, g_{12} and g_{22}.

As shown by the above example, one cannot predict the order of the system equivalent to a general partial differential equation, the latter depending strongly on the algebraic structure of the function F. However, one can calculate a *highest order* (denoted by N in the following) of the equivalent system of first order equations, counting the number of functions g_1, g_{11}, ... that can be introduced for a n-th order equation in a space of m dimensions. The number of partial derivatives of order k of f is given by:

$$\mu_k = \binom{m + k - 1}{k}. \quad (20.15)$$

The natural number (20.15) is the number of combinations with repetitions of k objects (the number of derivatives considered) from a collection of m (the directions along which derivatives are taken). It follows that the highest number of functions that can be introduced for Eq. (20.3), including f, is given by:

$$N = 1 + \sum_{k=1}^{n-1} \mu_k. \quad (20.16)$$

An idea on how the number N may vary with n and m is given in Table 20.1.

Table 20.1 Maximum numnber N of equations in a first order system equivalent to a partial differential equation of order n (columns) in a space of m dimensioni (rows)

	1	2	3	4	5	6	7	8	9	10
2	1	3	6	10	15	21	28	36	45	55
3	1	4	10	20	35	56	84	120	165	220
4	1	5	15	35	70	126	210	330	495	715
5	1	6	21	56	126	252	462	792	1287	2002

As in the case of the ordinary differential equations, one calls an *integral* of a partial differential equation a function f, that satisfies Eq. (20.3). In general, the integral of an equation of the type (20.3) is not unique. However, one is usually interested in determining an integral that satisfy additional conditions. In fact, Eq. (20.3) normally is not given in the whole space $\mathbb{R}^m$, but rather in an appropriate bounded domain $\Omega \subseteq \mathbb{R}^m$. Moreover, on the boundary of Ω, values of the function f and / or a suitable number of its derivatives (taken in the normal direction to $\partial\Omega$) are prescribed. This is known as a *differential problem*. In the next section some properties of particular differential problems will be analysed.

20.3 Classification of Linear Partial Differential Equations

Consider $m = 2$ ($x = (x, y)^\mathsf{T}$) and $n = p$ in Eq. (20.3), also assuming F to be *linear in f and in its derivatives*. Following the previous discussion, this equation is equivalent to a linear system of p first order partial differential equations. One can then analyse the nature of this equation directly considering the equivalent system, that is simpler since all the p equations in it are of first order.

The vector of the unknowns functions is defined as $u = (u_1, u_2, \ldots, u_p)^\mathsf{T}$ and the vector of the known terms is $w = (w_1, w_2, \ldots, w_p)^\mathsf{T}$. In addition, one has the matrices $a, b, c \in \mathbb{R}^p \times \mathbb{R}^p$ of the coefficients of the derivatives in x in y and the u coefficients.

State the following *Cauchy problem*:

$$\begin{cases} a(x) \cdot \partial_x u + b(x) \cdot \partial_y u + c(x) \cdot u = w(x) \\ u(x) = v(x) \quad \text{given at each } x \in \Gamma \end{cases} \tag{20.17}$$

where Γ is an arc of a regular curve belonging to the existence set of all x functions introduced above. The function $v : \mathbb{R}^2 \to \mathbb{R}^p$ is prescribed on the curve Γ. Moreover, it is very important to assume that at least one of the two matrices of the coefficients (a or b) has a nonzero determinant.

If one wanted to build a solution to the problem (20.17), at least in a neighborhood of the curve Γ (*i.e.*, an open subset in the (x, y)-plane containing Γ), one should start from the data on that curve and determine, from the equation in the problem (20.17), the vector u *at points exterior to the curve itself*. So, in order to carry out this task, *it must be possible to calculate from the equation in* (20.17) *the normal derivative* $\partial_v u$ at each point of the curve Γ. Denoting $\tau(x)$ and $v(x) = \tau^\perp(x)$ the tangent and normal unit vectors, respectively, to the curve Γ (the orientation is meaningless) at its point x, the tangent derivative of u is known from the data prescribed on the curve Γ in (20.17). Indeed:

$$\tau_x \partial_x u + \tau_y \partial_y u = \tau \cdot \nabla u = \tau \cdot \nabla v =: \frac{dv}{ds} = v'. \tag{20.18}$$

Make a system with Eq. (20.18) and the definition of normal derivative:

$$v_x \partial_x u + v_y \partial_y u = -\tau_y \partial_x u + \tau_x \partial_y u = \partial_\nu u$$

and get $\partial_x u$, $\partial_y u$:

$$\partial_x u = -\tau_y \partial_\nu u + \tau_x v' , \quad \partial_y u = +\tau_x \partial_\nu u + \tau_y v' . \tag{20.19}$$

Replacing derivatives (20.19), just obtained, in Eq. (20.17) one gets the following linear system in the normal derivatives of the unknown functions (components of u):

$$\left(-\tau_y a + \tau_x b\right) \cdot \partial_\nu u = -\left(\tau_x a + \tau_y b\right) \cdot v' - c \cdot v + w , \tag{20.20}$$

where, for the sake of brevity, the dependence on x of all terms has been omitted. The right-hand side of the system (20.20) is now defined, since the coefficients a, b and c and the constant term w are known, and v (and, therefore, its tangent derivative $\tau \cdot \nabla v = v'$) is prescribed on the curve Γ. The system (20.20) is compatible if and only if:

$$\det\left(-\tau_y a + \tau_x b\right) \neq 0 . \tag{20.21}$$

The condition (20.21) can then express a constraint on the direction tangent to Γ at its point x, which must be satisfied so that the differential problem (20.17) admit a solution in a neighborhood of Γ. Notice that the condition (20.21) also includes two special cases, corresponding to the choice of τ along the coordinate direction x ($\tau_y = 0$) or along y ($\tau_x = 0$). In the former case, the compatibility of the system is met only if $\det(b) \neq 0$, in the latter only if $\det(a) \neq 0$. The appearance of one of these two particular cases ($\tau_y = 0$ or $\tau_x = 0$) *should not be considered* in determining the *characteristics directions*, as it will be shown in the following.

The condition (20.21) is normally used in reverse. That is, given the coefficients a and b one searches at a given x the directions along which this condition is violated or, equivalently, the components x and y of a unit vector[2] τ such that:

$$\det\left(-\tau_y a + \tau_x b\right) = 0 . \tag{20.22}$$

Equation (20.22) is an algebraic equation of degree not higher than p in the ratio $\rho = \tau_x/\tau_y$ (if $\tau_y \neq 0$) or in the reciprocal ratio $1/\rho = \tau_y/\tau_x$ (if $\tau_x \neq 0$). The roots of Eq. (20.22) identify directions through x that are called *characteristic*.

Denote with $r \leq p$ the degree of the (20.22). The linear equation in (20.17) is then said *hyperbolic* if r roots of the algebraic Eq. (20.22) are real and distinct. If this condition is met and $r = p$, we speak of equation one says the equation is *strictly hyperbolic*. Instead, if the roots are not real, the equation is said *elliptical*.

---

[2] The fact that τ is a *unit vector* is not crucial in the following considerations. In fact, Eq. (20.22) does not define the modulus of τ. It would be enough to talk about the direction identified by τ, with no reference to the module of this vector.

In all other cases, the equation is said *parabolic*. From the qualitative point of view, hyperbolic equations describe phenomena of wave propagation, in which certain "signals" propagate simultaneously in several directions. A disturbance generated in a small region of the field will modify the solution only in a limited region (which is called *domain of influence* of the disturbance). Conversely, elliptic equations are often used in the search for the potential of vector fields. A disturbance generated in a small region of the field affects the whole solution, even very far from the area it originated. Finally, parabolic equations are typical of heat transfer and, more generally, of diffusive phenomena. A couple of examples are in order.

Example 20.1 Consider the third order linear equation:

$$x\partial^3_{xxx} f - y\partial^3_{xxy} f + \partial^2_{xy} f - \partial^2_{yy} f = 0 \qquad (20.23)$$

for $x \neq 0$. Introduced the three functions $g_{xx} := \partial^2_{xx} f, g_{xy} := \partial^2_{xy} f$ and $g_{yy} := \partial^2_{yy} f$, if one consider the derivative $\partial^3_{xxy} f$ as $\partial_x g_{xy}$, Eq. (20.23) is equivalent to the system:

$$\begin{cases} x\partial_x g_{xx} - y\partial_x g_{xy} + g_{xy} - g_{yy} = 0 \\ \partial_y g_{xx} - \partial_x g_{xy} = 0 \\ \partial_x g_{yy} - \partial_y g_{xy} = 0, \end{cases}$$

that can be written in the form (20.17) with the following matrices:

$$u = \begin{pmatrix} g_{xx} \\ g_{xy} \\ g_{yy} \end{pmatrix}, \quad a = \begin{pmatrix} x & -y & 0 \\ 0 & -1 & 0 \\ 0 & 0 & 1 \end{pmatrix}, \quad b = \begin{pmatrix} 0 & 0 & 0 \\ 1 & 0 & 0 \\ 0 & -1 & 0 \end{pmatrix}, \quad c = \begin{pmatrix} 0 & 1 & -1 \\ 0 & 0 & 0 \\ 0 & 0 & 0 \end{pmatrix}, \quad w = 0.$$

$$(20.24)$$

Observe that $\det(b) = 0$, so it is expected that at least one zero of the characteristic determinant be obtained at $\tau_y = 0$. In fact, write the determinant (20.22):

$$\det(-\tau_y a + \tau_x b) = \begin{vmatrix} -x\tau_y & y\tau_y & 0 \\ \tau_x & \tau_y & 0 \\ 0 & -\tau_x & -\tau_y \end{vmatrix} = \tau_y^2(y\tau_x + x\tau_y), \qquad (20.25)$$

and search for its zeros. Excluding the solution $\tau_y = 0$, from the previous equation one identifies only one characteristic direction at an angle $-\arctan(y/x)$ with respect to the same axis. Equation (20.23) turns out to be hyperbolic.

Finally, observe that the derivative $\partial^3_{xxy} f$ could also be seen as $\partial_y g_{xx}$. In such case the matrices a and b (20.24) would be transformed into the following ones:

$$a = \begin{pmatrix} x & 0 & 0 \\ 0 & -1 & 0 \\ 0 & 0 & 1 \end{pmatrix}, \quad b = \begin{pmatrix} -y & 0 & 0 \\ 1 & 0 & 0 \\ 0 & -1 & 0 \end{pmatrix},$$

but this would not change the value of the determinant (20.25), even though it takes the new form:

$$\begin{vmatrix} -x\tau_y - y\tau_x & 0 & 0 \\ \tau_x & \tau_y & 0 \\ 0 & -\tau_x & -\tau_y \end{vmatrix}.$$

Example 20.2 Consider the linear equation of third order:

$$\partial^3_{xxy} f - \partial^3_{xyy} f + x\partial^3_{xxx} f + \partial^2_{xx} f = 1, \tag{20.26}$$

for $x \neq 0$. Using the functions $g_{xx} := \partial^2_{xx} f$ e $g_{xy} := \partial^2_{xy} f$, Eq. (20.26) turns out to be equivalent to the system of *two* first order equations:

$$\begin{cases} x\partial_x g_{xx} + \partial_x g_{xy} - \partial_y g_{xy} + g_{xx} = 1 \\ \partial_y g_{xx} - \partial_x g_{xy} = 0, \end{cases}$$

that can be rewritten in the vector form (20.17), once the following matrices have been defined:

$$u = \begin{pmatrix} g_{xx} \\ g_{xy} \end{pmatrix}, \quad a = \begin{pmatrix} x & 1 \\ 0 & -1 \end{pmatrix}, \quad b = \begin{pmatrix} 0 & -1 \\ 1 & 0 \end{pmatrix}, \quad c = \begin{pmatrix} 1 & 0 \\ 0 & 0 \end{pmatrix}, \quad w = \begin{pmatrix} 1 \\ 0 \end{pmatrix}.$$

The determinant (20.22) in this case take the form:

$$\det\left(-\tau_y a + \tau_x b\right) = \begin{vmatrix} -x\tau_y & -\tau_y - \tau_x \\ \tau_x & \tau_y \end{vmatrix} = -x\tau_y^2 + \tau_x\tau_y + \tau_x^2,$$

having assumed $x \neq 0$, $\tau_x \neq 0$ and, at $\rho = \tau_y/\tau_x$ one finds the roots:

$$\rho_{1,2} = \frac{1 \pm \sqrt{4x + 1}}{2x}. \tag{20.27}$$

The roots (20.27) are not real for $x < -1/4$, conditions in which Eq. (20.26) turns out to be elliptic. At $x = -1/4$ the same equation is parabolic, whereas it becomes hyperbolic for $x > -1/4$.

20.4   Suggested Readings

A general introduction and the classification of partial differential equations is given in Lax and Terrell (2017, Chap. 9), Petrovsky (1954, Chap. I).

References

P. D. Lax and M. S. Terrell. *Multivariable Calculus with Applications*. Springer, Cham, CH, 2017.
I. G. Petrovsky. *Lectures on Partial Differential Equations*. Interscience, New York, 1954.

Chapter 21
Quasi-Linear Second Order Partial Differential Equations

In the past lecture it was discussed how can be classified a linear partial differential equation of any order, with unknown function $f : \mathbb{R}^2 \to \mathbb{R}$. Here the attention is focused on second order quasi-linear equations.

21.1 General Form of the Equations

Quasi-linear second order partial differential equations have the following form:

$$\alpha(\boldsymbol{x})\partial^2_{xx} f + \beta(\boldsymbol{x})\partial^2_{xy} f + \gamma(\boldsymbol{x})\partial^2_{yy} f = h(\boldsymbol{x}; f, \partial_x f, \partial_y f), \qquad (21.1)$$

being, as usual, $\boldsymbol{x} = (x, y)^{\mathrm{T}}$. Assume to analyse Eq. (21.1) at a point $\boldsymbol{x}$ such that $\alpha(\boldsymbol{x})$ and $\gamma(\boldsymbol{x})$ be nonvanishing. Notice that the form (21.1) is not the most general form of a second order quasi-linear equation, since the coefficients of the second derivatives of f (α, β and γ) are independent of f and of its first order derivatives.

Using the auxiliary functions $g_x := \partial_x f$, $g_y := \partial_y f$, Eq. (21.1) turns out to be equivalent to the system of three first order equations:

$$\begin{cases} \alpha\partial_x g_x + \beta\partial_y g_x + \gamma\partial_y g_y = h \\ \partial_x f - g_x = 0 \\ \partial_y g_x - \partial_x g_y = 0, \end{cases}$$

which suggests to choose as the vector of the unknowns:

$$\boldsymbol{u} = \begin{pmatrix} f \\ g_x \\ g_y \end{pmatrix}.$$

© The Author(s), under exclusive license to Springer Nature Switzerland AG 2024
G. Riccardi et al., *Multidimensional Differential and Integral Calculus*,
https://doi.org/10.1007/978-3-031-70326-3_21

To the above system are then associated the matrices of the coefficients of the derivatives with respect to x (a) and y (b):

$$a = \begin{pmatrix} 0 & \alpha & 0 \\ 1 & 0 & 0 \\ 0 & 0 & -1 \end{pmatrix}, \quad b = \begin{pmatrix} 0 & \beta & \gamma \\ 0 & 0 & 0 \\ 0 & 1 & 0 \end{pmatrix}.$$

Observe that $\det(a) \neq 0$ (if $\alpha \neq 0$), whereas $\det(b) = 0$, so that one can expect a characteristic direction along the x axis ($\tau_y = 0$). The characteristic directions are identified by the condition:

$$\det\left(-\tau_y a + \tau_x b\right) = \begin{vmatrix} 0 & -\alpha\tau_y + \beta\tau_x & \gamma\tau_x \\ -\tau_y & 0 & 0 \\ 0 & \tau_x & \tau_y \end{vmatrix}$$

$$= -\tau_y \left(\alpha\tau_y^2 - \beta\tau_x\tau_y + \gamma\tau_x^2\right) = 0. \tag{21.2}$$

Equation (21.2) shows that there exists a characteristic direction along the coordinate x axis ($\tau_y = 0$), whereas for $\tau_y \neq 0$ one find the two roots for $\rho = \tau_x/\tau_y$:

$$\rho_{1,2} = \frac{\beta \pm \sqrt{\beta^2 - 4\alpha\gamma}}{2\gamma}. \tag{21.3}$$

It follows that Eq. (21.1) is elliptic if $\beta^2 < 4\alpha\gamma$, parabolic when $\beta^2 = 4\alpha\gamma$ and, finally, strictly hyperbolic if $\beta^2 > 4\alpha\gamma$.

A partial differential problem for each of the three classes of equations just introduced will be analysed in the following, using Eq. (21.1) as typical equation. The elliptic case ($\beta^2 < 4\alpha\gamma$) is considered first.

21.2 An Example of Elliptic Problem

Consider the following partial differential problem:

$$\begin{cases} \nabla^2 f = g(x, y) & (x, y) \in (-\pi, +\pi) \times (0, h) \\ f(-\pi, y) = f(+\pi, y) & y \in (0, h) \\ f(x, 0) = p(x) & x \in (-\pi, +\pi) \\ \partial_y f(x, h) = q(x) & x \in (-\pi, +\pi) \end{cases} \tag{21.4}$$

where the equation is of the type (21.1) with $\alpha \equiv 1$, $\beta \equiv 0$ and $\gamma \equiv 1$ and, moreover, the right-hand side is independent of the unknown function (f) and on its first order derivatives. The equation is then elliptic in the whole plane: it is called *Poisson's equation* if it is actually $g \neq 0$, whereas if the known term g is identically zero, it is called *Laplace's equation*. The boundary conditions associated to the equation are of three different kinds: on the vertical sides of the rectangle $[-\pi, +\pi] \times [0, h]$ *periodicity conditions* are prescribed (what happens at a point $(-\pi, y)$ also happens at the corresponding point $(+\pi, y)$), whereas on the bottom side the unknown function is prescribed and on the top side its normal derivative. Of course, the known term g must satisfy the periodicity condition $g(-\pi, y) = g(+\pi, y)$, the laplacian of a periodic function being itself a periodic function. The problem (21.4) is called a *Poisson's problem*, with mixed boundary conditions.

In order to find the solution of the problem (21.4) it is appropriate to exploit two important properties of the partial differential equation and their boundary conditions. The linear nature of the equation allows to build its solution as a countable sum of simple solutions of the equation, up to satisfy the boundary conditions. This approach leads to exploit the mathematical properties of the series of functions, described in detail in lesson 8. Moreover, the periodicity of the boundary conditions allows to use trigonometric functions, as the Fourier's series, introduced in Sect. 8.5.

The periodic structure with respect to x of the boundary conditions suggests to search for the solution in the form of a Fourier series in the x variable. In fact, when the known term g, the data p and q, as well as the sought solution f, are expanded in Fourier series:

$$g(x, y) = \sum_{k=-\infty}^{+\infty} \psi_k(y)\, e^{ikx}$$

$$p(x) = \sum_{k=-\infty}^{+\infty} \chi_k\, e^{ikx}$$

$$q(x) = \sum_{k=-\infty}^{+\infty} \mu_k\, e^{ikx} \tag{21.5}$$

$$f(x, y) = \sum_{k=-\infty}^{+\infty} \phi_k(y)\, e^{ikx}$$

one finds that the periodicity conditions get automatically satisfied. It should be observed that the (complex) coefficients in the Fourier series (21.5) satisfy the conditions to result in a real sum, e.g. $\psi_{-k} = \overline{\psi_k}$ and analogue, where the bar denotes the conjugate. Using the expansions (21.5) in the problem (21.4) one obtains, for relative integer k, the ordinary differential problem:

$$\begin{cases} \dfrac{d^2\phi_k}{dy^2} - k^2\phi_k = \psi_k \ \ y \in (0, h) \\[2mm] \phi_k(0) = \chi_k \\[2mm] \dfrac{d\phi}{dy}(h) = \mu_k \end{cases} \tag{21.6}$$

The solution of the general problem (21.6) can be written as:

$$\phi_k(y) = \chi_k \cosh(ky)$$

$$+\frac{1}{\cosh(kh)} \left\{ \mu_k - k\chi_k \sinh(kh) - \int_0^h d\eta \cosh\left[k(h-\eta)\right] \psi_k(\eta) \right\} \frac{\sinh(ky)}{k}$$

$$+ \int_0^y d\eta \, \frac{\sinh\left[k(y-\eta)\right]}{k} \, \psi_k(\eta) \tag{21.7}$$

and it holds true for every relative integer k. In fact, for $k = 0$ the problem (21.6) has the solution:

$$\phi_0(y) = \chi_0 + \left[\mu_0 - \int_0^h d\eta \, \psi_0(\eta) \right] y + \int_0^y d\eta \, (y-\eta) \, \psi_k(\eta),$$

that can also be obtained from the general form (21.7) taking the limit as $k \to 0$ (remember that $\sinh(a\xi)/a \to \xi$ when $a \to 0$).

Once the Fourier coefficients (21.7) have been calculated, the solution f can be obtained summing the series in (21.5). Two examples are given in the following.

Example 21.1 Calculate the solution of the problem (21.4) for the following choice of the known term and of the boundary conditions:

$$g(x, y) = e^{-\alpha y} \sin x, \quad p(x) = \cos(2x), \quad q(x) = \sin(3x), \tag{21.8}$$

with the condition $|\alpha| \neq 1$.

First of all, write the Fourier coefficients of the functions (21.8), corresponding to the non negative wave numbers

$$\psi_{+1}(y) = -\frac{i}{2} e^{-\alpha y}, \quad \chi_{+2} = \frac{1}{2}, \quad \mu_{+3} = -\frac{i}{2},$$

whereas the homologous coefficients for the negative wave numbers can be obtained through the conjugate. Coefficients that do not appear are identically zero. Thus, observe that, due to the linear nature of the problem (21.4), the only non zero components of f are those corresponding to values $k = \pm1, \pm2$ and ±3. Using the solution formula (21.7) one finds the Fourier components:

$$\phi_{+1}(y) = \frac{1}{2i} \frac{1}{\alpha^2 - 1} \left[\left(\alpha e^{-\alpha h} + \sinh h \right) \frac{\sinh y}{\cosh h} + e^{-\alpha y} - \cosh y \right]$$

$$\phi_{+2}(y) = \frac{1}{2} \left[\cosh(2y) - \tanh(2h) \sinh(2y) \right]$$

$$\phi_{+3}(y) = \frac{1}{2i} \frac{1}{3} \frac{\sinh(3y)}{\cosh(3h)},$$

whence the solution follows:

$$f(x, y) = \frac{\sin x}{\alpha^2 - 1} \left[(\alpha e^{-\alpha h} + \sinh h) \frac{\sinh y}{\cosh h} + e^{-\alpha y} - \cosh y \right]$$

$$+ \cos(2x) \left[\cosh(2y) - \tanh(2h) \sinh(2y) \right] + \frac{\sin(3x) \sinh(3y)}{3 \cosh(3h)}.$$

(21.9)

Exercise 21.1 What happens if $\alpha = \pm 1$?

Example 21.2 Consider the solution of the problem (21.4) corresponding to the choice:

$$g(x, y) = \sum_{k=1}^{n} \sin(ky) \cos(kx), \quad p(x) = \sum_{k=1}^{n} \frac{\sin(kx)}{1 + k^2}, \quad q(x) = \sum_{k=1}^{n} \frac{\cos(kx)}{1 + k}.$$

(21.10)

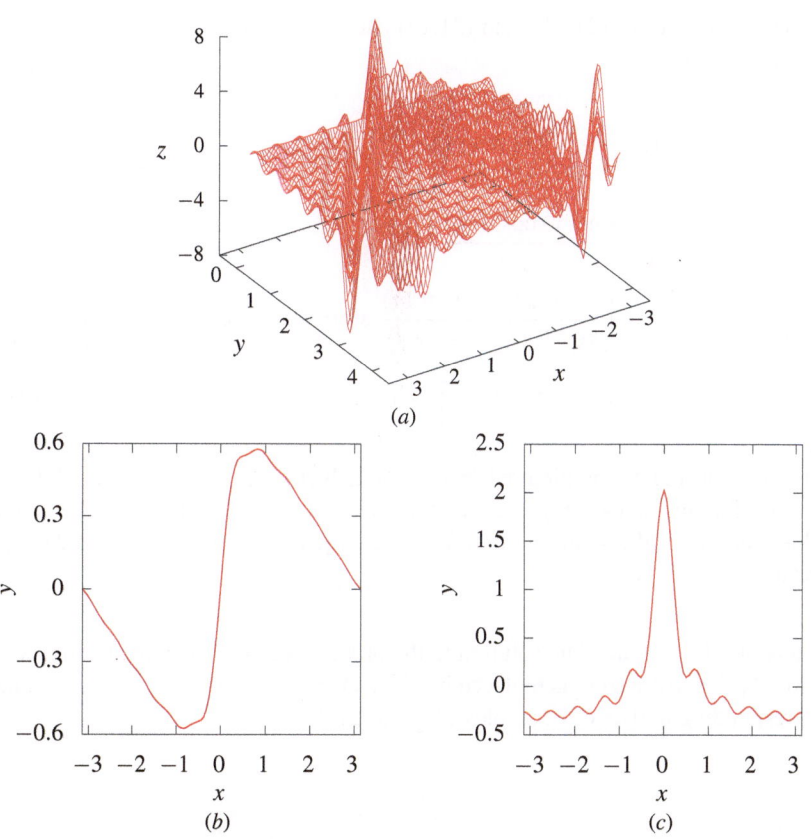

(a)

(b)

(c)

Fig. 21.1 Data of Example 21.1: functions g **a**, p **b** and q **c** (21.10) for $n = 10$, $h = 4$

To have an idea of the pattern of this data one can look at Fig. 21.1. Corresponding to the above data one finds the Fourier coefficients ($k = 1, 2, \ldots, n$):

$$\psi_k(y) = +\frac{1}{2}\sin(ky) \ k = +1, +2, \ldots \quad \psi_k(y) = -\frac{1}{2}\sin(ky) \ k = -1, -2, \ldots$$

$$\chi_k = +\frac{1}{2i}\frac{1}{1+k^2} \ k = +1, +2, \ldots \quad \chi_k = -\frac{1}{2i}\frac{1}{1+k^2} \ k = -1, -2, \ldots$$

$$\mu_k = \frac{1}{2}\frac{1}{1+k} \qquad k = +1, +2, \ldots \quad \mu_k = \frac{1}{2}\frac{1}{1-k} \qquad k = -1, -2, \ldots ,$$

$$(21.11)$$

and moreover, integrating by parts twice, one calculates the following integrals ($k \neq 0$):

$$I_k(y) = \int_0^y d\eta \, \cosh[k(y-\eta)]\sin(k\eta) = \frac{\cosh(ky) - \cos(ky)}{2k}$$

$$J_k(y) = \int_0^y d\eta \, \sinh[k(y-\eta)]\sin(k\eta) = \frac{\sinh(ky) - \sin(ky)}{2k}.$$

$$(21.12)$$

Based on definitions (21.11), and (21.12) one can rewrite the solution (21.7) in the present case as:

$$f(x, y) = \sum_{k=1}^{n} \left[\phi_k(y)e^{+ikx} + \phi_{-k}(y)e^{-ikx}\right]$$

$$= \sum_{k=1}^{n} \left\{ \frac{\sin(kx)\cosh(ky)}{k^2+1} \right.$$

$$+ \left[\frac{\cos(kx)}{k(k+1)} - \frac{\sinh(kh)}{k^2+1}\sin(kx) - \frac{1}{k}\cos(kx)I_k(h)\right]\frac{\sinh(ky)}{\cosh(kh)}$$

$$+ \left. \frac{1}{k}\cos(kx)J_k(y)\right\}.$$

$$(21.13)$$

The solution (21.13) is pictured in Fig. 21.2. Notice, in particular, the "filtering" action affecting g (see Fig. 21.1a): the function f results far smoother then g. This property of the solution of the Poisson's equation is widely used in the digital treatment of images.

Exercise 21.2 Deduce the solution to the problem (21.4), when the boundary condition $\partial_y f(x, h) = q(x)$ is replaced by $f(x, h) = q(x)$. Calculate the new solutions corresponding to data of Examples 21.1 and 21.2.

Fig. 21.2 Solution f (21.13) to the problem (21.4), with $n = 10$ and $h = 4$

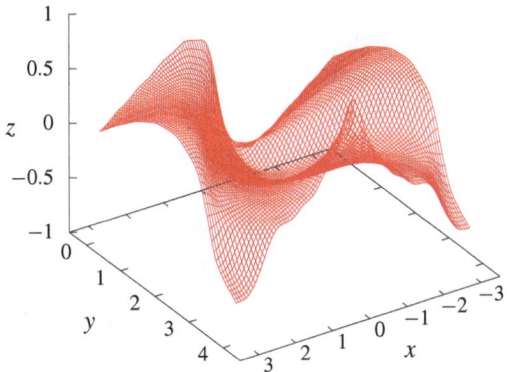

21.3   Why the Poisson's Equation is Very Important

After these brief considerations on the solution (in simple geometry) of the Poisson's problem, it is time to discuss some examples of Poisson's (or Laplace's) equation of physical interest. Normally, the Poisson's equation occurs when considering the divergence of a vector function of a vector variable f that can be written as the gradient of a scalar function φ of the same vector variable:

$$f = \nabla \varphi .\tag{21.14}$$

The function φ is a potential generated by f. As it is obvious by its definition (21.14), it is not unique. Often the divergence g of f has an important physical meaning. Thus, observe that from Eq. (21.14) it follows that:

$$g = \nabla \cdot f = \nabla \cdot \nabla \varphi = \nabla^2 \varphi ,\tag{21.15}$$

that is just a Poisson's equation.

One may ask what conditions must f satisfy so that (21.14) be valid. First, it is shown that asking for the validity of the relation (21.14) is equivalent to requiring that given two points x_0 and x in the space where f works, the integral on a curve $\mathscr{C}(x_0, x)$ starting at x_0 and ending at x:

$$\int_{\mathscr{C}(x_0,x)} dx' \cdot f(x')\tag{21.16}$$

be independent of the curve $\mathscr{C}(x_0, x)$. In fact, if the condition (21.14) holds, then:

$$\int_{\mathscr{C}(x_0,x)} dx' \cdot f(x') = \int_{\mathscr{C}(x_0,x)} dx' \cdot \nabla \varphi(x') = \int_{\mathscr{C}(x_0,x)} d\varphi = \varphi(x) - \varphi(x_0) ,$$

of course, independently of the path chosen to integrate between x_0 and x. Viceversa, suppose that the integral (21.16) proves to be independent of the integration path. Thus this integral defines a new function:

$$\phi(x) := \int_{\mathscr{C}(x_0,x)} dx' \cdot f(x'). \tag{21.17}$$

It is relatively easy to calculate the derivative of this function along the generic i-th coordinate direction e_i. Fix an arbitrary point $x^\star$, with the only constraint to choose it on the straight line through x parallel to i-th coordinate direction. To connect the two points x_0 and x, then choose an arbitrary curve $\mathscr{C}$ passing through $x^\star$, including arbitrary x_0 and $x^\star$ and straight between $x^\star$ and x. In this way the integral (21.17) is rewritten as follows:

$$\phi(x) = \int_{\mathscr{C}(x_0,x^\star)} dx' \cdot f(x') + \int_{\mathscr{C}(x^\star,x)} dx' \cdot f(x')$$

$$= \int_{\mathscr{C}(x_0,x^\star)} dx' \cdot f(x') + \int_0^{x_i - x_i^\star} d\xi \, f_i(x^\star + \xi e_i). \tag{21.18}$$

Using the differentiation rule for an integral whose endpoints depend on the differentiation variable, differentiate with respect to x_i the expression (21.18) of the function (21.17). In this way one obtains:

$$f_i(x) = \partial_{x_i} \phi(x). \tag{21.19}$$

The relation (21.19), holding true for all coordinate directions, proves that the function f is the gradient of the scalar function ϕ, that is (21.14). Observe also that the formula (21.17) provides an operational tool to calculate the function ϕ, given f. Furthermore, the functions ϕ (21.17) and φ (21.14) differ from each other only by constants, and then can be identified.

Thus, it has been shown that (21.14) and the path independence of the integral (21.16) are two equivalent conditions. The second condition has, however, deep implications. In fact, consider any *closed* and *oriented* curve $\mathscr{P}$. Fixed two arbitrary points x_0 and x on $\mathscr{P}$, this can be thought of as the union of two open curves: the arc $\mathscr{P}_+$, that is the part of $\mathscr{P}$ traveled from x_0 to x and the arc $\mathscr{P}_-$, traveled from x to x_0, following the positive direction on $\mathscr{P}$. By virtue of the path independence of the integral (21.16), one knows that:

$$\int_{\mathscr{P}_+(x_0,x)} dx' \cdot f(x') = -\int_{\mathscr{P}_-(x,x_0)} dx' \cdot f(x'),$$

that is:

$$\int_{\mathscr{P}} dx' \cdot f(x') = 0. \tag{21.20}$$

Then, if f can be expressed as the gradient of a scalar function, thus circulation on every closed curve $\mathscr{P}$ is zero.

If the curve $\mathscr{P}$ can be thought of as the contour of an arbitrary surface Σ all included in the field of existence of f ($\partial \Sigma = \mathscr{P}$), then one knows from the Stokes' theorem that the flux of the curl of f through this surface is zero, according to (21.20). Since the surface Σ remains arbitrary, this implies that the curl of f itself must be zero:

$$\nabla \times f \equiv 0. \tag{21.21}$$

However, it is not always possible to find a surface Σ entirely inside the existence field of f such that $\partial \Sigma = \mathscr{P}$. In fact, $\mathscr{P}$ may embrace a "hole" in the field of existence of f, which is called a *gap* in mathematical language. In this case, the reasoning here exposed has to be completely reviewed.

21.4 Suggested Readings

For linear equations, an overview of the main properties and classification is found in Weinberger (1995, Chap. II)

Reference

H. F. Weinberger. *A First Course in Partial Differential Equations*. Dover, New York, 1995.

Chapter 22
Parabolic and Hyperbolic Partial Differential Equations

This lecture briefly presents solutions for a parabolic and a hyperbolic equation, studying partial differential problems on domains of simple shape. As already seen in the previous lectures, parabolic equations generally govern diffusive problems, whereas hyperbolic equations appear when studying wave propagation phenomena.

22.1 An Example of Parabolic Problem

One searches for a function $f(x, y; t)$ of time t and two spatial variables x and y, that for any $(x, y; t) \in [-\pi, +\pi] \times [-\pi, +\pi] \times [0, +\infty)$ satisfies the problem:

$$\begin{cases} \partial_t f = \kappa \nabla^2 f + F \\ f(x, y; 0) = f_0(x, y) \quad \text{(prescribed)} \\ f(-\pi, y; t) = f(+\pi, y; t) \\ f(x, -\pi; t) = f(x, +\pi; t), \end{cases} \tag{22.1}$$

where κ is a given positive coefficient generally called *diffusion coefficient*.

Exercise 22.1 Determine the physical dimensions of κ. Do they depend on the physical dimensions of f?

The problem (22.1) is a classical one in the study of thermal diffusion (in this case f is the temperature) and it is composed by three sets of equations: a *field equation*, to be satisfied at any $t > 0$ and at every point (x, y) of the square $(-\pi, +\pi) \times (-\pi, +\pi)$, an *initial condition*, prescribing the values of the solution f at time $t = 0$, and conditions prescribing values of the solution f on the boundary of the square, called *boundary conditions*.

G. Riccardi et al., *Multidimensional Differential and Integral Calculus*,
https://doi.org/10.1007/978-3-031-70326-3_22

One can show that the equation in the problem (22.1) is parabolic. Some work is requested to extend to the case $m = 3$ arguments concerning the classification already shown in the case of a two-dimensional independent variable. To begin with, it must be reduced to a system of first order equations, introducing two new auxiliary functions: $g_x := \partial_x f$ and $g_y := \partial_y f$. This results in the following first order system in the unknown functions $(f, g_x, g_y)^\mathsf{T} =: \boldsymbol{u}$:

$$\begin{cases} \partial_t f - \kappa \partial_x g_x - \kappa \partial_y g_y = 0 \\ \partial_y g_x - \partial_x g_y \qquad\quad = 0 \\ \partial_x f - g_x \qquad\qquad\ = 0. \end{cases} \tag{22.2}$$

This is a linear system, of the form:

$$\boldsymbol{a}(\boldsymbol{x}) \cdot \partial_x \boldsymbol{u} + \boldsymbol{b}(\boldsymbol{x}) \cdot \partial_y \boldsymbol{u} + \boldsymbol{c}(\boldsymbol{x}) \cdot \partial_z \boldsymbol{u} = \boldsymbol{w}(\boldsymbol{x}), \tag{22.3}$$

where $\boldsymbol{x} = (x, y, z)$ and $\boldsymbol{a}, \boldsymbol{b}$ and $\boldsymbol{c}$ are three 3×3 matrices:

$$\boldsymbol{a} = \begin{pmatrix} 0 & -\kappa & 0 \\ 0 & 0 & -1 \\ 1 & 0 & 0 \end{pmatrix}, \quad \boldsymbol{b} = \begin{pmatrix} 0 & 0 & -\kappa \\ 0 & 1 & 0 \\ 0 & 0 & 0 \end{pmatrix}, \quad \boldsymbol{c} = \begin{pmatrix} 1 & 0 & 0 \\ 0 & 0 & 0 \\ 0 & 0 & 0 \end{pmatrix}. \tag{22.4}$$

In the case of Eq. (22.1) one sets t in the direction of z-coordinate. Since the differential Eq. (22.3) holds with $m = 3$, it is evident that one must set a differential problem prescribing $\boldsymbol{u}$ on a surface of the three-dimensional space, instead of a curve.

In the following, conditions are sought that enable to calculate the derivative of $\boldsymbol{u}$ normal to a given surface Σ in $\mathbb{R}^3$, based on the linear Eq. (22.3). Take a point $\boldsymbol{x} \in \Sigma$ and through it two directions $\boldsymbol{\tau}_{1,2}$ tangent to Σ at the same point, forming a base on the tangent plane. The normal direction is defined by means of the two angles θ and φ as follows:

$$\boldsymbol{v} \propto (\cos\theta \cos\varphi, \sin\theta \cos\varphi, \sin\varphi)^\mathsf{T}$$

using the vector product $\boldsymbol{\tau}_1 \times \boldsymbol{\tau}_2$. Call $\tau_{1,2}$ the moduli of vectors $\boldsymbol{\tau}_{1,2}$ and σ_{12} the scalar product $\boldsymbol{\tau}_1 \cdot \boldsymbol{\tau}_2$. Observe that orthogonality of $\boldsymbol{\tau}_1$ and $\boldsymbol{\tau}_2$ ($\sigma_{12} = 0$) is not requested, nor their normalization.

The two tangent derivatives of $\boldsymbol{u}$ in the directions $\boldsymbol{\tau}_{1,2}$ are denoted with $\boldsymbol{v}_{1,2}$:

$$\boldsymbol{v}_1 := \boldsymbol{\tau}_1 \cdot \nabla \boldsymbol{u}, \quad \boldsymbol{v}_2 := \boldsymbol{\tau}_2 \cdot \nabla \boldsymbol{u}. \tag{22.5}$$

Observe that these vectors are known, because of the knowledge of data on Σ. Then, calculate the three derivatives $\partial_x \boldsymbol{u}, \partial_y \boldsymbol{u}$ and $\partial_z \boldsymbol{u}$ as a function of the known quantities (22.5) and of the normal derivative $\boldsymbol{u}$ (unknown), solving the linear system:

$$\begin{cases} \tau_{1x}\partial_x u + \tau_{1y}\partial_y u + \tau_{1z}\partial_z u = v_1 \\ \tau_{2x}\partial_x u + \tau_{2y}\partial_y u + \tau_{2z}\partial_z u = v_2 \\ v_x\partial_x u + n_y\partial_y u + n_z\partial_z u = \partial_v u . \end{cases}$$

The determinant of the matrix of coefficients is $v^2 = |\mathbf{v}|^2$ and, solving by the method of Cramer, one obtains:

$$\begin{cases} v^2\,\partial_x u = v_x\partial_v u + p_{1x}v_1 + p_{2x}v_2 \\ v^2\,\partial_y u = n_y\partial_v u + p_{1y}v_1 + p_{2y}v_2 \\ v^2\,\partial_z u = v_z\partial_v u + p_{1z}v_1 + p_{2z}v_2 , \end{cases}$$

where, for the sake of brevity, the two vectors:

$$\mathbf{p}_1 = \tau_2^2\boldsymbol{\tau}_1 - \sigma_{12}\boldsymbol{\tau}_2 , \qquad \mathbf{p}_2 = \tau_1^2\boldsymbol{\tau}_2 - \sigma_{12}\boldsymbol{\tau}_1 .$$

have been introduced. Using the above expressions in Eq. (22.3) one obtains a system defining the normal derivative:

$$\left(v_x\mathbf{a} + v_y\mathbf{b} + v_z\mathbf{c} \right) \cdot \partial_v u = w' , \tag{22.6}$$

that is the analogue of the one obtained in the case $m = 2$. The term w' in Eq. (22.6) is given by:

$$w' = v^2 w - \left(p_{1x}\mathbf{a} \cdot \mathbf{v}_1 + p_{2x}\mathbf{a} \cdot \mathbf{v}_2 + p_{1y}\mathbf{b} \cdot \mathbf{v}_1 + p_{2y}\mathbf{b} \cdot \mathbf{v}_2 + p_{1z}\mathbf{c} \cdot \mathbf{v}_1 + p_{2z}\mathbf{c} \cdot \mathbf{v}_2 \right) .$$

The condition ensuring that the system (22.6) can be solved is turned into the requirement that the determinant of the matrix of coefficients be nonvanishing. Search for the plane attitudes that does not allow to obtain $\partial_v u$ from the system (22.6), requiring that the determinant of the matrix of coefficients built from arrays (22.4) be zero:

$$\begin{vmatrix} v_z & -\kappa v_x & -\kappa v_y \\ 0 & v_y & -v_x \\ v_x & 0 & 0 \end{vmatrix} = \kappa v_x \left(v_x^2 + v_y^2 \right) \propto \cos\theta \cos^3\varphi = 0 . \tag{22.7}$$

The condition (22.7) shows there are three characteristic roots coinciding at $\varphi = +\pi/2$: the equation in the problem (22.1) is thus parabolic, with characteristic surfaces given by planes $t = $ constant (that is, with $n_x = n_y = 0$). This means That the solution evolves in the time "direction": consequently, the integration of the differential problem (22.1) must proceed in time.

In the problem (22.1) it is required that the solution takes the same values on to parallel sides, that is *periodic* conditions along x and y are enforced. The functions $F(x, y; t)$, in the right-hand side of field equation, and $f_0(x, y)$ are known periodic functions, with period 2π along x and y (that is why they will be said *biperiodic*). Finally, the function f is assumed to be continuous with continuous first order time derivatives and second order space derivatives. The differential equation, as well as the boundary conditions in problem (22.1) are *linear*, that is depend linearly on the unknown function f and its derivatives. It follows that if $f_1(x, y; t)$ ed $f_2(x, y; t)$ are solutions of the problem (22.1), then every linear combination of them $\alpha_1 f_1(x, y; t) + \alpha_2 f_2(x, y; t)$ *with constant coefficients α_1 and α_2* is a solution of the same equation and satisfies the same boundary conditions. This suggests to search for a solution $f(x, y; t)$ of the problem (22.1) in the form of a Fourier series expansion in x and y:

$$f(x, y; t) = \sum_{p,q=-\infty}^{+\infty} c_{pq}(t)\, e^{i\,(px+qy)}, \qquad (22.8)$$

where the sum is taken on both p and q. Observe that the expansion (22.8) allows to automatically satisfy the conditions on the contour of the square, since all functions $\exp[i\,(px+qy)]$ are periodic with period 2π or submultiples of it.

Exercise 22.2 Having defined the lengths l_x (along x) and l_y (along y) and the grid of points $x_{kh} = (x_k, y_h) = ((k-1/2)l_x, (h-1/2)l_y)$ for k and h relative integers, write the Fourier series of the functions defined for $x \in (x_k - l_x/2, x_k + l_x/2) \times (y_h - l_y/2, y_h + l_y/2)$:

$$f_1(x) = \sin\left[6\pi\,(x - x_k)/l_x\right] \cos\left[4\pi\,(y - y_h)/l_y\right]$$

$$f_2(x) = \sin\left[2\pi\,(x - x_k)/l_x\right] \left[l_y^2/4 - (y - y_h)^2\right]$$

$$f_3(x) = \left[l_x^2/4 - (x - x_k)^2\right]\left[l_y^2/4 - (y - y_h)^2\right].$$

The partial derivatives appearing in Eq. (22.1) are as follows:

$$\partial_t f = \sum_{p,q=-\infty}^{+\infty} \frac{dc_{pq}}{dt}\, e^{i\,(px+qy)} \quad \text{and} \quad \nabla^2 f = -\sum_{p,q=-\infty}^{+\infty} (p^2 + q^2)\, c_{pq}\, e^{i\,(px+qy)}.$$

$$(22.9)$$

Moreover, since the two functions F and f_0 are given, it will be assumed to be known the Fourier coefficients of the expansions:

$$F(x, y; t) = \sum_{p,q=-\infty}^{+\infty} C_{pq}(t)\, e^{i\,(px+qy)} \quad \text{and} \quad f_0(x, y) = \sum_{p,q=-\infty}^{+\infty} c_{pq}^0\, e^{i\,(px+qy)}.$$

$$(22.10)$$

Using the derivatives (22.9) and the first of the expansions (22.10) in Eq. (22.1), one gets the infinitely many ordinary differential equations in time:

$$\frac{dc_{pq}}{dt} + \kappa\,(p^2 + q^2)\, c_{pq} = C_{pq},$$

each corresponding to a relative integer value of p and q. The above equations are solved starting from the initial conditions $c_{pq}(0) = c^0_{pq}$:

$$c_{pq}(t) = c^0_{pq}\, e^{-\kappa(p^2+q^2)t} + \int_0^t e^{-\kappa(p^2+q^2)(t-\tau)}\, C_{pq}(\tau)\, d\tau \,. \tag{22.11}$$

Using the solutions (22.11) (with $C_{pq}(t)$ given) one can calculate all coefficients of the expansion (22.8) at any time.

Suppose now that F is constant in time and has only one non zero (and constant) Fourier coefficient, say C_{lm}. The solutions (22.11) thus imply that as $t \to +\infty$ the solution $f(x, y; t)$ tends exponentially to the function $F(x, y)/\chi_{lm}$ with $\chi_{lm} = \kappa(l^2 + m^2)$, that is proportional to the forcing function. This asymptotic value is a decreasing function of κ: for increasing κ, the function f takes progressively smaller values, whereas "reaching" of the asymptotic solution requires progressively shorter times. Finally, consider the effect of a forcing function defined by a single Fourier coefficient $C_{lm}(t)$ being a sinusoidal function of time: $C_{lm}(t) = \sin(\omega t + \phi)$. The solution (22.11) for the corresponding Fourier coefficient of f is as follows:

$$rcl c_{lm}(t) = \underbrace{\left(c^0_{lm} + \frac{\omega \cos\phi - \chi_{lm} \sin\phi}{\chi^2_{lm} + \omega^2} \right) e^{-\chi_{lm}t}}_{\text{part in decay}} + \underbrace{\frac{\chi_{lm}}{\chi^2_{lm} + \omega^2} \sin(\omega t + \phi)}_{\text{component in phase}}$$

$$+ \underbrace{\frac{-\omega}{\chi^2_{lm} + \omega^2} \cos(\omega t + \phi)}_{\text{component in quadrature}} \,.$$

The forcing function causes a component in phase and a component in quadrature, along with damped part with time constant $1/\chi_{lm}$, rapidly decreasing with l ed m. Observe how the wave numbers included in χ_{lm} and the angular frequency ω of the forcing function play different roles in defining the amplitude of the components in phase and in quadrature.

22.2 An Example of a Hyperbolic Problem

Given a function $f : \mathbb{R}^2 \to \mathbb{R}$, consider now the one-dimensional *wave equation*:

$$\partial^2_{tt} f - c^2 \partial^2_{xx} f = 0 \,, \tag{22.12}$$

where c is the speed of propagation (taken as constant), known from arguments of different nature. Suppose f to be continuous, with continuous second order derivatives.

First of all it must be checked that Eq. (22.12) is hyperbolic. Introducing the vector of the independent variables $x = (x, t)^T$, the auxiliary functions $g_x := \partial_x f$ and $g_t := \partial_t f$ and the new vector of the unknowns $u = (g_x, g_t)^T$ one obtains the first order system:

$$\begin{cases} -c^2 \partial_x g_x + \partial_t g_t & = 0 \\ \partial_t g_x - \partial_x g_t & = 0, \end{cases}$$

and the associated matrices of the coefficients:

$$a = \begin{pmatrix} -c^2 & 0 \\ 0 & -1 \end{pmatrix}, \qquad b = \begin{pmatrix} 0 & 1 \\ 1 & 0 \end{pmatrix}.$$

The equation for the characteristic directions $\tau = (\tau_x, \tau_t)^T$ then follows:

$$|-\tau_t a + \tau_x b| = \begin{vmatrix} c^2 \tau_t & \tau_x \\ \tau_x & \tau_t \end{vmatrix} = c^2 \tau_t^2 - \tau_x^2 = 0, \tag{22.13}$$

which has two real and distinct solutions: $\tau_x = \pm c \tau_t$. Equation (22.12) is then strictly hyperbolic and the relation (22.13) shows that the characteristic curves are straight lines whose equations are $x \pm ct = $ constant.

Try to write Eq. (22.12) along the characteristic curves. Introduce the new coordinates $\xi = x + ct$ ed $\eta = x - ct$ and read $f[x(\xi, \eta), t(\xi, \eta)] := F(\xi, \eta)$, or, equivalently, $F[\xi(x, t), \eta(x, t)] = f(x, t)$. Equation (22.12), in the new variables ξ ed η, becomes ($\nabla^2 = \partial_{\xi\xi}^2 + \partial_{\eta\eta}^2$):

$$\underbrace{c^2 \left(\nabla^2 F - 2\partial_{\xi\eta}^2 F \right)}_{\partial_{tt}^2 f} - c^2 \underbrace{\left(\nabla^2 F + 2\partial_{\xi\eta}^2 F \right)}_{\partial_{xx}^2 f} = -4c^2 \partial_{\xi\eta}^2 F = 0.$$

Due to the presence of the mixed second derivative, the possible solutions of this equation are then confined to functions of the kind:

$$F(\xi, \eta) = p(\xi) + q(\eta), \tag{22.14}$$

where p and q are two arbitrary functions (continue along with their second order derivatives) of relevant variables. Coming back to the original vector variable $x = (x, t)^T$, the structure of the solution (22.14) to the wave equation becomes:

$$f(x, t) = F[\xi(x, t), \eta(x, t)] = p(x + ct) + q(x - ct). \tag{22.15}$$

Thus, Eq. (22.15) provides the general structure of the solution to the wave Eq. (22.12): the two functions p and q are then determined using additional conditions.

In most cases, Eq. (22.12) is part of a differential problem also including suitable initial conditions on f and its time derivative:

$$\begin{cases} \partial_{tt}^2 f - c^2 \partial_{xx}^2 f = 0 \\ f(x,0) = f_0(x), \quad \partial_t f(x,0) = c g_0(x). \end{cases} \tag{22.16}$$

Based on the structure (22.15) of the solution, determine the two functions p and q. Enforcing the two initial conditions one has:

$$\begin{cases} p(x) + q(x) = f_0(x) \\ p'(x) - q'(x) = g_0(x) \end{cases}$$

and integrating the second in x starting at an arbitrary point x_0 one finally gets the system:

$$\begin{cases} p(x) + q(x) = f_0(x) \\ p(x) - q(x) = \int_{x_0}^x d\sigma \, g_0(\sigma) + p(x_0) - q(x_0). \end{cases}$$

Solving the above system one obtains:

$$\begin{cases} p(x) = \dfrac{f_0(x)}{2} + \dfrac{1}{2} \int_{x_0}^x d\sigma \, g_0(\sigma) + \dfrac{p(x_0) - q(x_0)}{2} \\ q(x) = \dfrac{f_0(x)}{2} - \dfrac{1}{2} \int_{x_0}^x d\sigma \, g_0(\sigma) - \dfrac{p(x_0) - q(x_0)}{2} \end{cases} \tag{22.17}$$

and substituting in the general structure of the solution (22.15) one write the solution of the problem (22.16):

$$\begin{aligned} f(x,t) &= \frac{f_0(\xi) + f_0(\eta)}{2} + \frac{1}{2} \left[\int_{x_0}^\xi d\sigma \, g_0(\sigma) + \int_\eta^{x_0} d\sigma \, g_0(\sigma) \right] \\ &= \frac{f_0(x+ct) + f_0(x-ct)}{2} + \frac{1}{2} \int_{x-ct}^{x+ct} d\sigma \, g_0(\sigma). \end{aligned} \tag{22.18}$$

The relation (22.18) is called *d'Alembert's solution*,[1] because it was deduced by the famous eighteenth-century French mathematician. As it is shown by the formula (22.18), the solution at time $t > 0$ is given by the superposition of two "waves": one travels in the positive direction of x ($f_0(x - ct)/2$) and the other in the negative direction ($f_0(x + ct)/2$), replicating the initial data, but with half its amplitude. In addition, to these two waves the effect of the initial data on the $\partial_t f$ is superimposed.

Figure 22.1 shows two time histories corresponding to two different initial data. In the case (*a*) one has:

$$f_0(x) = \begin{cases} \sin(x) & -\pi < x < +\pi \\ 0 & x < -\pi, \quad x > +\pi \end{cases}, \qquad g(x) \equiv 0, \tag{22.19}$$

---

[1] https://mathshistory.st-andrews.ac.uk/Biographies/DAlembert/.

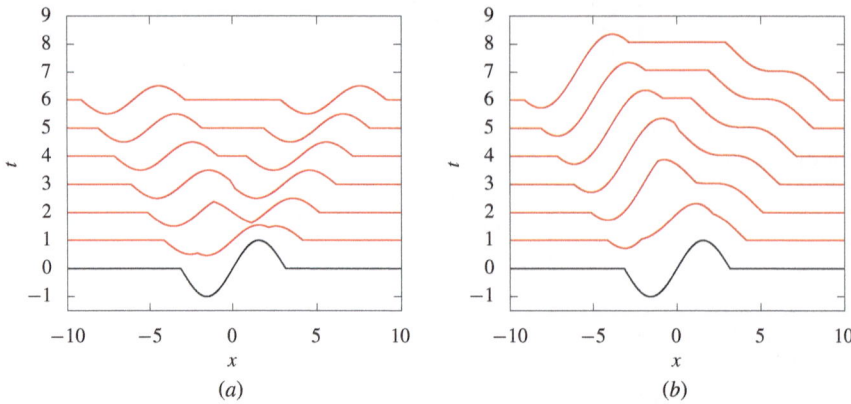

Fig. 22.1 Time histories corresponding to the choices (22.19) **a** and (22.20) **b**. The speed c is assumed unity. The graphs corresponding to times 0 (black line) and 1, 2, ..., 6 (red) are drawn overlapped to the straight lines $t = 0, t = 1, \ldots, t = 6$

whereas in the case (b) the original data f_0 and the new data g have been used:

$$g(x) = \begin{cases} \pi^2 - x^2 & -\pi < x < +\pi \\ 0 & x < -\pi, \quad x > +\pi . \end{cases} \qquad (22.20)$$

As it can be easily seen in (a), the waves traveling along the two families of characteristic lines sum up resulting in modified wave profiles. Between time 3 and time 4 the two waves separates and propagate undisturbed in the positive and negative direction of x. If also a derivative $\partial_t f$ at time 0 is present, the previous evolution is more complicated, since a flat term overlaps central region which becomes progressively large as time elapses.

As it can be easily deduced also from the evolutions just illustrated, since a wave travels on the straight line $x - ct = \eta$ and the other one on the straight line $x + ct = \xi$, at a fixed point of the (x, t) plane the solution depends only on the initial data lying between the values η and ξ of the abscissa x (see Fig. 22.2a). The region between the straight lines $x - ct = \eta$, $x + ct = \xi$ and the x axis is called *domain of influence* for the solution at point (x, t). Viceversa, the region of the (x, t) plane between the straight lines $x - ct = \eta$ and $x + ct = \xi$ and above them is called *domain of dependence* for the point (x, t). Given the data of the problem (22.16) on the line segment (a, b) of the real axis, the solution can be calculated only in the domain between that segment and the straight lines $x + ct = b$, $x - ct = a$, as shown in Fig. 22.2b.

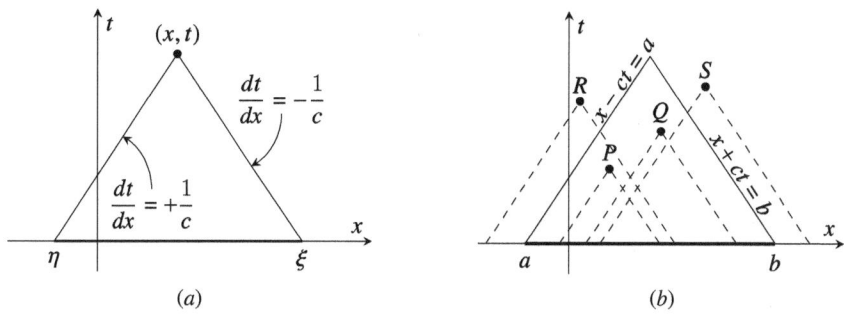

Fig. 22.2 Domain of dependence **a** of the point (x, t) and procedure of construction of the solution **b**. As far as the latter is concerned, if one wants to calculate the solution at points $P = (x_P, t_P)$ and $Q = (x_Q, t_Q)$, interior to the domain of influence, one can use the data in $x_P - ct_P$, $x_P + ct_P$ and $x_Q - ct_Q$, $x_Q + ct_Q$. On the other hand, if one wants to calculate the solution at points R and S it is necessary to know the data at points exterior to the interval (a, b)

22.3   Suggested Readings

Hyperbolic equations are extensively treated in Petrovsky (1954, Chap. II) and, for special problems in wave propagation, in Weinberger (1995, Chaps. I, IV, pp. 112–116; Chap. VI, pp. 152–155; Chap. X, pp. 333–337). An introduction to parabolic equations is given in Petrovsky (1954, Chap. IV) and Weinberger (1995, Chap. III, pp. 58–61; Chap. IV, pp. 92–95).

References

I. G. Petrovsky. *Lectures on Partial Differential Equations*. Interscience, New York, 1954.
H. F. Weinberger. *A First Course in Partial Differential Equations*. Dover, New York, 1995.

Chapter 23
Solution of Poisson's Equation

As mentioned in Sect. 2.2, Poisson's problems play important roles related to the introduction of potentials both in Electromagnetism and in Continuum Mechanics. Due to their importance, several methods for solving these problems have been proposed and used. The present lesson is dedicated to a particular class of methods, in which Green's functions are used.

23.1 Introduction

In this lecture some methods are briefly presented for the solution of Poisson problems, with Dirichlet[1] or Neumann[2] boundary conditions. It is worth to recall that if $f : \mathbb{R}^n \to \mathbb{R}$ has continuous second derivatives and $F : \mathbb{R}^n \to \mathbb{R}$ is continuous, the Poisson's equation is written:

$$\nabla^2 f = F \,, \tag{23.1}$$

where f is the unknown function and F the known term. Problems dealt with in this section will be posed in a bounded domain D of the space $\mathbb{R}^n$, having a regular boundary ∂D of finite measure.

As already mentioned, a first example of a partial differential problem involving a Poisson's equation that is frequently encountered in the applications is as follows:

$$\begin{cases} \nabla^2 f = F & \text{in } D \\ \partial_\nu f = g & \text{on } \partial D \,, \end{cases} \tag{23.2}$$

---

[1] https://mathshistory.st-andrews.ac.uk/Biographies/Dirichlet/.

[2] https://mathshistory.st-andrews.ac.uk/Biographies/Neumann/.

where $g(x)$ is a continuous function whose values are given at every $x \in \partial D$. In the problem (23.2) values of the derivative of function f normal to the boundary are prescribed on ∂D, and this is what is called a *Neumann* problem. Of course, it can be shown that the problem (23.2) identifies f within an arbitrary additive constant. However, a *compatibility condition* is requested to be satisfied, involving the known term F in the equation and the data g prescribed on the boundary, which is derived integrating on D both sides of the equation and using the Gauss' theorem:

$$\int_{\partial D} dS\, g = \int_{\partial D} dS\, \partial_\nu f = \int_D dV\, \nabla^2 f = \int_D dV\, F . \tag{23.3}$$

The problem where values of f are prescribed on ∂D (rather than those of its normal derivative, is written as:

$$\begin{cases} \nabla^2 f = F & \text{in } D \\ f = h & \text{on } \partial D , \end{cases} \tag{23.4}$$

where $h(x)$ is a continuous function with continuous seconde partial derivatives, given at every $x \in \partial D$. The problem (23.4) is called *Dirichlet* problem. It will be shown that it has a unique solution f.

23.2   Harmonic Functions

The case when Eq. (23.1) is a homogeneous one, that is $F = 0$, is studied first. A function $f : \mathbb{R}^n \to \mathbb{R}$ satisfying the Laplace's equation $\nabla^2 f = 0$ is said *harmonic* and has the following interesting property, which holds whatever the dimension n of the space where it operates.

Consider the mean value M of the function f on the surface of a sphere $B_r(x)$ of radius r, centered at an arbitrary point x:

$$M(r) = \frac{1}{|\partial B_r(x)|} \int_{\partial B_r(x)} dS(y)\, f(y) , \tag{23.5}$$

where $|\partial B_r(x)|$ is the area of the spherical surface, given by the product of a number c_n, depending on the dimension of the space, by r^{n-1}, i.e. $|\partial B_r(x)| = c_n\, r^{n-1}$. As it is shown at the left-hand side of Eq. (23.5), the mean value M is, in general, a function of the radius r of the sphere $B_r(x)$.

Equation (23.5) is now written in hyperspherical coordinates using the area element $dS(\varphi) = r^{n-1}\, \Phi(\varphi)\, d\varphi_1\, d\varphi_2 \cdot \ldots \cdot d\varphi_{n-1}$. Recall that the vector of angles φ belongs in this case to the bounded set $A \subset \mathbb{R}^{n-1}$ (its measure is bounded by $(2\pi)^{n-1}$). So, the mean value (23.5) becomes:

$$M(r) = \frac{1}{c_n} \int_A d\varphi_1\, d\varphi_2 \cdot \ldots \cdot d\varphi_{n-1}\, \Phi(\varphi)\, f[y(\varphi)] . \tag{23.6}$$

Denoting by $\boldsymbol{v}(\boldsymbol{\varphi})$ the outward unit normal vector to the surface $\partial B_r(\boldsymbol{x})$ at its point of hyperspherical coordinates $\boldsymbol{\varphi}$ and considering that this unit vector is aligned with the radius of the sphere $B_r(\boldsymbol{x})$ passing through the point of coordinates $\boldsymbol{\varphi}$, that is $\boldsymbol{y}(\boldsymbol{\varphi}) = \boldsymbol{x} + r\,\boldsymbol{v}(\boldsymbol{\varphi})$, one can take the derivative of both sides of the relation (23.6), thus obtaining:

$$\frac{dM}{dr}(r) = \frac{1}{c_n} \int_A d\varphi_1\, d\varphi_2 \cdot \ldots \cdot d\varphi_{n-1}\, \Phi(\boldsymbol{\varphi})\, \boldsymbol{v}(\boldsymbol{\varphi}) \cdot \nabla f[\boldsymbol{y}(\boldsymbol{\varphi})]. \qquad (23.7)$$

Now, multiplying again by r^{n-1} the numerator and the denominator of the right-hand side of (23.7) and considering the expression of the element of area dS:

$$\frac{dM}{dr}(r) = \frac{1}{|\partial B_r(\boldsymbol{x})|} \int_{\partial B_r(\boldsymbol{x})} dS(\boldsymbol{y})\, \boldsymbol{v}(\boldsymbol{y}) \cdot \nabla f(\boldsymbol{y}),$$

through the Gauss' theorem one gets:

$$\frac{dM}{dr}(r) = \frac{1}{|\partial B_r(\boldsymbol{x})|} \int_{B_r(\boldsymbol{x})} dV(\boldsymbol{y})\, \nabla^2 f(\boldsymbol{y}) = 0,$$

being f harmonic, so that $\nabla^2 f = 0$. So, it has been shown that the mean value $M(r)$ of a harmonic function on the surface of a sphere of radius r does not depend on the radius of the sphere $B_r(\boldsymbol{x})$. Thus, *a harmonic function f cannot have isolated maxima or minima in the interior of its domain of definition*, but achieve its maxima and minima on the boundary of it. This property is very important, leading to unexpected consequences in many fields.

Based on the above result, one can show that the solution of problem (23.4) is unique. In fact, suppose there exist two functions f_1 and f_2, both satisfying the problem (23.4). It follows that their difference $f_1 - f_2 =: \delta$ is harmonic in D and moreover it vanishes at every point on the surface ∂D. Then δ vanishes everywhere in D.

Finally, it must be emphasised the paramount importance of the harmonic functions in Complex Analysis, where it can be shown that a function of complex variable $z = x + iy$ (i is the imaginary unit) *differentiable in the complex sense* is composed by a real and an imaginary part, which are both harmonic functions of the real variables x and y.

23.3 Green's Functions

Consider now a particular harmonic function G from $\mathbb{R}^n - \{\boldsymbol{0}\}$, that is the space $\mathbb{R}^n$ deprived of the origin, to $\mathbb{R}$, depending on $|\boldsymbol{x}|$ only, and having the following property:

$$\int_{\partial B_r(\boldsymbol{0})} dS(\boldsymbol{x})\, \partial_v G(\boldsymbol{x}) \equiv 1, \qquad (23.8)$$

for each positive value of the radius r of the sphere $B_r(\mathbf{0})$, centered at the origin.
Two important examples of harmonic functions having the property (23.8) are:

$$G(\mathbf{x}) = \frac{1}{2\pi} \log|\mathbf{x}| \quad \text{for } n = 2, \qquad G(\mathbf{x}) = -\frac{1}{4\pi}\frac{1}{|\mathbf{x}|} \quad \text{for } n = 3. \quad (23.9)$$

For exercise, verify that these functions are harmonic with the property (23.8). A
function of this kind is called a *Green's function* of the ∇^2 operator in the space $\mathbb{R}^n$.

A convention is now introduced. Notice first that if the condition (23.8) is satisfied:
$dG/dr \propto r^{-n+1}$, since $|\partial B_r(\mathbf{0})| \propto r^{n-1}$. As a consequence, for the laplacian of G one
has: $\nabla^2 G \sim d^2 G/dr^2 \propto r^{-n}$ and then the function $\nabla^2 G$ is not integrable at $\mathbf{x} = \mathbf{0}$,
nor at infinity. Nevertheless, the condition (23.8) will often be rewritten in the form
of a volume integral, using the Green's formulae:

$$\int_{B_r(\mathbf{0})} dV(\mathbf{x}) \, \nabla^2 G(\mathbf{x}) \equiv 1, \qquad (23.10)$$

true for any r. It must be kept in mind that the formula (23.10) *is only symbolic*,
being $\nabla^2 G$ not integrable at $\mathbf{0}$.

Suppose now to have a function f from $\mathbb{R}^n$ to $\mathbb{R}$ that is continuous with continuous
first partial derivatives in a neighbourhood of $\mathbf{x} = \mathbf{0}$. Calculate the integral of the
product $[f(\mathbf{y}) - f(\mathbf{0})]\nabla^2 G(\mathbf{y})$ on the sphere $B_r(\mathbf{0})$. To this end, subdivide $B_r(\mathbf{0})$
into a small sphere $B_\rho(\mathbf{0})$, with $\rho < r$, and the spherical crown that is the complement
of $B_\rho(\mathbf{0})$ in $B_r(\mathbf{0})$. The integral extending to the crown described above certainly
vanishes, since $\nabla^2 G(\mathbf{x})$ is zero outside the origin. Consequently:

$$\int_{B_r(\mathbf{0})} dV(\mathbf{y}) \left[f(\mathbf{y}) - f(\mathbf{0}) \right] \nabla^2 G(\mathbf{y}) = \int_{B_\rho(\mathbf{0})} dV(\mathbf{y}) \left[f(\mathbf{y}) - f(\mathbf{0}) \right] \nabla^2 G(\mathbf{y}),$$
$$(23.11)$$

for any $\rho < r$. The function $f(\mathbf{y}) - f(\mathbf{0})$ is infinitesimal as $\mathbf{y} \to \mathbf{0}$, being f continuous, so that $[f(\mathbf{y}) - f(\mathbf{0})]\nabla^2 G(\mathbf{y})$ is integrable at $\mathbf{0}$, although the function $\nabla^2 G(\mathbf{y})$
is not. Finally, notice that the integral (23.11) vanishes. In fact, since (23.11) holds
for arbitrary $\rho \in (0, r)$, it must be true as $\rho \to 0^+$. The limit as $\rho \to 0^+$ of the
right-hand side of (23.11) can be calculated through the Gauss' theorem:

$$\int_{B_\rho(\mathbf{0})} dV(\mathbf{y}) \left[f(\mathbf{y}) - f(\mathbf{0}) \right] \nabla^2 G(\mathbf{y}) = \int_{\partial B_\rho(\mathbf{0})} dS(\mathbf{y}) \left[f(\mathbf{y}) - f(\mathbf{0}) \right] \partial_\nu G(\mathbf{y})$$
$$- \int_{B_\rho(\mathbf{0})} dV(\mathbf{y}) \, \nabla f(\mathbf{y}) \cdot \nabla G(\mathbf{y}),$$

whence also that limit turns out to be zero. In fact, $\partial_\nu G$ on the surface of the sphere
$B_\rho(\mathbf{0})$ is proportional to ρ^{-n+1}, so that $[f(\mathbf{y}) - f(\mathbf{0})]\partial_\nu G(\mathbf{y})$ on $\partial B_\rho(\mathbf{0})$ diverges
less rapidly than ρ^{-n+1} as $\rho \to 0^+$, and it follows that, taking such limit, the first
integral vanishes. Since ∇G is integrable in $B_r(\mathbf{0})$ and $|\nabla f|$ is bounded, the second
integral also goes to zero. Using this result one can write:

$$0 = \int_{B_r(0)} dV(x) [f(y) - f(0)] \nabla^2 G(y)$$

$$= \int_{\partial B_r(0)} dS(y) [f(y) - f(0)] \partial_\nu G(y) - \int_{B_r(0)} dV(y) \nabla f(y) \cdot \nabla G(y)$$

$$= \int_{B_r(0)} dV(y) f(y) \nabla^2 G(y) - f(0),$$

where the definition (23.8) of G has been used. It must be emphasised again that the sum of the two integrals in the second line does make sense, whereas rewriting it in terms of just the volume integral in the third line is understood as merely symbolic. Then one has:

$$\int_{B_r(0)} dV(y) f(y) \nabla^2 G(y) = f(0),$$

which means that the integral of a continuous function with continuous partial derivatives at the origin multiplied by the Laplacian of G provides the value of the function at the origin. Repeating the argument above at a point x, different from the origin of $\mathbb{R}^n$, one obtains the relation at $x \in D$:

$$\int_D dV(y) f(y) \nabla_y^2 G(x - y) = f(x), \qquad (23.12)$$

where ∇_y^2 is the Laplacian calculated taking derivatives with respect to the variable y. The fundamental property (23.12) of the Green's function of the Laplacian is often used to develop solution methods for the problems of Neumann (23.2) and Dirichlet (23.4).

23.4   Green's Function Method

Return now to the solution of Neumann (23.2) and Dirichlet (23.4) problems, to illustrate a method, widely used in the applications, based on an *integral representation* of the solution. Consider two functions $p : \mathbb{R}^n \to \mathbb{R}$ and $q : \mathbb{R}^n \to \mathbb{R}$ differentiable at least twice. Then, the *Green's second identity*:

$$p \nabla^2 q - q \nabla^2 p \equiv \nabla \cdot (p \nabla q - q \nabla p), \qquad (23.13)$$

holds, as it is easily obtained through direct calculation. All derivatives in the identity (23.13) are taken with respect to a a variable denoted with y. Consider a point x in the interior of the domain D and rewrite the identity (23.13), selecting as the function p the unknown function f and as the function q the Green's function G of the Laplace operator in $\mathbb{R}^n$ (23.9) evaluated at point $x - y$. Integrating both sides of identity (23.13) on the domain D with the functions p and q specified as above, one gets the relation:

$$\int_D dV(y) \left[f(y) \nabla_y^2 G(x - y) - G(x - y) \nabla^2 f(y) \right] =$$

$$= \int_{\partial D} dS(y) \left[f(y) \partial_{\nu(y)} G(x - y) - G(x - y) \partial_\nu f(y) \right] . \quad (23.14)$$

Using the relation (23.12) in the first integral at the left-hand side and replacing $\nabla^2 f$ with the known term F of the Poisson's Eq. (23.1) in the second integral at the left-hand side, one gets at point x *interior* to the domain D:

$$f(x) = \int_D dV(y) \, G(x - y) \, F(y)$$

$$+ \int_{\partial D} dS(y) \left[f(y) \partial_{\nu(y)} G(x - y) - G(x - y) \partial_{\nu(y)} f(y) \right] . \quad (23.15)$$

From the relation (23.15) it follows that f is known at interior points of the domain D, once the values of f and its normal derivative $\partial_\nu f$ are given on ∂D.

Exercise 23.1 Why adding a constant to f does not modify the relation (23.15)?

The relation (23.15) does not hold at points $x \in \partial D$. In fact the relation (23.12) holds at interior points of D, that is at x such that there exists a spherical neighbourhood $B_r(x) \subset D$ of it: $\nabla_y^2 G(x - y)$ is identically zero outside such a neighbourhood. For any point $x \in \partial D$, no spherical neighbourhood of it exist, however small is selected its radius r, that is interior to the domain D. However, if the boundary at x is regular enough to allow the definition of the tangent plane, then, for r sufficiently small, a half neighbourhood $B_r(x)$ belongs to D and the remaining half will be exterior to it. Given the above conditions, one can show that the relation (23.12) takes the following slightly different form:

$$\int_D dV(y) \, f(y) \nabla_y^2 G(x - y) = \frac{1}{2} f(x) , \quad (23.16)$$

as it could be expected. Thus, from the relation (23.14) one obtains the following equation:

$$\frac{1}{2} f(x) - \int_{\partial D} dS(y) \, f(y) \, \partial_{\nu(y)} G(x - y) =$$

$$= \int_D dV(y) \, G(x - y) \, F(y) - \int_{\partial D} dS(y) \, G(x - y) \, g(y) , \quad (23.17)$$

whose terms including the function f moved to the left-hand side, leaving at the right-hand side the term with the normal derivative. Equation (23.17) is an *integral equation* in the unknown function f on the surface ∂D, whose solution can often be easily found by numerical approximation. Once the values of f on the boundary of the domain D are known (at least in approximated form), (23.15) provides the solution of the Neumann problem (23.2) at every interior point.

The solution of the Dirichlet problem (23.4) through the Green's function is somewhat more complicated. Only a procedure working for domains D of very simple shape (half-space, circle, ...) will be shown here. Suppose to be able to find a function $G^\star$ of the two variables x and y that is harmonic at the interior point y of D and has the property that at any interior point x of D and at any point y of the boundary of D the following relation holds:

$$G^\star(x; y) = G(x - y). \qquad (23.18)$$

A couple of examples for $n = 2$ are analysed in the following. If D is the half-plane $x_2 > 0$ one can consider the function:

$$G^\star(x_1, x_2; y_1, y_2) = \frac{1}{4\pi} \log\left[(x_1 - y_1)^2 + (x_2 + y_2)^2\right]$$

that is harmonic at y and has the property (23.18) along the boundary of D ($y_2 = 0$). On the other hand, when D is a circle of radius R centered at the origin one can takes the function:

$$G^\star(x, y) = \frac{1}{2\pi}\left(\log\left|\frac{R^2}{|x|^2}x - y\right| + \log|x| - \log R\right),$$

that is harmonic at the interior point y of D and has the property (23.18).

Applying the Green's second identity (23.13) to the functions f and $G^\star$ and recalling the property (23.18) one has:

$$0 = \int_D dV(y)\, G^\star(x; y)\, F(y)$$

$$+ \int_{\partial D} dS(y)\left[f(y)\, \partial_{\nu(y)}G^\star(x; y) - G(x - y)\, \partial_{\nu(y)}f(y)\right] \quad (23.19)$$

that, subtracted to (23.15) allows to cancel the contribution given by the normal derivative of f:

$$f(x) = \int_D dV(y)\left[G(x - y) - G^\star(x; y)\right] F(y)$$

$$+ \int_{\partial D} dS(y)\, f(y)\, \partial_{\nu(y)}\left[G(x - y) - G^\star(x; y)\right]. \quad (23.20)$$

The relation (23.20) suggests to introduce a *Green's function* G_D *of the domain* D defined by the following formula:

$$G_D(x; y) := G(x - y) - G^\star(x; y), \qquad (23.21)$$

that is harmonic at the interior point y of D and vanishes on the boundary of this domain. Using the function (23.21) the integral relation (23.20) can be simply rewritten:

$$f(x) = \int_D dV(y)\, G_D(x; y) F(y) + \int_{\partial D} dS(y)\, G_D(x; y)\, h(y)\,, \qquad (23.22)$$

and allows to solve the Dirichlet problem (23.4).

23.5 Suggested Readings

A general overview of elliptic equations is found in Petrovsky (1954, Chap. III), and, specifically, in Weinberger (1995, Chap. III, pp. 48–52; Chap. IV, pp. 95–112; Chap. VI, pp. 146–152; Chap. VII, pp. 194–197; Chap. VIII, pp. 236–267) for the Laplace's equation, and in Weinberger (1995, Chap. VI, pp. 155–158; Chap. VII, pp. 197–200) for the Poisson's equation.

References

I. G. Petrovsky. *Lectures on Partial Differential Equations*. Interscience, New York, 1954.
H. F. Weinberger. *A First Course in Partial Differential Equations*. Dover, New York, 1995.

Index